Finite Element Analysis

WILEY SERIES IN COMPUTATIONAL MECHANICS

Series Advisors:

René de Borst
Perumal Nithiarasu
Tayfun E. Tezduyar
Genki Yagawa
Tarek Zohdi

Finite Element Analysis

Method, Verification and Validation

Second Edition

Barna Szabó
Washington University in St. Louis

Ivo Babuška
The University of Texas at Austin

Registered Office
John Wiley & Sons, Inc., 111 River Street, Hoboken, NJ 07030, USA

Editorial Office
111 River Street, Hoboken, NJ 07030, USA

For details of our global editorial offices, customer services, and more information about Wiley products visit us at www.wiley.com.

Wiley also publishes its books in a variety of electronic formats and by print-on-demand. Some content that appears in standard print versions of this book may not be available in other formats.

Library of Congress Cataloging-in-Publication Data

Names: Szabó, B. A. (Barna Aladar), 1935- author. | Babuška, Ivo, author.
Title: Finite element analysis : method, verification and validation /
    Barna Szabó, Ivo Babuška.
Description: Second edition. | Hoboken, NJ : Wiley, 2021. | Series: Wiley
    series in computational mechanics | Includes bibliographical references
    and index.
Identifiers: LCCN 2021005389 (print) | LCCN 2021005390 (ebook) | ISBN
    9781119426424 (cloth) | ISBN 9781119426387 (adobe pdf) | ISBN
    9781119426462 (epub) | ISBN 9781119426479 (obook)
Subjects: LCSH: Finite element method.
Classification: LCC TA347.F5 S98 2021 (print) | LCC TA347.F5 (ebook) |
    DDC 620.001/51825–dc23
LC record available at https://lccn.loc.gov/2021005389
LC ebook record available at https://lccn.loc.gov/2021005390

Cover Design: Wiley
Cover Images: © Courtesy of Barna Szabó

Set in 9.5/12.5pt STIXTwoText by SPi Global, Chennai, India

10 9 8 7 6 5 4 3 2 1

Contents

Preface to the second edition

The first edition of this book, published in 1991, focused on the conceptual and algorithmic development of the finite element method from the perspective of solution verification, that is, estimation and control of the errors of approximation in terms of the quantities of interest. Since that time the importance of solution verification became widely recognized. It is a key constituent of predictive computational science, the branch of computational science concerned with the prediction of physical events.

Predictive computational science embraces the formulation of mathematical models, definition of the quantities of interest, code and solution verification, definition of statistical sub-models, calibration and validation of models, and forecasting physical events with quantified uncertainty. The second edition covers the main conceptual and algorithmic aspects of predictive computational science pertinent to solid mechanics. The formulation and application of design rules for mechanical and structural components subjected to cyclic loading are used for illustration.

Another objective in writing the first edition was to make some fundamentally important results of research in the field of applied mathematics accessible to the engineering community. Speaking generally, engineers and mathematicians view the finite element method very differently. Engineers see the method as a way to construct a numerical problem the solution of which is expected to provide quantitative information about the response of some physical system, for example a structural shell, to some form of excitation, such as the application of loads. Their view tends to be element-oriented: They tend to believe that sufficiently clever formulation of elements can overcome the various shortcomings of the method.

Mathematicians, on the other hand, view the finite element method as a method for approximating the exact solution of differential equations cast in a variational form. Mathematicians focus on a priori and a posteriori error estimation and error control. In the 1970s adaptive procedures were developed for the construction of sequences of finite element mesh such that the corresponding solutions converged to the exact solution in energy norm at optimal or nearly optimal rates. An alternative way of achieving convergence in energy norm through increasing the polynomial degrees of the elements on a fixed mesh was proven in 1981. The possibility of achieving exponential rates of convergence in energy norm for an important class of problems, that includes the problem of elasticity, was proven and demonstrated in 1984. This required the construction of sequences of finite element mesh and optimal assignment of polynomial degrees.

Superconvergent methods of extraction of certain quantities of interest (such as the stress intensity factor) from finite element solutions were developed by 1984.

These developments were fundamentally important milestones in a journey toward the emergence of predictive computational science. Our primary objective in publishing this second edition is to provide engineering analysts and software developers a comprehensive account of the conceptual and algorithmic aspects of verification, validation and uncertainty quantification illustrated by examples.

Quantification of uncertainty involves the application of methods of data analysis. A brief introduction to the fundamentals of data analysis is presented in this second edition.

We recommend this book to students, engineers and analysts who seek Professional Simulation Engineer (PSE) certification.

We would like to thank Dr. Ricardo Actis for many useful discussions, advice and assistance provided over many years; Dr. Börje Andersson for providing valuable convergence data relating to the solution of an interesting model problem of elasticity, and Professor Raul Tempone for guidance in connection with the application of data analysis procedures.

Barna Szabó and Ivo Babuška

Preface to the first edition

There are many books on the finite element method today. It seems appropriate therefore to say a few words about why this book was written. A brief look at the approximately 30-year history of the finite element method will not only help explain the reasons but will also serve to put the main points in perspective.

Systematic development of the finite element method for use as an analytical tool in engineering decision-making processes is usually traced to a paper published in 1956.[1] Demand for efficient and reliable numerical methods has been the key motivating factor for the development of the finite element method. The demand was greatly amplified by the needs of the space program in the United States during the 1960s. A great deal was invested into the development of finite element analysis technology in that period.

Early development of the finite element method was performed entirely by engineers. The names of Argyris, Clough, Fraeijs de Veubeke, Gallagher, Irons, Martin, Melosh, Pian, and Zienkiewicz come to mind in this connection.[2]

Development of the finite element method through the 1960s was based on intuitive reasoning, analogies with naturally discrete systems, such as structural frames, and numerical experimentation. The errors of discretization were controlled by uniform or nearly uniform refinement of the finite element mesh. Mathematical analysis of the finite element method begun in the late 1960s. Error estimation techniques were investigated during the 1970s. Adaptive mesh refinement procedures designed to reduce the errors of discretization with improved efficiency received a great deal of attention in this period.[3]

Numerical experiments conducted in the mid-1970s indicated that the use of polynomials of progressively increasing degree on a fixed finite element mesh can be much more advantageous than uniform or nearly uniform mesh refinement.[4] To distinguish between reducing errors of discretization by mesh refinement and the alternative approach, based on increasing the degree of the polynomial basis functions, the labels *h-version* and *p-version* gained currency for the following reason: Usually the symbol h is used to represent the size of the finite elements. Convergence occurs when the size of the largest element (h_{max}) is progressively reduced. Hence the name: h-version. The polynomial degree of elements is usually denoted by the symbol p. Convergence occurs when

1 Turner MJ, Clough RW, Martin HC and Topp LJ, Stiffness and deflection analysis of complex structures. *Journal of Aeronautical Sciences* **23**(9), 805–824, 1956.

2 Williamson CF, Jr. A History of the Finite Element Method to the Middle 1960s. Doctoral Dissertation, Boston University, 1976.

3 Babuška I and Rheinboldt WC. Adaptive approaches and reliability estimations in finite element analysis. *Computer Methods in Applied Mechanics and Engineering*, **17/18**, 519–540, 1979.

4 Szabó BA and Mehta AK. p-convergent finite element approximations in fracture mechanics. *Int. J. Num. Meth. Engng.*, **12**. 551–560, 1978.

the lowest polynomial degree p_{min} is progressively increased. Hence the name: p-version. The h- and p-versions are just special applications of the finite element method which, at least in principle, allows changing the finite element mesh concurrently with increasing the polynomial degree of elements. This general approach is usually called the *hp-version* of the finite element method. The theoretical basis for the p-version was established by 1981.[5] The understanding of how to combine mesh refinement with p-extensions most effectively was achieved by the mid-1980s.[6]

The p-version was developed primarily for applications in solid mechanics. A closely related recent development for applications primarily in fluid mechanics is the *spectral element method*.[7]

The 1980s brought another important development: superconvergent methods for the extraction of engineering data from finite element solutions were developed and demonstrated. At the time of writing good understanding exists regarding how finite element discretizations should be designed and how engineering data should be extracted from finite element solutions for optimal reliability and efficiency.

The knowledge which allows construction of advanced finite element computer programs, that work efficiently and reliably for very large classes of problems, and all admissible data that characterize those problems, is well established today. It must be borne in mind, however, that finite element solutions and engineering data extracted from them can be valuable only if they serve the purposes of making correct engineering decisions.

Our purpose in writing this book is to introduce the finite element method to engineers and engineering students in the context of the engineering decision-making process. Basic engineering and mathematical concepts are the starting points. Key theoretical results are summarized and illustrated by examples. The focus is on the developments in finite element analysis technology during the 1980s and their impact on reliability, quality assurance procedures in finite element computations, and performance. The principles that guide the construction of mathematical models are described and illustrated by examples.

---

5 Babuška I, Szabó B and Katz IN. The p-version of the finite element method. *SIAM J. Numer. Anal.*, **18**, 515–545, 1981.
6 Babuška I. The p- and hp-versions of the finite element method. The state of the art. In: DL Dwoyer, MY Hussaini and RG Voigt, editors, *Finite Elements: Theory and Applications*, Springer-Verlag New York, Inc., 1988.
7 Maday Y and Patera AT. Spectral element methods for the incompressible Navier-Stokes equations. In: AK Noor and JT Oden, editors. *State-of-the-Art Surveys on Computational Mechanics*, American Society of Mechanical Engineers, New York, 1989.

Preface

This book, *Finite Element Analysis: Method, Verification and Validation, 2nd Edition*, is written by two well-recognized, leading experts on finite element analysis. The first edition, published in 1991, was a landmark contribution in finite element methods, and has now been updated and expanded to include the increasingly important topic of error estimation, validation – the process to ascertain that the mathematical/numerical model meets acceptance criteria – and verification – the process for acceptability of the approximate solution and computed data. The systematic treatment of formulation, verification and validation procedures is a distinguishing feature of this book and sets it apart from other texts on finite elements. It encapsulates contemporary research on proper model selection and control of modeling errors. Unique features of the book are accessibility and readability for students and researchers alike, providing guidance on modeling, simulation and implementation issues. It is an essential, self-contained book for any person who wishes to fully master the finite element method.

About the companion website

This book is accompanied by a companion website:

www.wiley.com/go/szabo/finite_element_analysis

The website includes solutions manual, PowerPoint slides for instructors, and a link to finite element software

1

Introduction to the finite element method

This book covers the fundamentals of the finite element method in the context of numerical simulation with specific reference to the simulation of the response of structural and mechanical components to mechanical and thermal loads.

We begin with the question: what is the meaning of the term "simulation"? By its dictionary definition, simulation is the imitative representation of the functioning of one system or process by means of the functioning of another. For instance, the membrane analogy introduced by Prandtl[1] in 1903 made it possible to find the shearing stresses in bars of arbitrary cross-section, loaded by a twisting moment, through mapping the deflected shape of a thin elastic membrane. In other words, the distribution and magnitude of shearing stress in a twisted bar can be simulated by the deflected shape of an elastic membrane.

The membrane analogy exists because two unrelated phenomena can be modeled by the same partial differential equation. The physical meaning associated with the coefficients of the differential equation depends on which problem is being solved. However, the solution of one is proportional to the solution of the other: At corresponding points the shearing stress in a bar, subjected to a twisting moment, is oriented in the direction of the tangent to the contour lines of a deflected thin membrane and its magnitude is proportional to the slope of the membrane. Furthermore, the volume enclosed by the deflected membrane is proportional to the twisting moment.

In the pre-computer years the membrane analogy provided practical means for estimating shearing stresses in prismatic bars. This involved cutting the shape of the cross-section out of sheet metal or a wood panel, covering the hole with a thin elastic membrane, applying pressure to the membrane and mapping the contours of the deflected membrane. In present-day practice both problems would be formulated as mathematical problems which would then be solved by a numerical method, most likely by the finite element method.

There are many other useful analogies. For example, the same differential equations simulate the response of assemblies of mechanical components, such as linear spring-mass-viscous damper systems and assemblies of electrical components, such as capacitors, inductors and resistors. This has been exploited by the use of analogue computers. Obviously, it is much easier to build and manipulate electrical circuitry than mechanical assemblies. In present-day practice both simulation problems would be formulated as mathematical problems which would be solved by a numerical method.

At the heart of simulation of aspects of physical reality is a mathematical problem cast in a generalized form[2]. The solution of the mathematical problem is approximated by a numerical method,

---

1 Ludwig Prandtl 1875–1953.
2 The generalized form is also called variational form or weak form.

Finite Element Analysis: Method, Verification and Validation, Second Edition. Barna Szabó and Ivo Babuška.
© 2021 John Wiley & Sons, Inc. Published 2021 by John Wiley & Sons, Inc.
Companion Website: www.wiley.com/go/szabo/finite_element_analysis

such as the finite element method, which is the subject of this book. The quantities of interest (QoI) are extracted from the approximate solution. The errors of approximation in the QoI depend on how the mathematical problem was discretized[3] and how the QoI were extracted from the numerical solution. When the errors of approximation are larger than what is considered acceptable then the discretization has to be changed either by an automated adaptive process or by action of the analyst.

Estimation and control of numerical errors are fundamentally important in numerical simulation. Consider, for example, the problem of design certification. Design rules are typically stated in the form

$$F_{max} \leq F_{all} \tag{1.1}$$

where $F_{max} > 0$ (resp. $F_{all} > 0$) is the maximum (resp. allowable) value of a quantity of interest, for example the first principal stress. Since in numerical simulation only an approximation to F_{max} is available, denoted by F_{num}, it is necessary to know the size of the numerical error τ:

$$|F_{max} - F_{num}| \leq \tau F_{max}. \tag{1.2}$$

In design and design certification the worst case scenario has to be considered, which is underestimation of F_{max}, that is,

$$F_{num} = (1 - \tau)F_{max}. \tag{1.3}$$

Therefore it has to be shown that

$$F_{num} \leq (1 - \tau)F_{all}. \tag{1.4}$$

Without a reliable estimate of the size of the numerical error it is not possible to certify design and, furthermore, numerical errors penalize design by lowering the allowable value, as indicated by eq. (1.4). Generally speaking, it is far more economical to ensure that τ is small than to accept the consequences of decreased allowable values.

We distinguish between finite element modeling and numerical simulation. As explained in greater detail in Chapter 5, finite element modeling evolved well before the theoretical basis of numerical simulation was developed. In finite element modeling a numerical problem is formulated by assembling elements from a library of finite elements that contains intuitively constructed beam, plate, shell, solid elements of various description. The numerical problem so created may not correspond to a well defined mathematical problem and therefore a solution may not even exist. For that reason it is not possible to speak of errors of approximation. Nevertheless, finite element modeling is widely practiced with success in some cases but with disappointing results in others. Such practice should be regarded as a practice of art, guided by intuition and experience, rather than a scientific activity. This is because practitioners of finite element modeling have to balance two kinds of very large errors: (a) conceptual errors in the formulation and (b) approximation errors in the numerical solution of an improperly posed mathematical problem.

In numerical simulation, on the other hand, the formulation of mathematical models is treated separately from their numerical solution. A mathematical model should be understood to be a precise statement of an idea of physical reality that permits the prediction of the occurrence, or probability of occurrence, of physical events, given certain data. The intuitive aspects of simulation are confined to the formulation of mathematical models whereas their numerical solution involves the application of well established procedures of applied mathematics. Separation of mathematical

---

3 The term "discretization" refers to processes by which approximating functions are defined. The most widely used discretizations will be described and illustrated by examples in this and subsequent chapters.

models from their numerical solution makes separate treatment of errors associated with the formulation of mathematical models and their numerical approximation possible. Errors associated with the formulation of mathematical models are called model form errors. Errors associated with the numerical solution of mathematical problems are called errors of approximation or errors of discretization. In the early papers and books on the finite element method no such distinction was made.

In this chapter we introduce the finite element method as a method by which the exact solution of a mathematical problem, cast in a generalized form, can be approximated. We also introduce the relevant mathematical concepts, terminology and notation in the simplest possible setting. Generalization of these concepts to two- and three-dimensional problems will be discussed in subsequent chapters.

We first consider the formulation of a second order ordinary differential equation without reference to any physical interpretation. This is to underline that once a mathematical problem was formulated, the approximation process is independent from why the mathematical problem was formulated. This important point is often missed by engineering users of legacy finite element codes because the formulation and approximation of mathematical problems is mixed in finite element libraries.

We show that the exact solution of the generalized formulation is unique. Approximation of the exact solution by the finite element method is described and various discretization strategies are explored. Efficient methods for the computation of QoIs and a posteriori error estimation are described. This chapter serves as a foundation for subsequent chapters.

We would like to assure engineering students who are not yet familiar with the concepts and notation of that branch of applied mathematics on which the finite element method is based that their investment of time and effort to master the contents of this chapter will prove to be highly rewarding.

1.1 An introductory problem

We introduce the finite element method through approximating the exact solution of the following second order ordinary differential equation

$$-(\kappa u')' + cu = f \quad \text{on the closed interval } \bar{I} = [0 \leq x \leq \ell] \tag{1.5}$$

with the boundary conditions

$$u(0) = u(\ell) = 0 \tag{1.6}$$

where the prime indicates differentiation with respect to x. It is assumed that $0 < \alpha \leq \kappa(x) \leq \beta < \infty$ where α and β are real numbers, $\kappa' < \infty$ on $\bar{I}$, $c \geq 0$ and $f = f(x)$ are defined such that the indicated operations are meaningful on I. For example, the indicated operations would not be meaningful if $(\kappa u')'$, c or f would not be finite in one or more points on the interval $0 \leq x \leq \ell$. The function f is called a forcing function.

We seek an approximation to u in the form:

$$u_n = \sum_{j=1}^{n} a_j \varphi_j(x), \quad \varphi_j(0) = \varphi_j(\ell) = 0 \text{ for all } j \tag{1.7}$$

where $\varphi_j(x)$ are fixed functions, called basis functions, and a_j are the coefficients of the basis functions to be determined. Note that the basis functions satisfy the zero boundary conditions.

Let us find a_j such that the integral $\mathcal{I}$ defined by

$$\mathcal{I} = \frac{1}{2}\int_0^{\ell}\left(\kappa(u' - u_n')^2 + c(u - u_n)^2\right)\,dx \tag{1.8}$$

is minimum. While there are other plausible criteria for selecting a_j, we will see that this criterion is fundamentally important in the finite element method. Differentiating $\mathcal{I}$ with respect to a_i and letting the derivative equal to zero, we have:

$$\frac{d\mathcal{I}}{da_i} = \int_0^{\ell}\left(\kappa(u' - u_n')\varphi_i' + c(u - u_n)\varphi_i\right)\,dx = 0, \quad i = 1, 2, \ldots, n. \tag{1.9}$$

Using the product rule: $(\kappa u'\varphi_i)' = (\kappa u')'\varphi_i + \kappa u'\varphi_i'$ we write

$$\int_0^{\ell}\kappa u'\varphi_i'\,dx = \int_0^{\ell}\left((\kappa u'\varphi_i)' - (\kappa u')'\varphi_i\right)\,dx$$

$$= \underbrace{(\kappa u'\varphi_i)_{x=\ell}}_{=0} - \underbrace{(\kappa u'\varphi_i)_{x=0}}_{=0} - \int_0^{\ell}(\kappa u')'\varphi_i\,dx. \tag{1.10}$$

The underbraced terms vanish on account of the boundary conditions, see eq. (1.7). On substituting this expression into eq. (1.9), we get

$$\int_0^{\ell}\underbrace{(-(\kappa u')' + cu)}_{=f(x)}\varphi_i\,dx - \int_0^{\ell}(\kappa u_n'\varphi_i' + cu_n\varphi_i)\,dx = 0$$

which will be written as

$$\int_0^{\ell}(\kappa u_n'\varphi_i' + cu_n\varphi_i)\,dx = \int_0^{\ell}f\varphi_i\,dx, \quad i = 1, 2, \ldots, n. \tag{1.11}$$

We define

$$k_{ij} = \int_0^{\ell}\kappa\varphi_i'\varphi_j'\,dx, \quad m_{ij} = \int_0^{\ell}c\varphi_i\varphi_j\,dx, \quad r_i = \int_0^{\ell}f\varphi_i\,dx \tag{1.12}$$

and write eq. (1.11) in the following form

$$\sum_{j=1}^{n}(k_{ij} + m_{ij})a_j = r_i, \quad i = 1, 2, \ldots, n \tag{1.13}$$

which represents n simultaneous equations in n unknowns. It is usually written in matrix form:

$$([K] + [M])\{a\} = \{r\}. \tag{1.14}$$

On solving these equations we find an approximation u_n to the exact solution u in the sense that u_n minimizes the integral $\mathcal{I}$.

Example 1.1   Let $\kappa = 1, c = 1, \ell = 2$ and

$$f = \sin(\pi x/\ell) + \sin(2\pi x/\ell).$$

With these data the exact solution of eq. (1.5) is

$$u = \frac{1}{1 + \pi^2/\ell^2}\sin(\pi x/\ell) + \frac{1}{1 + 4\pi^2/\ell^2}\sin(2\pi x/\ell).$$

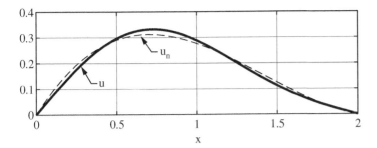

Figure 1.1 Exact and approximate solutions for the problem in Example 1.1.

We seek an approximation to u in the form:

$$u_n = u_2 = a_1 x(\ell - x) + a_2 x(\ell - x)^2.$$

On computing the elements of $[K]$, $[M]$ and $\{r\}$ we get

$$[K] = \begin{bmatrix} 2.6667 & 2.6667 \\ 2.6667 & 4.2667 \end{bmatrix} \quad [M] = \begin{bmatrix} 1.0667 & 1.0667 \\ 1.0667 & 1.2190 \end{bmatrix} \quad \{r\} = \begin{Bmatrix} 1.0320 \\ 1.4191 \end{Bmatrix}.$$

The solution of this problem is $a_1 = 0.0556$, $a_2 = 0.2209$. These coefficients, together with the basis functions, define the approximate solution u_n. The exact and approximate solutions are shown in Fig. 1.1.

The choice of basis functions

By definition, a set of functions $\varphi_j(x)$, $(j = 1, 2, \ldots, n)$ are linearly independent if

$$\sum_{j=1}^{n} a_j \varphi_j(x) = 0$$

implies that $a_j = 0$ for $j = 1, 2, \ldots, n$. It is left to the reader to show that if the basis functions are linearly independent then matrix $[M]$ is invertible.

Given a set of linearly independent functions $\varphi_j(x)$, $(j = 1, 2, \ldots, n)$, the set of functions that can be written as

$$u_n = \sum_{j=1}^{n} a_j \varphi_j(x)$$

is called the span and $\varphi_j(x)$ are basis functions of S.

We could have defined other polynomial basis functions, for example;

$$u_n = \sum_{i=1}^{n} c_i \psi_i(x) , \qquad \psi_i(x) = x^i(\ell - x). \tag{1.15}$$

When one set of basis functions $\{\varphi\} = \{\varphi_1 \ \varphi_2 \ \cdots \ \varphi_n\}^T$ can be written in terms of another set $\{\psi\} = \{\psi_1 \ \psi_2 \ \cdots \ \psi_n\}^T$ in the form:

$$\{\psi\} = [B]\{\varphi\} \tag{1.16}$$

where $[B]$ is an invertible matrix of constant coefficients then both sets of basis functions are said to have the same span. The following exercise demonstrates that the approximate solution depends on the span, not on the choice of basis functions.

Exercise 1.1 Solve the problem of Example 1.1 using the basis functions $\varphi_1 = x(\ell - x)$, $\varphi_2 = x^2(\ell - x)$ and show that the resulting approximate solution is identical to the approximate solution obtained in Example 1.1. The span of the basis functions in this exercise and in Example 1.1 is the same: It is the set of polynomials of degree less than or equal to 3 that vanish in the points $x = 0$ and $x = \ell$.

Summary of the main points

1. The definition of the integral $\mathcal{I}$ by eq. (1.8) made it possible to find an approximation to the exact solution u of eq. (1.5) without knowing u.
2. A formulation cannot be meaningful unless all indicated operations are defined. In the case of eq. (1.5) this means that $(\kappa u')'$ and cu are finite on the interval $0 \le x \le \ell$. In the case of eq. (1.11) the integral

$$\int_0^\ell \left(\kappa (u')^2 + cu^2 \right) \, dx$$

   must be finite which is a much less stringent condition. In other words, eq. (1.8) is meaningful for a larger set of functions u than eq. (1.5) is. Equation (1.5) is the strong form, whereas eq. (1.11) is the generalized or weak form of the same differential equation. When the solution of eq. (1.5) exists then u_n converges to that solution in the sense that the limit of the integral $\mathcal{I}$ is zero.
3. The error $e = u - u_n$ depends on the span and not on the choice of basis functions.

1.2 Generalized formulation

We have seen in the foregoing discussion that it is possible to approximate the exact solution u of eq. (1.5) without knowing u when $u(0) = u(\ell) = 0$. In this section the formulation is outlined for other boundary conditions.

The generalized formulation outlined in this section is the most widely implemented formulation; however, it is only one of several possible formulations. It has the properties of stability and consistency. For a discussion on the requirements of stability and consistency in numerical approximation we refer to [5].

1.2.1 The exact solution

If eq. (1.5) holds then for an arbitrary function $v = v(x)$, subject only to the restriction that all of the operations indicated in the following are properly defined, we have

$$\int_0^\ell \left((-\kappa u')' + cu - f \right) v \, dx = 0. \tag{1.17}$$

Using the product rule; $(\kappa u' v)' = (\kappa u')' v + \kappa u' v'$ we get

$$\int_0^\ell (-\kappa u')' v \, dx = -(\kappa u' v)_{x=\ell} + (\kappa u' v)_{x=0} + \int_0^\ell \kappa u' v' \, dx$$

therefore eq. (1.17) is transformed to:

$$\int_0^\ell (\kappa u' v' + cuv) \, dx = \int_0^\ell fv \, dx + (\kappa u' v)_{x=\ell} - (\kappa u' v)_{x=0}. \tag{1.18}$$

We introduce the following notation:

$$B(u, v) \overset{\text{def}}{=} \int_0^\ell (\kappa u' v' + cuv) \, dx \tag{1.19}$$

where $B(u, v)$ is a bilinear form. A bilinear form has the property that it is linear with respect to each of its two arguments. The properties of bilinear forms are listed Section A.1.3 of Appendix A. We define the linear form:

$$F(v) \overset{\text{def}}{=} \int_0^\ell fv \, dx + (\kappa u' v)_{x=\ell} - (\kappa u' v)_{x=0}. \tag{1.20}$$

The forcing function $f(x)$ may be a sum of forcing functions: $f(x) = f_1(x) + f_2(x) + \ldots$, some or all of which may be the Dirac delta function[4] multiplied by a constant. For example if $f_k(x) = F_0\delta(x_0)$ then

$$\int_0^\ell f_k(x)v \, dx = \int_0^\ell F_0\delta(x_0)v \, dx = F_0 v(x_0). \tag{1.21}$$

The properties of linear forms are listed in Section A.1.2. Note that $F_0 v(x_0)$ in eq. (1.21) is a linear form only if v is continuous and bounded.

The definitions of $B(u, v)$ and $F(v)$ are modified depending on the boundary conditions. Before proceeding further we need the following definitions.

1. The *energy norm* is defined by

$$\|u\|_{E(I)} \overset{\text{def}}{=} \sqrt{\frac{1}{2}B(u, u)} \tag{1.22}$$

   where I represents the open interval $I = \{x \mid 0 < x < \ell\}$. This notation should be understood to mean that $x \in I$ if and only if x satisfies the condition to the right of the bar ($\mid$). This notation may be shortened to $I = (0, \ell)$, or more generally $I = (a, b)$ where $b > a$ are real numbers. If the interval includes both boundary points then the interval is a closed interval denoted by $\bar{I} \overset{\text{def}}{=} [0, \ell]$.
   We have seen in the introductory example that the error is minimized in energy norm, that is, $\|u - u_n\|_{E(I)}^2$, equivalently $\|u - u_n\|_{E(I)}$ is minimum. The square root is introduced so that $\|\alpha u\|_{E(I)} = |\alpha| \|u\|_{E(I)}$ (where α is a constant) holds. This is one of the definitive properties of norms listed in Section A.1.1.

2. The energy space, denoted by $E(I)$, is the set of all functions u defined on I that satisfy the following condition:

$$E(I) \overset{\text{def}}{=} \{u \mid \|u\|_{E(I)} < \infty\}. \tag{1.23}$$

   Since infinitely many linearly independent functions satisfy this condition, the energy space is infinite-dimensional.

3. The *trial space*, denoted by $\tilde{E}(I)$, is a subspace of $E(I)$. When boundary conditions are prescribed on u, such as $u(0) = \hat{u}_0$ and/or $u(\ell) = \hat{u}_\ell$, then the functions that lie in $\tilde{E}(I)$ satisfy those boundary conditions. Note that when $\hat{u}_0 \neq 0$ and/or $\hat{u}_\ell \neq 0$ then $\tilde{E}(I)$ is not a linear space. This is because the condition stated under item 1 in Section A.1.1 is not satisfied. When u is prescribed on a boundary then that boundary condition is called an essential boundary condition. If no essential boundary conditions are prescribed on u then $\tilde{E}(I) = E(I)$.

---

4 See Definition A.5 in the appendix.

4. The *test space*, denoted by $E^0(I)$, is a subspace of $E(I)$. When boundary conditions are prescribed on u, such as $u(0) = \hat{u}_0$ and/or $u(\ell) = \hat{u}_\ell$ then the functions that lie in $E^0(I)$ are zero in those boundary points.

If no boundary conditions are prescribed on u then $\tilde{E}(I) = E^0(I) = E(I)$. If $u(0) = \hat{u}_0$ is prescribed and $u(\ell)$ is not known then

$$\tilde{E}(I) \stackrel{\text{def}}{=} \{u \mid u \in E(I),\ u(0) = \hat{u}_0\} \tag{1.24}$$

$$E^0(I) \stackrel{\text{def}}{=} \{u \mid u \in E(I),\ u(0) = 0\}. \tag{1.25}$$

If $u(0)$ is not known and $u(\ell) = \hat{u}_\ell$ is prescribed then

$$\tilde{E}(I) \stackrel{\text{def}}{=} \{u \mid u \in E(I),\ u(\ell) = \hat{u}_\ell\} \tag{1.26}$$

$$E^0(I) \stackrel{\text{def}}{=} \{u \mid u \in E(I),\ u(\ell) = 0\}. \tag{1.27}$$

If $u(0) = \hat{u}_0$ and $u(\ell) = \hat{u}_\ell$ are prescribed then

$$\tilde{E}(I) \stackrel{\text{def}}{=} \{u \mid u \in E(I),\ u(0) = \hat{u}_0,\ u(\ell) = \hat{u}_\ell\} \tag{1.28}$$

$$E^0(I) \stackrel{\text{def}}{=} \{u \mid u \in E(I),\ u(0) = 0,\ u(\ell) = 0\}. \tag{1.29}$$

We are now in a position to describe the generalized formulation for various boundary conditions in a concise manner;

1. When u is prescribed on a boundary then the boundary condition is called essential or Dirichlet[5] boundary condition. Let us assume that u is prescribed on both boundary points. In this case we write $u = \bar{u} + u^\star$ where $\bar{u} \in E^0(I)$ is the function to be approximated and $u^\star \in \tilde{E}$ is an arbitrary fixed function that satisfies the boundary conditions. Substituting $\bar{u} + u^\star$ for u in eq. (1.18) we have:

$$\underbrace{\int_0^\ell (\kappa \bar{u}' v' + c \bar{u} v)\, dx}_{B(\bar{u}, v)} = \underbrace{\int_0^\ell f v\, dx - \int_0^\ell (\kappa (u^\star)' v' + c u^\star v)\, dx}_{F(v)} \tag{1.30}$$

and the generalized formulation is stated as follows: "Find $\bar{u} \in E^0(I)$ such that $B(\bar{u}, v) = F(v)$ for all $v \in E^0(I)$" where $E^0(I)$ is defined by eq. (1.29). Note that $u \in \tilde{E}(I)$ is independent of the choice of $u^\star$. Essential boundary conditions are enforced by restriction on the space of admissible functions.

2. When $\kappa u' = F$ is prescribed on a boundary then the boundary condition is called Neumann[6] boundary condition. Assume that $u(0) = \hat{u}_0$ and $(\kappa u')_{x=\ell} = F_\ell$ are prescribed. In this case

$$\underbrace{\int_0^\ell (\kappa \bar{u}' v' + c \bar{u} v)\, dx}_{B(\bar{u}, v)} = \underbrace{\int_0^\ell f v\, dx + F_\ell v(\ell) - \int_0^\ell (\kappa (u^\star)' v' + c u^\star v)\, dx}_{F(v)} \tag{1.31}$$

and the generalized formulation is: "Find $\bar{u} \in E^0(I)$ such that $B(\bar{u}, v) = F(v)$ for all $v \in E^0(I)$" where $E^0(I)$ is defined by eq. (1.25).

---

5 Peter Gustav Lejeune Dirichlet 1805–1859.
6 Carl Gottfried Neumann 1832–1925.

An important special case is when $c = 0$ and $(\kappa u')_{x=0} = F_0$ and $(\kappa u')_{x=\ell} = F_\ell$ are prescribed. In this case:

$$\underbrace{\int_0^\ell \kappa u' v' \, dx}_{B(u,v)} = \underbrace{\int_0^\ell f v \, dx - F_0 v(0) + F_\ell v(\ell)}_{F(v)} \tag{1.32}$$

and the generalized formulation is "Find $u \in E(I)$ such that $B(u, v) = F(v)$ for all $v \in E(I)$ where $E(I)$ is defined by eq. (1.23)." Since the left-hand side is zero for $v = C$ (constant) the specified data must satisfy the condition

$$\int_0^\ell f \, dx - F_0 + F_\ell = 0. \tag{1.33}$$

3. When $(\kappa u')_{x=0} = k_0(u(0) - \delta_0)$ and/or $(\kappa u')_{x=\ell} = k_\ell(\delta_\ell - u(\ell))$, where $k_0 > 0$, $k_\ell > 0$, δ_0 and δ_ℓ are given real numbers, is prescribed on a boundary then the boundary condition is called a Robin[7] boundary condition. Assume, for example, that $(\kappa u')_{x=0} = k_0(u(0) - \delta_0)$ and $(\kappa u')_{x=\ell} = F_\ell$ are prescribed. In that case

$$\underbrace{\int_0^\ell (\kappa u' v' + cuv) \, dx + k_0 u(0)v(0)}_{B(u,v)} = \underbrace{\int_0^\ell f v \, dx + F_\ell v(\ell) - k_0 \delta_0 v(0)}_{F(v)} \tag{1.34}$$

and the generalized formulation is: "Find $u \in E(I)$ such that $B(u, v) = F(v)$ for all $v \in E(I)$ where $E(I)$ is defined by eq. (1.23)."

These boundary conditions may be prescribed in any combination. The Neumann and Robin boundary conditions are called natural boundary conditions. Natural boundary conditions cannot be enforced by restriction. This is illustrated in Exercise 1.3.

The generalized formulation is stated as follows: "Find $u_{EX} \in X$ such that $B(u_{EX}, v) = F(v)$ for all $v \in Y$". The space X is called the trial space, the space Y is called the test space. We will use this notation with the understanding that the definitions of X, Y, $B(u, v)$ and $F(v)$ depend on the boundary conditions. It is essential for analysts to understand and be able to precisely state the generalized formulation for any set of boundary conditions.

Under frequently occurring special conditions the mathematical problem can be formulated on a subdomain and the solution extended to the full domain by symmetry, antisymmetry or periodicity. The symmetric, antisymmetric and periodic boundary conditions will be discussed in Chapter 2.

Theorem 1.1 The solution of the generalized formulation is unique in the energy space. The proof is by contradiction: Assume that there are two solutions u_1 and u_2 in $X \subset E(I)$ that satisfy

$$B(u_1, v) = F(v) \quad \text{for all } v \in Y$$
$$B(u_2, v) = F(v) \quad \text{for all } v \in Y.$$

Using property 1 of bilinear forms stated in the appendix, Section A.1.3, we have

$$B(u_1 - u_2, v) = 0 \quad \text{for all } v \in Y.$$

Selecting $v = u_1 - u_2$ we have $B(u_1 - u_2, u_1 - u_2) \equiv 2\|u_1 - u_2\|_{E(I)}^2 = 0$. That is, $u_1 = u_2$ in energy space. Observe that when $c = 0$ and $u_1 = u_2 + C$ where C is an arbitrary constant, then $\|u_1 - u_2\|_{E(I)} = 0$.

---

7 Victor Gustave Robin (1855–1897).

Summary of the main points

The exact solution of the generalized formulation u_{EX} is called the generalized solution or weak solution whereas the solution that satisfies equation (1.5) is called the strong solution. The generalized formulation has the following important properties:

1. The exact solution, denoted by u_{EX}, exists for all data that satisfy the conditions $0 < \alpha \leq \kappa(x) \leq \beta < \infty$ where α and β are real numbers, $0 \leq c(x) < \infty$ and f is such that $F(v)$ satisfies the definitive properties of linear forms listed in Section A.1.2 for all $v \in E(I)$. Note that κ, c and f can be discontinuous functions.
2. The exact solution is unique in the energy space, see Theorem 1.1.
3. If the data are sufficiently smooth for the strong solution to exist then the strong and weak solutions are the same.
4. This formulation makes it possible to find approximations to u_{EX} with arbitrary accuracy. This will be addressed in detail in subsequent sections.

Exercise 1.2 Assume that $u(0) = \hat{u}_0$ and $(\kappa u')_{x=\ell} = k_\ell(\delta_\ell - u(\ell))$ are given. State the generalized formulation.

Exercise 1.3 Consider the sequence of functions $u_n(x) \in E(I)$

$$u_n(x) = \begin{cases} -x + (2\ell/n + b) & \text{for} \quad 0 \leq x \leq \ell/n \\ x + b & \text{for} \quad \ell/n < x \leq \ell \end{cases}$$

illustrated in Fig. 1.2. Show that $u_n(x)$ converges to $u(x) = x + b$ in the space $E(I)$ as $n \to \infty$. For the definition of convergence refer to Section A.2 in the appendix.

This exercise illustrates that restriction imposed on u' (or higher derivatives of u) at the boundaries will not impose a restriction on $E(I)$. Therefore natural boundary conditions cannot be enforced by restriction. Whereas all functions in $E(I)$ are continuous and bounded, the derivatives do not have to be continuous or bounded.

Exercise 1.4 Show that $F(v)$ defined on $E(I)$ by eq. (1.20) satisfies the properties of linear forms listed in Section A.1.2 if f is square integrable on I. This is a sufficient but not necessary condition for $F(v)$ to be a linear form.

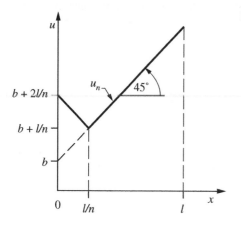

Figure 1.2 Exercise 1.3: The function $u_n(x)$.

Remark 1.1 $F(v)$ defined on $E(I)$ by eq. (1.20) satisfies the properties of linear forms listed in Section A.1.2 if the following inequality is satisfied:

$$\int_0^\ell fv \, dx < \infty \quad \text{for all } v \in E(I).$$ (1.35)

1.2.2 The principle of minimum potential energy

Theorem 1.2 The function $u \in \tilde{E}(I)$ that satisfies $B(u, v) = F(v)$ for all $v \in E^0(I)$ minimizes the quadratic functional[8] $\pi(u)$, called the potential energy;

$$\pi(u) \stackrel{\text{def}}{=} \frac{1}{2}B(u, u) - F(u)$$ (1.36)

on the space $\tilde{E}(I)$.

Proof: For any $v \in E^0(I)$, $\|v\|_E \neq 0$ we have:

$$\pi(u + v) = \frac{1}{2}B(u + v, u + v) - F(u + v)$$

$$= \frac{1}{2}B(u, u) + B(u, v) + \frac{1}{2}B(v, v) - F(u) - F(v)$$

$$= \pi(u) + \underbrace{B(u, v) - F(v)}_{0} + \frac{1}{2}B(v, v)$$ (1.37)

where $B(v, v) > 0$ unless $\|v\|_{E(I)} = 0$. Therefore any admissible nonzero perturbation of u will increase $\pi(u)$.

This important theorem, called the theorem or principle of minimum potential energy, will be used in Chapter 7 as our starting point in the formulation of mathematical models for beams, plates and shells.

Given the potential energy and the space of admissible functions, it is possible to determine the strong form. This is illustrated by the following example.

Example 1.2 Let us determine the strong form corresponding to the potential energy defined by

$$\pi(u) = \frac{1}{2}\int_0^\ell \left(\kappa(u')^2 + cu^2\right) \, dx + \frac{1}{2}k_0 u^2(0) - \int_0^\ell fu \, dx - k_0\delta_0 u(0)$$ (1.38)

with $\tilde{E}(I) = \{u \mid u \in E(I), \; u(\ell) = \hat{u}_\ell\}$.

Since u minimizes $\pi(u)$, any perturbation of u by $v \in E^0(I)$ will increase $\pi(u)$. Therefore $\pi(u + \epsilon v)$ is minimum at $\epsilon = 0$ and hence

$$\left.\frac{d\pi(u + \epsilon v)}{d\epsilon}\right|_{\epsilon=0} = 0.$$ (1.39)

Therefore we have

$$\int_0^\ell \left(\kappa u'v' + cuv\right) \, dx - \int_0^\ell fv \, dx + \underbrace{k_0 u(0)v(0) - k_0\delta_0 v(0)}_{0} = 0$$ (1.40)

where the last two terms are zero because $v \in E^0(I)$. Integrating the first term by parts,

$$\int_0^\ell \kappa u'v' \, dx = \underbrace{\kappa u'(\ell)v(\ell) - \kappa u'(0)v(0)}_{0} - \int_0^\ell (\kappa u')'v \, dx$$

---

8 A functional is a real-valued function defined on a space of functions or vectors.

and, substituting this into eq. (1.40), we get

$$\int_0^\ell \left(-(\kappa u')' + cu - f \right) v \, dx = 0. \tag{1.41}$$

Since this holds for all $v \in E^0(I)$, the bracketed expression must be zero. In other words, the solution of the differential equation

$$-(\kappa u')' + cu = f, \quad (\kappa u')_{x=0} = k_0(u(0) - \delta_0), \quad u(\ell) = \hat{u}_\ell \tag{1.42}$$

minimizes the potential energy defined by eq. (1.38). This is the strong form of the problem.

Remark 1.2 The procedure in Example 1.2 is used in the calculus of variations for identifying the differential equation, known as the Euler[9]-Lagrange[10] equation, the solution of which maximizes or minimizes a functional. In this example the solution minimizes the potential energy on the space $\tilde{E}(I)$.

Remark 1.3 Whereas the strain energy is always positive, the potential energy may be positive, negative or zero.

1.3 Approximate solutions

The trial and test spaces defined in the preceding section are infinite-dimensional, that is, they span infinitely many linearly independent functions. To find an approximate solution, we construct finite-dimensional subspaces denoted, respectively, by $S \subset X$, $V \subset Y$ and seek the function $u \in S$ that satisfies $B(u, v) = F(v)$ for all $v \in V$. Let us return to the introductory example described in Section 1.1 and define

$$u = u_n = \sum_{j=1}^n a_j \varphi_j, \quad v = v_n = \sum_{i=1}^n b_i \varphi_i$$

where φ_i ($i = 1, 2, \ldots n$) are basis functions. Using the definitions of k_{ij} and m_{ij} given in eq. (1.12), we write the bilinear form as

$$B(u, v) \equiv \int_0^\ell (\kappa u' v' + cuv) \, dx = \sum_{i=1}^n \sum_{j=1}^n (k_{ij} + m_{ij}) a_j b_i$$

$$= \{b\}^T ([K] + [M]) \{a\}. \tag{1.43}$$

Similarly,

$$F(v) \equiv \int_0^\ell fv \, dx = \sum_{i=1}^n b_i r_i = \{b\}^T \{r\} \tag{1.44}$$

where r_i is defined in eq. (1.12). Therefore we can write $B(u, v) - F(v) = 0$ in the following form:

$$\{b\}^T (([K] + [M])\{a\} - \{r\}) = 0. \tag{1.45}$$

Since this must hold for any choice of $\{b\}$, it follows that

$$([K] + [M])\{a\} = \{r\} \tag{1.46}$$

---

9 Leonhard Euler 1707–1783.
10 Joseph-Louis Lagrange 1736–1813.

which is the same system of linear equations we needed to solve when minimizing the integral $\mathcal{I}$, see eq. (1.14). Of course, this is not a coincidence. The solution of the generalized problem: "Find $u_n \in S$ such that $B(u_n, v) = F(v)$ for all $v \in V$", minimizes the error in the energy norm. See Theorem 1.4.

Theorem 1.3 The error e defined by $e = u - u_n$ satisfies $B(e, v) = 0$ for all $v \in S^0(I)$. This result follows directly from

$$B(u, v) = F(v) \quad \text{for all } v \in S^0(I)$$

$$B(u_n, v) = F(v) \quad \text{for all } v \in S^0(I).$$

Subtracting the second equation from the first we have,

$$B(u - u_n, v) \equiv B(e, v) = 0 \quad \text{for all } v \in S^0(I). \tag{1.47}$$

This equation is known as the Galerkin[11] orthogonality condition.

Theorem 1.4 If $u_n \in S^0(I)$ satisfies $B(u_n, v) = F(v)$ for all $v \in S^0(I)$ then u_n minimizes the error $u_{EX} - u_n$ in energy norm where u_{EX} is the exact solution:

$$\left\| u_{EX} - u_n \right\|_{E(I)} = \min_{u \in \tilde{S}} \left\| u_{EX} - u \right\|_{E(I)}. \tag{1.48}$$

Proof: Let $e = u - u_n$ and let v be an arbitrary function in $S^0(I)$. Then

$$\left\| e + v \right\|_{E(I)}^2 \equiv \frac{1}{2} B(e + v, e + v) = \frac{1}{2} B(e, e) + B(e, v) + \frac{1}{2} B(v, v).$$

The first term on the right is $\left\| e \right\|_{E(I)}^2$, the second term is zero on account of Theorem 1.3, the third term is positive for any $v \neq 0$ in $S^0(I)$. Therefore $\left\| e \right\|_{E(I)}$ is minimum.

Theorem 1.4 states that the error depends on the exact solution of the problem u_{EX} and the definition of the trial space $\tilde{S}(I)$.

The finite element method is a flexible and powerful method for constructing trial spaces. The basic algorithmic structure of the finite element method is outlined in the following sections.

1.3.1 The standard polynomial space

The standard polynomial space of degree p, denoted by $S^p(I_{st})$, is spanned by the monomials $1, \xi, \xi^2, \ldots, \xi^p$ defined on the standard element

$$I_{st} = \{\xi \mid -1 < \xi < 1\}. \tag{1.49}$$

The choice of basis functions is guided by considerations of implementation, keeping the condition number of the coefficient matrices small, and personal preferences. For the symmetric positive-definite matrices considered here the condition number C is the largest eigenvalue divided by the smallest. The number of digits lost in solving a linear problem is roughly equal to $\log_{10} C$. Characterizing the condition number as being large or small should be understood in this context. In the finite element method the condition number depends on the choice of the basis functions and the mesh.

The standard polynomial basis functions, called shape functions, can be defined in various ways. We will consider shape functions based on Lagrange polynomials and Legendre[12] polynomials. We will use the same notation for both types of shape function.

---

11 Boris Grigoryevich Galerkin 1871–1945.
12 Adrien-Marie Legendre 1752–1833.

Lagrange shape functions

Lagrange shape functions of degree p are constructed by partitioning I_{st} into p sub-intervals. The length of the sub-intervals is typically $2/p$ but the lengths may vary. The node points are $\xi_1 = -1$, $\xi_2 = 1$ and $-1 < \xi_3 < \xi_4 < \cdots < \xi_{p+1} < 1$. The ith shape function is unity in the ith node point and is zero in the other node points:

$$N_i(\xi) = \prod_{\substack{k=1 \\ k \neq i}}^{p+1} \frac{\xi - \xi_k}{\xi_i - \xi_k}, \quad i = 1, 2, \ldots, p+1, \quad \xi \in I_{st}. \tag{1.50}$$

These shape functions have the following important properties:

$$N_i(\xi_j) = \begin{cases} 1 & \text{if } i = j \\ 0 & \text{if } i \neq j \end{cases} \quad \text{and} \quad \sum_{i=1}^{p+1} N_i(\xi) = 1. \tag{1.51}$$

For example, for $p = 2$ the equally spaced node points are $\xi_1 = -1, \xi_2 = 1, \xi_3 = 0$. The corresponding Lagrange shape functions are illustrated in Fig. 1.3.

Exercise 1.5   Sketch the Lagrange shape functions for $p = 3$.

Legendre shape functions

For $p = 1$ we have

$$N_1 = \frac{1 - \xi}{2}, \quad N_2 = \frac{1 + \xi}{2}. \tag{1.52}$$

For $p \geq 2$ we define the shape functions as follows:

$$N_i(\xi) = \sqrt{\frac{2i - 3}{2}} \int_{-1}^{\xi} P_{i-2}(t)\, dt \quad i = 3, 4, \ldots, p+1 \tag{1.53}$$

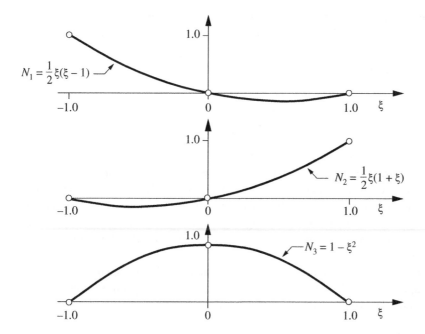

Figure 1.3   Lagrange shape functions in one dimension, $p = 2$.

where $P_i(t)$ are the Legendre polynomials. The definition of Legendre polynomials is given in Appendix D. These shape functions have the following important properties:

1. Orthogonality. For $i, j \geq 3$:

$$\int_{-1}^{+1} \frac{dN_i}{d\xi} \frac{dN_j}{d\xi} \, d\xi = \begin{cases} 1 & \text{if } i = j \\ 0 & \text{if } i \neq j. \end{cases} \tag{1.54}$$

   This property follows directly from the orthogonality of Legendre polynomials, see eq. (D.13) in the appendix.
2. The set of shape functions of degree p is a subset of the set of shape functions of degree $p + 1$. Shape functions that have this property are called hierarchic shape functions.
3. These shape functions vanish at the endpoints of I_{st}: $N_i(-1) = N_i(+1) = 0$ for $i \geq 3$.

The first five hierarchic shape functions are shown in Fig. 1.4. Observe that all roots lie in I_{st}. Additional shape functions, up to $p = 8$, can be found in the appendix, Section D.1.

Exercise 1.6 Show that for the hierarchic shape functions, defined by eq. (1.53), $N_i(-1) = N_i(+1) = 0$ for $i \geq 3$.

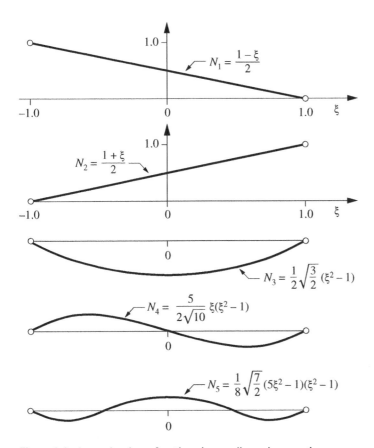

Figure 1.4 Legendre shape functions in one dimension, $p = 4$.

Exercise 1.7 Show that the hierarchic shape functions defined by eq. (1.53) can be written in the form:

$$N_i(\xi) = \frac{1}{\sqrt{2(2i-3)}} \left(P_{i-1}(\xi) - P_{i-3}(\xi) \right) \qquad i = 3, 4, \dots \tag{1.55}$$

Hint: note that $P_n(1) = 1$ for all n and use equations (D.10) and (D.12) in Appendix D.

1.3.2 Finite element spaces in one dimension

We are now in a position to provide a precise definition of finite element spaces in one dimension.

The domain $I = \{x \mid 0 < x < \ell\}$ is partitioned into M non-overlapping intervals called finite elements. A partition, called finite element mesh, is denoted by Δ. Thus $M = M(\Delta)$. The boundary points of the elements are the node points. The coordinates of the node points, sorted in ascending order, are denoted by x_i, $(i = 1, 2, \dots, M+1)$ where $x_1 = 0$ and $x_{M+1} = \ell$. The kth element I_k has the boundary points x_k and x_{k+1}, that is, $I_k = \{x \mid x_k < x < x_{k+1}\}$.

Various approaches are used for the construction of sequences of finite element mesh. We will consider four types of mesh design:

1. A mesh is uniform if all elements have the same size. On the interval $I = (0, \ell)$ the node points are located as follows:

$$x_k = (k-1)\ell / M(\Delta) \quad \text{for } k = 1, 2, 3, \dots, M(\Delta) + 1.$$

2. A sequence of meshes Δ_K ($K = 1, 2, \dots$) is quasiuniform if there exist positive constants C_1, C_2, independent of K, such that

$$C_1 \le \frac{\ell_{\max}^{(K)}}{\ell_{\min}^{(K)}} \le C_2, \quad K = 1, 2, \dots \tag{1.56}$$

where $\ell_{\max}^{(K)}$ (resp. $\ell_{\min}^{(K)}$) is the length of the largest (resp. smallest) element in mesh Δ_K. In two and three dimensions ℓ_k is defined as the diameter of the kth element, meaning the diameter of the smallest circle or sphere that envelopes the element. For example, a sequence of quasi-uniform meshes would be generated in one dimension if, starting from an arbitrary mesh, the elements would be successively halved.

3. A mesh is geometrically graded toward the point $x = 0$ on the interval $0 < x < \ell$ if the node points are located as follows:

$$x_k = \begin{cases} 0 & \text{for } k = 1 \\ q^{M(\Delta)+1-k}\ell & \text{for } k = 2, 3, \dots, M(\Delta) + 1 \end{cases} \tag{1.57}$$

where $0 < q < 1$ is called grading factor or common factor. These are called geometric meshes.

4. A mesh is a radical mesh if on the interval $0 < x < \ell$ the node points are located by

$$x_k = \left(\frac{k-1}{M(\Delta)} \right)^{\theta} \ell, \quad \theta > 1, \quad k = 1, 2, \dots, M(\Delta) + 1. \tag{1.58}$$

The question of which of these schemes is to be preferred in a particular application can be answered on the basis of a priori information concerning the regularity of the exact solution and aspects of implementation. Practical considerations that should guide the choice of the finite element mesh will be discussed in Section 1.5.2.

When the exact solution has one or more terms like $|x - x_0|^\alpha$, and $\alpha > 1/2$ is a fractional number, then the ideal mesh is a geometrically graded mesh and the polynomial degrees are assigned in such a way that the smallest elements are assigned the lowest polynomial degree, the largest elements the highest. The optimal grading factor is $q = (\sqrt{2} - 1)^2 \approx 0.17$ which is independent of α. The assigned polynomial degrees should increase at a rate of approximately 0.4 [45].

The ideal meshes are radical meshes when the same polynomial degree is assigned to each element. The optimal value of θ depends on p and α:

$$\theta = \frac{p + 1/2}{\alpha - 1/2 + (n-1)/2} \tag{1.59}$$

where n is the number of spatial dimensions. For a detailed analysis of discretization schemes in one dimension see reference [45].

The relationship between the kth element of the mesh and the standard element I_{st} is defined by the mapping function

$$x = Q_k(\xi) = \frac{1 - \xi}{2} x_k + \frac{1 + \xi}{2} x_{k+1}, \quad \xi \in I_{st}. \tag{1.60}$$

A finite element space S is a set of functions characterized by Δ, the assigned polynomial degrees $p_k \geq 1$ and the mapping functions $Q_k(\xi)$, $k = 1, 2, \ldots, M(\Delta)$. Specifically;

$$S = S(I, \Delta, \mathbf{p}, \mathbf{Q}) = \{u \mid u \in E(I), \ u(Q_k(\xi)) \in S^{p_k}(I_{st}), \ k = 1, 2, \ldots, M(\Delta)\} \tag{1.61}$$

where $\mathbf{p}$ and $\mathbf{Q}$ represent, respectively, the arrays of the assigned polynomial degrees and the mapping functions. This should be understood to mean that $u \in S$ if and only if u satisfies the conditions on the right of the vertical bar (|). The first condition $u \in E(I)$ is that u must lie in the energy space. In one dimension this implies that u must be continuous on I. The expression $u(Q_k(\xi)) \in S^{p_k}(I_{st})$ indicates that on element I_k the function $u(x)$ is mapped from the standard polynomial space $S^{p_k}(I_{st})$.

The finite element test space, denoted by $S^0(I)$, is defined by the intersection $S^0(I) = S(I) \cap E(I)$, that is, $u \in S^0(I)$ is zero in those boundary points where essential boundary conditions are prescribed. The number of basis functions that span $S^0(I)$ is called the number of degrees of freedom.

The process by which the number of degrees of freedom is progressively increased by mesh refinement, with the polynomial degree fixed, is called h-extension and its implementation the h-version of the finite element method. The process by which the number of degrees of freedom is progressively increased by increasing the polynomial degree of elements, while keeping the mesh fixed, is called p-extension and its implementation the p-version of the finite element method. The process by which the number of degrees of freedom is progressively increased by concurrently refining the mesh and increasing the polynomial degrees of elements is called hp-extension and its implementation the hp-version of the finite element method.

Remark 1.4 It will be explained in Chapter 5 that the separate naming of the h, p and hp versions is related to the evolution of the finite element method rather than its theoretical foundations.

1.3.3 Computation of the coefficient matrices

The coefficient matrices are computed element by element. The numbering of the coefficients is based on the numbering of the standard shape functions, the indices range from 1 through p_{k+1}. This numbering will have to be reconciled with the requirement that each basis function must be continuous on I and must have an unique identifying number. This will be discussed separately.

Computation of the stiffness matrix

The first term of the bilinear form in eq. (1.43) is computed as a sum of integrals over the elements

$$\int_0^{\ell} \kappa(x) u_n' v_n' \, dx = \sum_{k=1}^{M(\Delta)} \int_{x_k}^{x_{k+1}} \kappa(x) u_n' v_n' \, dx. \tag{1.62}$$

We will be concerned with the evaluation of the integral on the kth element:

$$\int_{x_k}^{x_{k+1}} \kappa(x) u_n' v_n' \, dx = \int_{x_k}^{x_{k+1}} \kappa(x) \left(\sum_{j=1}^{p_k+1} a_j \frac{dN_j}{dx} \right) \left(\sum_{i=1}^{p_k+1} b_i \frac{dN_i}{dx} \right) dx.$$

The shape functions N_i are defined on the standard domain I_{st}. Referring to the mapping function given by eq. (1.60), we have

$$dx = \frac{x_{k+1} - x_k}{2} d\xi \equiv \frac{\ell_k}{2} \, d\xi \tag{1.63}$$

where $\ell_k \stackrel{\text{def}}{=} x_{k+1} - x_k$ is the length of the kth element. Also,

$$\frac{d}{dx} = \frac{d}{d\xi} \frac{d\xi}{dx} = \frac{2}{x_{k+1} - x_k} \frac{d}{d\xi} \equiv \frac{2}{\ell_k} \frac{d}{d\xi}.$$

Therefore

$$\int_{x_k}^{x_{k+1}} \kappa(x) u_n' v_n' \, dx = \frac{2}{\ell_k} \int_{-1}^{+1} \kappa(Q_k(\xi)) \left(\sum_{j=1}^{p_k+1} a_j \frac{dN_j}{d\xi} \right) \left(\sum_{i=1}^{p_k+1} b_i \frac{dN_i}{d\xi} \right) d\xi.$$

We define

$$k_{ij}^{(k)} = \frac{2}{\ell_k} \int_{-1}^{+1} \kappa(Q_k(\xi)) \frac{dN_i}{d\xi} \frac{dN_j}{d\xi} \, d\xi \tag{1.64}$$

and write

$$\int_{x_k}^{x_{k+1}} \kappa(x) u_n' v_n' \, dx = \sum_{i=1}^{p_k+1} \sum_{j=1}^{p_k+1} k_{ij}^{(k)} a_j b_i \equiv \{b\}^T [K^{(k)}] \{a\}. \tag{1.65}$$

The terms of the stiffness matrix $k_{ij}^{(k)}$ depend on the the mapping, the definition of the shape functions and the function $\kappa(x)$. The matrix $[K^{(k)}]$ is called the element stiffness matrix. Observe that $k_{ij}^{(k)} = k_{ji}^{(k)}$, that is, $[K^{(k)}]$ is symmetric. This follows directly from the symmetry of $B(u, v)$ and the fact that the same basis functions are used for u_n and v_n.

In the finite element method the integrals are evaluated by numerical methods. Numerical integration is discussed in Appendix E. In the important special case when $\kappa(x) = \kappa_k$ is constant on I_k, it is possible to compute $[K^{(k)}]$ once and for all. This is illustrated by the following example.

Example 1.3 When $\kappa(x) = \kappa_k$ is constant on I_k and the Legendre shape functions are used then, with the exception of the first two rows and columns, the element stiffness matrix is perfectly diagonal:

$$[K^{(k)}] = \frac{2\kappa_k}{\ell_k}
\begin{bmatrix}
1/2 & 1/2 & 0 & 0 & \cdots & 0 \\
 & 1/2 & 0 & 0 & & 0 \\
 & & 1 & 0 & & 0 \\
 & & & 1 & & 0 \\
 & \text{(sym.)} & & & \ddots & \vdots \\
 & & & & & 1
\end{bmatrix}. \tag{1.66}$$

Exercise 1.8 Assume that $\kappa(x) = \kappa_k$ is constant on I_k. Using the Lagrange shape functions displayed in Fig. 1.3 for $p = 2$, compute $k_{11}^{(k)}$ and $k_{13}^{(k)}$ in terms of κ_k and ℓ_k.

Computation of the Gram matrix

The second term of the bilinear form is also computed as a sum of integrals over the elements:

$$\int_0^\ell c(x) u_n v_n \, dx = \sum_{k=1}^{M(\Delta)} \int_{x_k}^{x_{k+1}} c(x) u_n v_n \, dx. \tag{1.67}$$

We will be concerned with evaluation of the integral

$$\int_{x_k}^{x_{k+1}} c(x) u_n v_n \, dx = \int_{x_k}^{x_{k+1}} c(x) \left(\sum_{j=1}^{p_k+1} a_j N_j \right) \left(\sum_{i=1}^{p_k+1} b_i N_i \right) \, dx$$

$$= \frac{\ell_k}{2} \int_{-1}^{+1} c(Q_k(\xi)) \left(\sum_{j=1}^{p_k+1} a_j N_j \right) \left(\sum_{i=1}^{p_k+1} b_i N_i \right) \, d\xi.$$

Defining:

$$m_{ij}^{(k)} = \frac{\ell_k}{2} \int_{-1}^{1} c(Q_k(\xi)) N_i N_j \, d\xi \tag{1.68}$$

the following expression is obtained:

$$\int_{x_k}^{x_{k+1}} c(x) u_n v_n \, dx = \sum_{i=1}^{p_k+1} \sum_{j=1}^{p_k+1} m_{ij}^{(k)} a_j b_i = \{b\}^T [M^{(k)}] \{a\} \tag{1.69}$$

where $\{a\} = \{a_1 \ a_2 \ \ldots \ a_{p_k+1}\}^T$, $\{b\}^T = \{b_1 \ b_2 \ \ldots \ b_{p_k+1}\}$ and

$$[M^{(k)}] = \begin{bmatrix} m_{11}^{(k)} & m_{12}^{(k)} & \cdots & m_{1,p_k+1} \\ m_{21}^{(k)} & m_{22}^{(k)} & \cdots & m_{2,p_k+1} \\ \vdots & & \ddots & \vdots \\ m_{p_k+1,1}^{(k)} & m_{p_k+1,2}^{(k)} & \cdots & m_{p_k+1,p_k+1} \end{bmatrix}.$$

The terms of the coefficient matrix $m_{ij}^{(k)}$ are computable from the mapping, the definition of the shape functions and the function $c(x)$. The matrix $[M^{(k)}]$ is called the element-level Gram matrix[13] or the element-level mass matrix. Observe that $[M^{(k)}]$ is symmetric. In the important special case where $c(x) = c_k$ is constant on I_k it is possible to compute $[M^{(k)}]$ once and for all. This is illustrated by the following example.

Example 1.4   When $c(x) = c_k$ is constant on I_k and the Legendre shape functions are used then the element-level Gram matrix is strongly diagonal. For example, for $p_k = 5$ the Gram matrix is:

$$[M^{(k)}] = \frac{c_k \ell_k}{2} \begin{bmatrix} 2/3 & 1/3 & -1/\sqrt{6} & 1/3\sqrt{10} & 0 & 0 \\ & 2/3 & -1/\sqrt{6} & -1/3\sqrt{10} & 0 & 0 \\ & & 2/5 & 0 & -1/5\sqrt{21} & 0 \\ & & & 2/21 & 0 & -1/7\sqrt{45} \\ & \text{(sym.)} & & & 2/45 & 0 \\ & & & & & 2/77 \end{bmatrix} \tag{1.70}$$

---

13 Jörgen Pedersen Gram 1850–1916.

Remark 1.5 For $p_k \geq 2$ a simple closed form expression can be obtained for the diagonal terms and the off-diagonal terms. Using eq. (1.55) it can be shown that:

$$
\begin{aligned}
m_{ii}^{(k)} &= \frac{c_k \ell_k}{2} \frac{1}{2(2i-3)} \int_{-1}^{+1} (P_{i-1}(\xi) - P_{i-3}(\xi))^2 \, d\xi \\
&= \frac{c_k \ell_k}{2} \frac{2}{(2i-1)(2i-5)}, \qquad i \geq 3
\end{aligned}
\tag{1.71}
$$

and all off-diagonal terms are zero for $i \geq 3$, with the exceptions:

$$
m_{i,i+2}^{(k)} = m_{i+2,i}^{(k)} = -\frac{c_k \ell_k}{2} \frac{1}{(2i-1)\sqrt{(2i-3)(2i+1)}}, \qquad i \geq 3.
\tag{1.72}
$$

Remark 1.6 It has been proposed to make the Gram matrix perfectly diagonal by using Lagrange shape functions of degree p with the node points coincident with the Lobatto points. Therefore $N_i(\xi_j) = \delta_{ij}$ where δ_{ij} is the Kronecker delta[14]. Then, using $p+1$ Lobatto points, we get:

$$
m_{ij}^{(k)} = \frac{c_k \ell_k}{2} \int_{-1}^{1} N_i N_j \, d\xi \approx \frac{c_k \ell_k}{2} w_i \delta_{ij}
$$

where w_i is the weight of the ith Lobatto point. There is an integration error associated with this term because the integrand is a polynomial of degree $2p$. To evaluate this integral exactly $n \geq (2p+3)/2$ Lobatto points would be required (see Appendix E), whereas only $p+1$ Lobatto points are used. Throughout this book we will be concerned with errors of approximation that can be controlled by the design of mesh and the assignment of polynomial degrees. We will assume that the errors of integration and errors in mapping are negligibly small in comparison with the errors of discretization.

Exercise 1.9 Assume that $c(x) = c_k$ is constant on I_k. Using the Lagrange shape functions of degree $p = 3$, with the nodes located in the Lobatto points, compute $m_{33}^{(k)}$ numerically using 4 Lobatto points. Determine the relative error of the numerically integrated term. Refer to Remark 1.6 and Appendix E.

Exercise 1.10 Assume that $c(x) = c_k$ is constant on I_k. Using the Lagrange shape functions of degree $p = 2$, compute $m_{11}^{(k)}$ and $m_{13}^{(k)}$ in terms of c_k and ℓ_k.

1.3.4 Computation of the right hand side vector

Computation of the right hand side vector involves evaluation of the functional $F(v)$, usually by numerical means. In particular, we write:

$$
F(v_n) = \int_0^\ell f(x) v_n \, dx = \sum_{k=1}^{M(\Delta)} \int_{x_k}^{x_{k+1}} f(x) v_n \, dx.
\tag{1.73}
$$

The element-level integral is computed from the definition of v_n on I_k:

$$
\int_{x_k}^{x_{k+1}} f(x) v_n \, dx = \frac{\ell_k}{2} \int_{-1}^{+1} f(Q_k(\xi)) \left(\sum_{i=1}^{p_{k+1}} b_i^{(k)} N_i \right) d\xi = \sum_{i=1}^{p_{k+1}} b_i^{(k)} r_i^{(k)}
\tag{1.74}
$$

---

14 The definition of δ_{ij} is given by eq. (2.1).

where

$$r_i^{(k)} \stackrel{\text{def}}{=} \frac{\ell_k}{2} \int_{-1}^{+1} f(Q_k(\xi)) N_i(\xi) \, d\xi \tag{1.75}$$

which is computed from the given data and the shape functions.

Example 1.5 Let us assume that $f(x)$ is a linear function on I_k. In this case $f(x)$ can be written as

$$f(x) = \frac{1-\xi}{2} f(x_k) + \frac{1+\xi}{2} f(x_{k+1}) = f(x_k) N_1(\xi) + f(x_{k+1}) N_2(\xi).$$

Using the Legendre shape functions we have:

$$r_1^{(k)} = f(x_k) \frac{\ell_k}{2} \int_{-1}^{+1} N_1^2 \, d\xi + f(x_{k+1}) \frac{\ell_k}{2} \int_{-1}^{+1} N_1 N_2 \, d\xi = \frac{\ell_k}{6} \left(2f(x_k) + f(x_{k+1}) \right)$$

$$r_2^{(k)} = f(x_k) \frac{\ell_k}{2} \int_{-1}^{+1} N_1 N_2 \, d\xi + f(x_{k+1}) \frac{\ell_k}{2} \int_{-1}^{+1} N_2^2 \, d\xi = \frac{\ell_k}{6} \left(f(x_k) + 2f(x_{k+1}) \right)$$

$$r_3^{(k)} = f(x_k) \frac{\ell_k}{2} \int_{-1}^{+1} N_1 N_3 \, d\xi + f(x_{k+1}) \frac{\ell_k}{2} \int_{-1}^{+1} N_2 N_3 \, d\xi$$

$$= -\frac{\ell_k}{6} \sqrt{\frac{3}{2}} \left(f(x_k) + f(x_{k+1}) \right).$$

Exercise 1.11 Assume that $f(x)$ is a linear function on I_k. Using the Legendre shape functions compute $r_4^{(k)}$ and show that $r_i^{(k)} = 0$ for $i > 4$. Hint: Make use of eq. (1.55).

Exercise 1.12 Let

$$f(x) = f_k \sin \frac{x - x_k}{\ell_k} \pi, \quad x \in I_k$$

where f_k is a constant. Compute $r_5^{(k)}$ numerically in terms of f_k and ℓ_k using 3, 4 and 5 Gauss points. See Appendix E. Use the Legendre basis functions.

Exercise 1.13 Assume that $f(x)$ is a linear function on I. Using the Lagrange shape functions for $p = 2$, compute $r_1^{(k)}$.

1.3.5 Assembly

Having computed the coefficient matrices and right hand side vectors for each element, it is necessary to form the coefficient matrix and right hand side vector for the entire mesh. This process, called assembly, executes the summation in equations (1.62), (1.67) and (1.73). The local and global numbering of variables is reconciled in the assembly process. The algorithm is illustrated by the following example.

Example 1.6 Consider the three-element mesh shown in Fig. 1.5. The polynomial degrees $p_1 = 2$, $p_2 = 1$, $p_3 = 3$ are assigned to elements 1, 2, 3 respectively. The basis functions shown in Fig. 1.5 are composed of the mapped Legendre shape functions. For instance, the basis function $\varphi_2(x)$ is composed of the mapped shape function N_2 from element 1 and the mapped shape function N_1 from element 2. This basis function is zero over element 3. Basis function $\varphi_6(x)$ is the mapped shape function N_3 from element 3. This basis function is zero over elements 1 and 2.

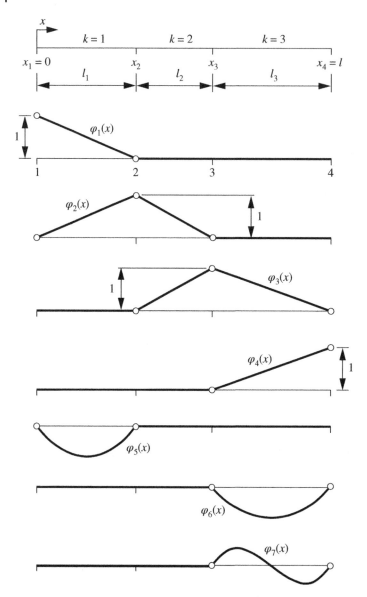

Figure 1.5 Typical finite element basis functions in one dimension.

Table 1.1 Local and global numbering in Example 1.6.

Numbering	Element number							
	1			**2**		**3**		
local	1	2	3	1	2	1	2	3	4
global	1	2	5	2	3	3	4	6	7

Each basis function is assigned a unique number, called a global number, and this number is associated with those element numbers and the shape function numbers from which the basis function is composed. The global and local numbers in this example are indicated in Table 1.1.

We denote $c_{ij}^{(k)} = k_{ij}^{(k)} + m_{ij}^{(k)}$ and, using equations (1.62) and (1.67), write $B(u_n, v_n)$ in the following form:

$$B(u_n, v_n) = \begin{matrix} & \begin{matrix} a_1 & a_2 & a_5 \end{matrix} \\ \begin{matrix} b_1 \\ b_2 \\ b_5 \end{matrix} & \begin{pmatrix} c_{11}^{(1)} & c_{12}^{(1)} & c_{13}^{(1)} \\ c_{21}^{(1)} & c_{22}^{(1)} & c_{23}^{(1)} \\ c_{31}^{(1)} & c_{32}^{(1)} & c_{33}^{(1)} \end{pmatrix} \end{matrix} + \begin{matrix} & \begin{matrix} a_2 & a_3 \end{matrix} \\ \begin{matrix} b_2 \\ b_3 \end{matrix} & \begin{pmatrix} c_{11}^{(2)} & c_{12}^{(2)} \\ c_{21}^{(2)} & c_{22}^{(2)} \end{pmatrix} \end{matrix}$$

$$+ \begin{matrix} & \begin{matrix} a_3 & a_4 & a_6 & a_7 \end{matrix} \\ \begin{matrix} b_3 \\ b_4 \\ b_6 \\ b_7 \end{matrix} & \begin{pmatrix} c_{11}^{(3)} & c_{12}^{(3)} & c_{13}^{(3)} & c_{14}^{(3)} \\ c_{21}^{(3)} & c_{22}^{(3)} & c_{23}^{(3)} & c_{24}^{(3)} \\ c_{31}^{(3)} & c_{32}^{(3)} & c_{33}^{(3)} & c_{34}^{(3)} \\ c_{41}^{(3)} & c_{42}^{(3)} & c_{43}^{(3)} & c_{44}^{(3)} \end{pmatrix} \end{matrix}$$

where the elements within the brackets are in the local numbering system whereas the coefficients a_j and b_i outside of the brackets are in the global system. The superscripts indicate the element numbers. The terms multiplied by $a_j b_i$ are summed to obtain the elements of the assembled coefficient matrix which will be denoted by c_{ij}. For example,

$$c_{11} = c_{11}^{(1)}, \quad c_{22} = c_{22}^{(1)} + c_{11}^{(2)}, \quad c_{33} = c_{22}^{(2)} + c_{11}^{(3)}, \quad c_{77} = c_{44}^{(3)}.$$

Assuming that the boundary conditions do not include Dirichlet conditions, the bilinear form can be written in terms of the 7×7 coefficient matrix as:

$$B(u_n, v_n) = \sum_{j=1}^{7} \sum_{i=1}^{7} c_{ij} a_j b_i = \{b_1 \ b_2 \ \cdots \ b_7\} \begin{bmatrix} c_{11} & c_{12} & \cdots & c_{17} \\ c_{21} & c_{22} & & c_{27} \\ \vdots & & & \vdots \\ c_{71} & c_{72} & \cdots & c_{77} \end{bmatrix} \begin{Bmatrix} a_1 \\ a_2 \\ \vdots \\ a_7 \end{Bmatrix}$$

$$\equiv \{b\}^T [C] \{a\}. \tag{1.76}$$

The treatment of Dirichlet conditions will be discussed separately in the next section.

The assembly of the right hand side vector from the element-level right hand side vectors is analogous to the procedure just described. Referring to eq. (1.73) we write $F(v_n)$ in the following form:

$$F(v_n) = \{b_1 \ b_2 \ b_5\} \begin{Bmatrix} r_1^{(1)} \\ r_2^{(1)} \\ r_3^{(1)} \end{Bmatrix} + \{b_2 \ b_3\} \begin{Bmatrix} r_1^{(2)} \\ r_2^{(2)} \end{Bmatrix} + \{b_3 \ b_4 \ b_6 \ b_7\} \begin{Bmatrix} r_1^{(3)} \\ r_2^{(3)} \\ r_3^{(3)} \\ r_4^{(3)} \end{Bmatrix}$$

$$= \{b_1 \ b_2 \ \cdots \ b_7\} \begin{Bmatrix} r_1 \\ r_2 \\ \vdots \\ r_7 \end{Bmatrix} \equiv \{b\}^T \{r\}$$

where $r_1 = r_1^{(1)}, r_2 = r_2^{(1)} + r_1^{(2)}, r_3 = r_2^{(2)} + r_1^{(3)}$, etc.

1.3.6 Condensation

Each element has $p - 1$ internal basis functions. Those elements of the coefficient matrix which are associated with the internal basis functions can be eliminated at the element level. This process is called condensation.

Let us partition the coefficient matrix and right hand side vector of a finite element with $p \geq 2$ such that

$$\begin{bmatrix} \mathbf{C}_{11} & \mathbf{C}_{12} \\ \mathbf{C}_{21} & \mathbf{C}_{22} \end{bmatrix} \begin{Bmatrix} \mathbf{a}_1 \\ \mathbf{a}_2 \end{Bmatrix} = \begin{Bmatrix} \mathbf{r}_1 \\ \mathbf{r}_2 \end{Bmatrix}$$

where the $\mathbf{a}_1 = \{a_1\ a_2\}^T$ and $\mathbf{a}_2 = \{a_3\ a_4\ \cdots\ a_{p+1}\}^T$. The coefficient matrix is symmetric therefore $\mathbf{C}_{21} = \mathbf{C}_{12}^T$. Using

$$\mathbf{a}_2 = -\mathbf{C}_{22}^{-1}\mathbf{C}_{21}\mathbf{a}_1 + \mathbf{C}_{22}^{-1}\mathbf{r}_2 \tag{1.77}$$

we get

$$\underbrace{\left(\mathbf{C}_{11} - \mathbf{C}_{12}\mathbf{C}_{22}^{-1}\mathbf{C}_{21}\right)\mathbf{a}_1}_{\text{Condensed}[C]} = \underbrace{\mathbf{r}_1 - \mathbf{C}_{12}\mathbf{C}_{22}^{-1}\mathbf{r}_2}_{\text{Condensed}\{r\}}. \tag{1.78}$$

The condensed stiffness matrices and load vectors are assembled and the Dirichlet boundary conditions are enforced as described in the following section. Upon solving the assembled system of equations the coefficients of the internal basis functions are computed from eq. (1.77) for each element.

1.3.7 Enforcement of Dirichlet boundary conditions

When Dirichlet conditions are specified on either or both boundary points then $u \in \tilde{S}(I)$ is split into two functions; a function $\bar{u} \in S^0(I)$ and an arbitrary specific function from $\tilde{S}(I)$, denoted by $u^\star$. We then seek $\bar{u} \in S^0(I)$ such that

$$\underbrace{\int_0^\ell (\kappa\bar{u}'v' + c\bar{u}v)\,dx}_{B(\bar{u},v)} = \underbrace{\int_0^\ell fv\,dx - \int_0^\ell (\kappa(u^\star)'v' + cu^\star v)\,dx}_{F(v)} \tag{1.79}$$

for all $v \in S^0(I)$. Observe that the solution $u = \bar{u} + u^\star$ is independent of the choice of $u^\star$.

We denote the global numbers of the basis functions that are unity at $x = 0$ and $x = \ell$ by K and L respectively. For instance, in Example 1.6 $K = 1$ and $L = 4$. It is advantageous to define $u^\star$ in terms of $\varphi_K(x)$ and $\varphi_L(x)$:

$$u^\star = \hat{u}_0\varphi_K(x) + \hat{u}_\ell\varphi_L(x) \tag{1.80}$$

as indicated in Fig. 1.6. When Dirichlet boundary condition is prescribed on only one of the boundary points then this expression is modified to include the term corresponding to that point only.

On substituting eq. (1.80) into eq. (1.79) the second term on the right-hand side of eq. (1.79) can be written as

$$\int_0^\ell (\kappa(u^\star)'v' + cu^\star v)\,dx = \sum_{i=1}^{N_u} b_i(c_{iK} + c_{iL})$$

where N_u is the number of unconstrained equations, that is, the number of equations prior to enforcement of the Dirichlet boundary conditions. (For instance, in Example 1.6 $N_u = 7$.) The coefficients c_{iK}, c_{iL} are elements of the assembled coefficient matrix.

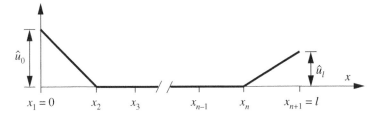

Figure 1.6 Recommended choice of the function $u^\star$ in one dimension.

Since $v \in S^0(I)$, we have $b_K = b_L = 0$ and therefore the Kth and Lth rows of matrix $[C]$ are multiplied by zero and can be deleted. The Kth and Lth columns of matrix $[C]$ are multiplied by $\hat{u}_0$ and $\hat{u}_\ell$ respectively, summed and the resulting vector is transferred to the right-hand side. The resulting coefficient matrix has the dimension N which is N_u minus the number of Dirichlet boundary conditions. The number N is called the number of degrees of freedom. It is the maximum number of linearly independent functions in $S^0(I)$.

Remark 1.7 In order to avoid having to renumber the coefficient matrix once the rows and columns corresponding to φ_K and φ_L were eliminated, all elements in the Kth and Lth rows and columns can be set to zero, with the exception of the diagonal elements, which are set to unity. The corresponding elements on the right hand side vector are set to $\hat{u}_0$ and $\hat{u}_\ell$. This is illustrated by the following example.

Example 1.7 Consider the problem

$$-u'' + 4u = 0, \quad u(0) = 1, \; u(1) = 2$$

the exact solution of which is

$$u = \frac{\exp(2) - 2}{\exp(2) - \exp(-2)} \exp(-2x) + \frac{2 - \exp(-2)}{\exp(2) - \exp(-2)} \exp(2x).$$

Using five elements of equal length on the interval $I = (0, 1)$ and $p = 1$ assigned to each element, find the finite element solution for this problem.

Referring to equations (1.66) and (1.70), the element-level coefficient matrix for each element is

$$\left[C^{(k)} \right] = \begin{bmatrix} 79/15 & -73/15 \\ -73/15 & 79/15 \end{bmatrix}, \quad k = 1, 2, \dots 5$$

where we used $\kappa_k = 1$, $c_k = 4$, $\ell_k = 1/5$. The assembled unconstrained coefficient matrix is:

$$[C] = \begin{bmatrix} 79/15 & -73/15 & 0 & 0 & 0 & 0 \\ -73/15 & 158/15 & -73/15 & 0 & 0 & 0 \\ 0 & -73/15 & 158/15 & -73/15 & 0 & 0 \\ 0 & 0 & -73/15 & 158/15 & -73/15 & 0 \\ 0 & 0 & 0 & -73/15 & 158/15 & -73/15 \\ 0 & 0 & 0 & 0 & -73/15 & 79/15 \end{bmatrix}.$$

Upon enforcement of the Dirichlet conditions the system of equations is

$$[C] = \begin{bmatrix} 158/15 & -73/15 & 0 & 0 \\ -73/15 & 158/15 & -73/15 & 0 \\ 0 & -73/15 & 158/15 & -73/15 \\ 0 & 0 & -73/15 & 158/15 \end{bmatrix} \begin{Bmatrix} a_2 \\ a_3 \\ a_4 \\ a_5 \end{Bmatrix} = \begin{Bmatrix} 73/13 \\ 0 \\ 0 \\ 146/15 \end{Bmatrix}$$

alternatively:

$$[C] = \begin{bmatrix} 1 & 0 & 0 & 0 & 0 & 0 \\ 0 & 158/15 & -73/15 & 0 & 0 & 0 \\ 0 & -73/15 & 158/15 & -73/15 & 0 & 0 \\ 0 & 0 & -73/15 & 158/15 & -73/15 & 0 \\ 0 & 0 & 0 & -73/15 & 158/15 & 0 \\ 0 & 0 & 0 & 0 & 0 & 1 \end{bmatrix} \begin{Bmatrix} a_1 \\ a_2 \\ a_3 \\ a_4 \\ a_5 \\ a_6 \end{Bmatrix} = \begin{Bmatrix} 1 \\ 73/15 \\ 0 \\ 0 \\ 146/15 \\ 2 \end{Bmatrix}$$

where the first and sixth equations are placeholders for the boundary conditions $a_1 = 1$, $a_6 = 2$. The solution is:

$$\{a\} = \{1.0000 \; 0.8784 \; 0.9012 \; 1.0722 \; 1.4194 \; 2.0000\}^T.$$

Exercise 1.14 Solve the problem in Example 1.7 with the boundary conditions $u(0) = 1$, $u'(1) = 3.6$.

Exercise 1.15 Solve the problem in Example 1.7 with the boundary conditions $u'(0) = -1$, $u(1) = 2$.

1.4 Post-solution operations

Following assembly of the coefficient matrix and enforcement of the essential boundary conditions (when applicable) the resulting system of simultaneous equations is solved by one of several methods designed to exploit the symmetry and sparsity of the coefficient matrix. The solvers are classified into two broad categories; direct and iterative solvers. Optimal choice of a solver in a particular application is based on consideration of the size of the problem and the available computational resources.

At the end of the solution process the finite element solution is available in the form

$$u_{FE} = \sum_{j=1}^{N_u} a_j \varphi_j(x) \tag{1.81}$$

where the indices reference the global numbering and N_u is the number of degrees of freedom plus the number of Dirichlet conditions.

The basis functions are decomposed into their constituent shape functions and the element-level solution records are created in the local numbering convention. Therefore the finite element solution on the kth element is available in the following form:

$$u_{FE}^{(k)} = \sum_{j=1}^{p_k+1} a_j^{(k)} N_j(\xi). \tag{1.82}$$

1.4.1 Computation of the quantities of interest

The computation of typical engineering quantities of interest (QoI) by direct and indirect methods is outlined in this section.

Computation of $u_{FE}(x_0)$
Direct computation of u_{FE} in the point $x = x_0$ involves a search to identify the element I_k in which point x_0 lies and, using the inverse of the mapping function defined by eq. (1.60), the standard

coordinate $\xi_0 \in I_{st}$ corresponding to x_0 is determined:

$$\xi_0 = Q_k^{-1}(x_0) = \frac{2x_0 - x_k - x_{k+1}}{x_{k+1} - x_k} \tag{1.83}$$

and $u_{FE}(x_0)$ is computed from

$$u_{FE}(x_0) = \sum_{j=1}^{p_k+1} a_j^{(k)} N_j(\xi_0). \tag{1.84}$$

Direct computation of $u'_{FE}(x_0)$
Direct computation of u'_{FE} in the point x_0 involves the computation of the corresponding standard coordinate $\xi_0 \in I_{st}$ using eq. (1.83) and evaluating the following expression:

$$\left(\frac{du_{FE}}{dx}\right)_{x=x_0} = \frac{2}{\ell_k}\left(\frac{du_{FE}}{d\xi}\right)_{\xi=\xi_0} = \frac{2}{\ell_k}\sum_{j=1}^{p_k+1} a_j^{(k)}\left(\frac{dN_j}{d\xi}\right)_{\xi=\xi_0} \tag{1.85}$$

where $\ell_k \overset{\text{def}}{=} x_{k+1} - x_k$. The computation of the higher derivatives is analogous.

Remark 1.8 When plotting quantities of interest such as the functions $u_{FE}(x)$ and $u'_{FE}(x)$, the data for the plotting routine are generated by subdividing the standard element into n intervals of equal length, n being the desired resolution. The QoIs corresponding to the grid-points are evaluated. This process does not involve inverse mapping. In node points information is provided from the two elements that share that node. If the computed QoI is discontinuous then the discontinuity will be visible at the nodes unless the plotting algorithm automatically averages the QoIs.

Indirect computation of $u'_{FE}(x_0)$ in node points
The first derivative in node points can be determined indirectly from the generalized formulation. For example, to compute the first derivative at node x_k from the finite element solution, we select $v = N_1(Q_k^{-1}(x))$ and use

$$\int_{x_k}^{x_{k+1}} \left(\kappa u'_{FE} v' + c u_{FE} v\right)\,dx = \int_{x_k}^{x_{k+1}} f v\,dx + \left[\kappa u'_{FE} v\right]_{x=x_{k+1}} - \left[\kappa u'_{FE} v\right]_{x=x_k}. \tag{1.86}$$

Test functions used in post-solution operations for the computation of a functional are called extraction functions. Here $v = N_1(Q_k^{-1}(x))$ is an extraction function for the functional $-\left[\kappa u'_{FE}\right]_{x=x_k}$. This is because $v(x_k) = 1$ and $v(x_{k+1}) = 0$ and hence

$$-\left[\kappa u'_{FE}\right]_{x=x_k} = \int_{x_k}^{x_{k+1}} \left(\kappa u'_{FE} v' + c u_{FE} v\right)\,dx - \int_{x_k}^{x_{k+1}} f v\,dx$$

$$= \sum_{j=1}^{p_k+1} c_{1j}^{(k)} a_j^{(k)} - r_1^{(k)} \tag{1.87}$$

where, by definition; $c_{ij}^{(k)} = k_{ij}^{(k)} + m_{ij}^{(k)}$.

Example 1.8 Let us find $u'_{FE}(1)$ for the problem in Example 1.7 by the direct and indirect methods. In this case the exact solution is known from which we have $u'_{EX}(1) = 3.5978$. By direct computation:

$$u'_{FE}(1) = \frac{2}{\ell_5}\left(\frac{du_{FE}}{d\xi}\right)_{\xi=1} = 5(a_6 - a_5) = 2.9028 \quad (19.32\% \text{ error})$$

and by indirect computation:

$$u'_{FE}(1) = -\frac{73}{15}a_5 + \frac{79}{15}a_6 = 3.6254 \quad (0.77\% \text{ error}).$$

Example 1.9 The following example illustrates that the indirect method can be used for obtaining the QoI efficiently and accurately even when the discretization was very poorly chosen. We will consider the problem

$$\int_0^\ell u'v' \, dx = \int_0^\ell \delta(x - \overline{x})v \, dx = v(\overline{x}), \quad u(0) = u(\ell) = 0$$

where δ is the delta function, see Definition A.5 in the appendix. Let us be interested in finding the approximate value of $u'(0)$. The data are $\ell = 1$ and $\overline{x} = 1/4$. We will use one finite element and $p = 2, 3, \dots$ This is a poorly chosen discretization because the derivatives of u are discontinuous in the point $x = \overline{x}$, whereas all derivatives of the shape functions are continuous. The proper discretization would have been to use two or more finite elements with a node point in $x = \overline{x}$. Then the exact solution would be obtained at $p = 1$.

If we use the Legendre shape functions then the coefficient matrix displayed in eq. (1.66) will be perfectly diagonal. The first two rows and columns will be zero on account of the boundary conditions and the diagonal term will be 2. Referring to eq. (1.75) the right hand side vector will be

$$r_i = N_i(\overline{\xi}) \quad \text{where} \quad \overline{\xi} = Q^{-1}(\overline{x}) = -1/2.$$

Therefore the coefficients of the shape functions can be written as $a_i = r_{i+2}/2$ $(i = 1, 2, \dots, p - 1)$ where the variables are renumbered through shifting the indices to account for the boundary conditions: $a_1 = a_2 = 0$. Hence

$$u_{FE} = \frac{1}{2} \sum_{i=1}^{p-1} N_{i+2}(\overline{\xi}) N_{i+2}(\xi)$$

and the QoI is:

$$u'_{FE}(0) = \sum_{i=1}^{p-1} N_{i+2}(\overline{\xi}) \frac{dN_{i+2}}{d\xi} \big|_{\xi=-1}.$$

From the definition of N_i in eq. (1.53) we have

$$\frac{dN_{i+2}}{d\xi}\bigg|_{\xi=-1} = \sqrt{\frac{2i+1}{2}} P_i(-1) = \sqrt{\frac{2i+1}{2}}(-1)^i$$

and the QoI can be written as

$$u'_{FE}(0) = \sum_{i=1}^{p-1} N_{i+2}(\overline{\xi}) \sqrt{\frac{2i+1}{2}}(-1)^i = \frac{1}{2} \sum_{i=1}^{p-1}(-1)^i(P_{i+1}(\overline{\xi}) - P_{i-1}(\overline{\xi}))$$

where we made use of eq. (1.55). The relationships between the polynomial degree ranging from 2 to 100 and the corresponding values of the QoI computed by the direct method are displayed in Fig. 1.7. It is seen that convergence to the exact value $u'_{EX}(0) = 0.75$ is very slow.

The indirect method is based on eq. (1.18) which, applied to this example, takes the form

$$\int_0^1 u'v' \, dx = \int_0^1 \delta(\overline{x})v \, dx + (u'v)_{x=1} - (u'v)_{x=0}.$$

Selecting $v = 1 - x$ and rearranging the terms we get

$$u'(0) = v(\overline{x}) + \int_0^1 u' \, dx = v(\overline{x}) = 0.75$$

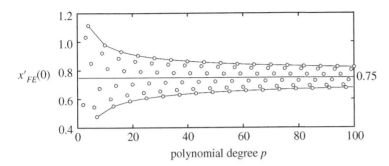

Figure 1.7 Example 1.9. Values of $u'_{FE}(0)$ computed by the direct method.

which is the exact solution. The choice $v = 1 - x$ was exceptionally fortuitous because it happens to be the Green's function (also known as the influence function) for $u'(0)$. Therefore the extracted value is independent of the solution $u \in E^0(I)$.

Let us choose $v = 1 - x^2$ for the extraction function. In this case

$$u'(0) = v(\bar{x}) - \int_0^1 u'v'\,dx = \frac{15}{16} + 2\int_0^1 u'x\,dx.$$

Substituting u'_{FE} for u':

$$\int_0^1 u'_{FE}\,x\,dx = \sum_{i=1}^{p-1} \frac{N_{i+2}(\bar{\xi})}{2}\sqrt{\frac{2i+1}{2}} \int_{-1}^1 P_i(\xi)\frac{1+\xi}{2}\,d\xi$$

$$= \frac{1}{4}\sum_{i=1}^{p-1} N_{i+2}(\bar{\xi})\sqrt{\frac{2i+1}{2}} \int_{-1}^1 P_i(\xi)(P_0(\xi) + P_1(\xi))\,d\xi = -\frac{3}{32}.$$

Taking the orthogonality of the Legendre polynomials (see eq. (D.13)) into account, the sum has to be evaluated only for $p = 2$. The extracted value of $u'_{FE}(0)$ for $p \geq 2$ is $u'_{FE}(0) = 0.5156\,(31.25\%$ error).

An explanation of why the extraction method is much more efficient than direct computation is given in Section 1.5.4.

Exercise 1.16 Find $u'_{FE}(0)$ for the problem in Example 1.7 by the direct and indirect methods. Compute the relative errors.

Exercise 1.17 For the problem in Example 1.9 let $v = 1 - x^3$ be the extraction function. Calculate the extracted value of $u'_{FE}(0)$ for $p \geq 3$.

Nodal forces
The vector of nodal forces associated with element k, denoted by $\{f^{(k)}\}$, is defined as follows:

$$\{f^{(k)}\} = [K^{(k)}]\{a^{(k)}\} - \{\bar{r}^{(k)}\} \qquad k = 1, 2, \dots, M(\Delta) \tag{1.88}$$

where $[K^{(k)}]$ is the stiffness matrix, $\{a^{(k)}\}$ is the solution vector and $\{\bar{r}^{(k)}\}$ is the load vector corresponding to traction forces, concentrated forces and thermal loads acting on element k.

The sign convention for nodal forces is different from the sign convention for the bar force: Whereas the bar force is positive when tensile, a nodal force is positive when acting in the direction of the positive coordinate axis.

Exercise 1.18 Assume that hierarchic basis functions based on Legendre polynomials are used. Show that when κ is constant and $c = 0$ on I_k then

$$f_1^{(k)} + f_2^{(k)} = r_1^{(k)} + r_2^{(k)}$$

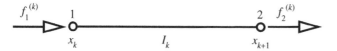

Figure 1.8 Exercise 1.18. Notation.

independently of the polynomial degree p_k. For sign convention refer to Fig. 1.8. Consider both thermal and traction loads. This exercise demonstrates that nodal forces are in equilibrium independently of the finite element solution. Therefore equilibrium of nodal forces is not an indicator of the quality of finite element solutions.

1.5 Estimation of error in energy norm

We have seen that the finite element solution minimizes the error in energy norm in the sense of eq. (1.48). It is natural therefore to use the energy norm as a measure of the error of approximation. There are two types of error estimators: (a) A priori estimators that establish the asymptotic rate of convergence of a discretization scheme, given information about the regularity (smoothness) of the exact solution and (b) a posteriori estimators that provide estimates of the error in energy norm for the finite element solution of a particular problem.

There is a very substantial body of work in the mathematical literature on the a priori estimation of the rate of convergence, given a quantitative measure of the regularity of the exact solution and a sequence of discretizations. The underlying theory is outside of the scope of this book; however, understanding the main results is important for practitioners of finite element analysis. For details we refer to [28, 45, 70, 84].

1.5.1 Regularity

Let us consider problems the exact solution of which has the functional form

$$u_{EX} = x^\alpha \varphi(x), \quad \alpha > 1/2, \quad x \in I = (0, \ell) \tag{1.89}$$

where $\varphi(x)$ is an analytic or piecewise analytic function, see Definition A.1 in the appendix. Our motivation for considering functions in this form is that this family of functions models the singular behavior of solutions of linear elliptic boundary value problems near vertices in polygonal and polyhedral domains. For u_{EX} to be in the energy space, its first derivative must be square integrable on I. Therefore

$$\int_0^\ell x^{2(\alpha-1)} \, dx > 0$$

from which it follows that α must be greater than $1/2$.

In the following we will see that when α is not an integer then the degree of difficulty associated with approximating u_{EX} by the finite element method is related to the size of $(\alpha - 1/2) > 0$. The smaller $(\alpha - 1/2)$ is, the more difficult it is to approximate u_{EX}.

If α is a fractional number then the measure of regularity used in the mathematical literature is the maximum number of square integrable derivatives, with the notion of derivative generalized to fractional numbers. See sections A.2.3 and A.2.4 in the appendix. For our purposes it is sufficient to remember that if u_{EX} has the functional form of eq. (1.89), and α is not an integer, then u_{EX} lies in the Sobolev space $H^{\alpha+1/2-\epsilon}(I)$ where $\epsilon > 0$ is arbitrarily small. This means that α must be larger than $1/2$ for the first derivative of u_{EX} to be square integrable. See, for example, [59].

If α is an integer then u_{EX} is an analytic or piecewise analytic function and the measure of regularity is the size of the derivatives of u_{EX}. Analogous definitions apply to two and three dimensions.

Remark 1.9 The kth derivative of a function $f(x)$ is a local property of $f(x)$ only when k is an integer. This is not the case for non-integer derivatives.

1.5.2 A priori estimation of the rate of convergence

Analysts are called upon to choose discretization schemes for particular problems. A sound choice of discretization is based on a priori information on the regularity of the exact solution. If we know that the exact solution lies in Sobolev space $H^k(I)$ then it is possible to say how fast the error in energy norm will approach zero as the number of degrees of freedom is increased, given a scheme by which a sequence of discretizations is generated. Index k can be inferred or estimated from the input data κ, c and f.

We define

$$h = \max_j \ell_j/\ell, \quad j = 1, 2, \ldots M(\Delta) \tag{1.90}$$

where ℓ_j is the length of the jth element, ℓ is the size of the of the solution domain $I = (1, \ell)$. This is generalized to two and three dimensions where ℓ is the diameter of the domain and ℓ_j is the diameter of the jth element. In this context diameter means the diameter of the smallest circle in one and two dimensions, or sphere in three dimensions, that contains the element or domain. In two and three dimensions the solution domain is denoted by Ω.

The a priori estimate of the relative error in energy norm for $u_{EX} \in H^k(\Omega)$, quasiuniform meshes and polynomial degree p is

$$(e_r)_E \overset{\text{def}}{=} \frac{\|u_{EX} - u_{FE}\|_{E(\Omega)}}{\|u_{EX}\|_{E(\Omega)}} \leq \begin{cases} C(k)\dfrac{h^{k-1}}{p^{k-1}}\|u_{EX}\|_{H^k(\Omega)} & \text{for } k - 1 \leq p \\[3mm] C(k)\dfrac{h^p}{p^{k-1}}\|u_{EX}\|_{H^{p+1}(\Omega)} & \text{for } k - 1 > p \end{cases} \tag{1.91}$$

where $E(\Omega)$ is the energy norm, k is typically a fractional number and $C(k)$ is a positive constant that depends on k but not on h or p. This inequality gives the upper bound for the asymptotic rate of convergence of the relative error in energy norm as $h \to 0$ or $p \to \infty$ [22]. This estimate holds for one, two and three dimensions. For one and two dimensions lower bounds were proven in [13, 24] and [46] and it was shown that when singularities are located in vertex points then the rate of convergence of the p-version is twice the rate of convergence of the h-version when both are expressed in terms of the number of degrees of freedom. It is reasonable to assume that analogous results can be proven for three dimensions; however, no proofs are available at present.

We will find it convenient to write the relative error in energy norm in the following form

$$(e_r)_E \leq \frac{C}{N^\beta} \tag{1.92}$$

where N is the number of degrees of freedom and C and β are positive constants, β is called the algebraic rate of convergence. In one dimension $N \propto 1/h$ for the h-version and $N \propto p$ for the p-version. Therefore for $k - 1 < p$ we have $\beta = k - 1$. However, for the important special case when the solution has the functional form of eq. (1.89) or, more generally, has a term like $u = |x - x_0|^\lambda$ and $x_0 \in \bar{I}$ is a nodal point then $\beta = 2(k-1)$ for the p-version: The rate of p-convergence is twice that of h-convergence [22, 84].

When the exact solution is an analytic function then $u_{EX} \in H^\infty(\Omega)$ and the asymptotic rate of convergence is exponential:

$$(e_r)_E \leq \frac{C}{\exp(\gamma N^\theta)} \tag{1.93}$$

where C, γ and θ are positive constants, independent of N. In one dimension $\theta \geq 1/2$, in two dimensions $\theta \geq 1/3$, in three dimensions $\theta \geq 1/5$, see [10].

When the exact solution is a piecewise analytic function then eq. (1.93) still holds provided that the boundary points of analytic functions are nodal points, or more generally, lie on the boundaries of finite elements.

The relationship between the error $e = u_{EX} - u_{FE}$ measured in energy norm and the error in potential energy is established by the following theorem.

Theorem 1.5

$$\|e\|_E^2 = \|u_{EX} - u_{FE}\|_{E(I)}^2 = \pi(u_{FE}) - \pi(u_{EX}). \tag{1.94}$$

Proof: Writing $e = u_{EX} - u_{FE}$ and noting that $e \in E^0(I)$, from the definition of $\pi(u_{FE})$ we have:

$$\begin{aligned}
\pi(u_{FE}) &= \pi(u_{EX} - e) = \frac{1}{2}B(u_{EX} - e, u_{EX} - e) - F(u_{EX} - e) \\
&= \frac{1}{2}B(u_{EX}, u_{EX}) - F(u_{EX}) \underbrace{-B(u_{EX}, e) + F(e)}_{0} + \frac{1}{2}B(e, e) \\
&= \pi(u_{EX}) + \|e\|_{E(I)}^2.
\end{aligned}$$

Remark 1.10 Consider the problem given by eq. (1.5) and assume that κ and c are constants. In this case the smoothness of u depends only on the smoothness of f: If $f \in C^k(I)$ then $u \in C^{k+2}(I)$ for any $k \geq 0$. Similarly, if $f \in H^k(I)$ then $u \in H^{k+2}(I)$ for any $k \geq 0$. This is known as the shift theorem. More generally, the smoothness of u depends on the smoothness of κ, c and F. For a precise statement and proof of the shift theorem we refer to [21].

Remark 1.11 An introductory discussion on how a priori estimates are obtained under the assumption that the second derivative of the exact solution is bounded can be found in Appendix B.

1.5.3 A posteriori estimation of error

The goal of finite element computations is to estimate certain quantities of interest (QoIs) such as, for example, the maximum and minimum values of u or u' on $I = (0, \ell)$. Since finite element solutions are approximations to an exact solution, it is not sufficient to report the value of a QoI computed from the finite element solution. It is also necessary to provide an estimate of the relative error in the QoI, or present evidence that the relative error in the QoI is not greater than an acceptable value.

In this section we will use the a priori estimates described in Section 1.5.2 to obtain a posteriori estimates of error in energy norm. It is possible to obtain very accurate estimates for a large class of problems which includes most problems of practical interest.

Error estimation based on extrapolation

For most practical problems the estimate (1.92) is sufficiently sharp so that the less than or equal sign ($\leq$) can be replaced by the approximately equal sign ($\approx$) and this a priori estimate can be used in an a posteriori fashion.

The computed values of the potential energy corresponding to a sequence of finite element spaces $S_1 \subset S_2 \subset \cdots S_n$ can be used for estimating the error in energy norm by extrapolation. Sequences of finite element spaces that have this property are called hierarchic sequences. By Theorem 1.5 and eq. (1.92) we have:

$$\pi(u_{FE}) - \pi(u_{EX}) \approx \frac{C^2}{N^{2\beta}} \tag{1.95}$$

where $C \stackrel{\text{def}}{=} C\|u_{EX}\|_{E(I)}$. There are three unknowns: $\pi(u_{EX})$, C and β. Assume that we have a sequence of solutions corresponding to the hierarchic sequence of finite element spaces $S_{i-2} \subset S_{i-1} \subset S_i$. Let us denote the corresponding computed potential energy values by π_{i-2}, π_{i-1}, π_i and the degrees of freedom by N_{i-2}, N_{i-1}, N_i. We will denote the estimate for $\pi(u_{EX})$ by π_∞. With this notation we have:

$$\pi_i - \pi_\infty \approx \frac{C^2}{N_p^{2\beta}} \tag{1.96}$$

$$\pi_{i-1} - \pi_\infty \approx \frac{C^2}{N_{i-1}^{2\beta}}. \tag{1.97}$$

On dividing eq. (1.96) with eq. (1.97) and taking the logarithm we get

$$\log \frac{\pi_i - \pi_\infty}{\pi_{i-1} - \pi_\infty} \approx 2\beta \log \frac{N_{i-1}}{N_i} \tag{1.98}$$

and, repeating with $i-1$ substituted for i, it is possible to eliminate 2β to obtain:

$$\frac{\pi_i - \pi_\infty}{\pi_{i-1} - \pi_\infty} \approx \left(\frac{\pi_{i-1} - \pi_\infty}{\pi_{i-2} - \pi_\infty} \right)^Q \tag{1.99}$$

where

$$Q = \log \frac{N_{i-1}}{N_i} \left(\log \frac{N_{i-2}}{N_{i-1}} \right)^{-1}.$$

Equation (1.99) can be solved for π_∞ to obtain an estimate for the exact value of the potential energy.

The relative error in energy norm corresponding to the ith finite element solution in the sequence is estimated from

$$e_i \approx \left(\frac{\pi_i - \pi_\infty}{|\pi_\infty|} \right)^{1/2}. \tag{1.100}$$

Usually the percent relative error is reported. This estimator has been tested against the known exact solution of many problems of various smoothness. The results have shown that it works well for a wide range of problems, including most problems of practical interest; however, it cannot be guaranteed to work well for all conceivable problems. For example, this method would fail if the exact solution would happen to be energy-orthogonal to all basis functions associated with (say) odd values of i.

Remark 1.12 From equation (1.92) we get

$$\log (e_r)_E \approx \log C - \beta \log N. \tag{1.101}$$

On plotting $(e_r)_E$ vs. N on log-log scale a straight line with the slope $-\beta$ will be seen for sufficiently large N. The estimated value of β, corresponding to the ith solution in the sequence, is denoted by β_i. It is computed from eq. (1.98):

$$\beta_i = \frac{1}{2} \frac{\log(\pi_i - \pi) - \log(\pi_{i-1} - \pi)}{\log N_{i-1} - \log N_i}. \tag{1.102}$$

Examples

The properties of the finite element solution with reference to a family of model problems is discussed in the following. The problems are stated as follows: Find $u_{FE} \in S^0(I)$ such that

$$\int_0^\ell \left(\kappa u'_{FE} v' + c u_{FE} v \right) \, dx = F(v) \quad \text{for all } v \in S^0(I) \tag{1.103}$$

where κ and c are constants and $F(v)$ is defined such that the exact solution is:

$$u_{EX} = x^\alpha(\ell - x), \quad \text{on } I = (0, \ell), \quad \alpha > 1/2. \tag{1.104}$$

As explained in Section 1.5.1, when α is not an integer, the case considered in the following, then this solution lies in the space $H^{\alpha+1/2-\epsilon}(I)$. Therefore the asymptotic rate of h-convergence on uniform meshes, predicted by eq. (1.92), is $\beta = \alpha - 1/2$ and the asymptotic rate of p-convergence on a fixed mesh is $\beta = 2\alpha - 1$.

We selected this problem because it is representative of the singular part of the exact solutions of two- and three-dimensional elliptic boundary value problems.

Referring to Theorem 1.3, we have $B(u_{EX} - u_{FE}, v) = 0$ for all $v \in S^0(I)$ therefore $F(v) = B(u_{EX}, v)$. Consequently for the kth element the load vector in the local numbering convention is:

$$r_i^{(k)} = \int_{x_k}^{x_{k+1}} \left(\kappa u'_{EX} \varphi'_i + c u_{EX} \varphi_i \right) \, dx, \quad i = 1, 2, \dots, p_k + 1 \tag{1.105}$$

where by definition $\varphi_i(Q_k(\xi)) = N_i(\xi)$.

When $1/2 < \alpha < 1$ then the first derivative of u_{EX} is infinity in the point $x = 0$. To avoid having u'_{EX} in the integrand, the first term in eq. (1.105) is integrated by parts:

$$\int_{x_k}^{x_{k+1}} \kappa u'_{EX} \varphi'_i \, dx = \left(\kappa u_{EX} \varphi'_i \right)_{x_k}^{x_{k+1}} - \int_{x_k}^{x_{k+1}} \kappa u_{EX} \varphi''_i \, dx.$$

Since $\varphi''_i = 0$ for $i = 1$ and $i = 2$, we have:

$$r_1^{(k)} = -\frac{1}{\ell_k} \left(\kappa u_{EX} \right)_{x=x_{k+1}} + \frac{1}{\ell_k} \left(\kappa u_{EX} \right)_{x=x_k} + \frac{\ell_k}{2} \int_{-1}^1 (c u_{EX})_{x=Q_k(\xi)} N_1 \, d\xi$$

$$r_2^{(k)} = \frac{1}{\ell_k} \left(\kappa u_{EX} \right)_{x=x_{k+1}} - \frac{1}{\ell_k} \left(\kappa u_{EX} \right)_{x=x_k} + \frac{\ell_k}{2} \int_{-1}^1 (c u_{EX})_{x=Q_k(\xi)} N_2 \, d\xi$$

and for $i \geq 3$ we have:

$$r_i^{(k)} = \sqrt{\frac{2i-3}{2}} \frac{2}{\ell_k} \left((\kappa u_{EX})_{x=x_{k+1}} - (-1)^i (\kappa u_{EX})_{x=x_k} - \int_{-1}^1 (\kappa u_{EX})_{x=Q_k(\xi)} \frac{dP_{i-2}}{d\xi} \, d\xi \right)$$

$$+ \frac{\ell_k}{2} \int_{-1}^1 (c u_{EX})_{x=Q_k(\xi)} N_i \, d\xi \tag{1.106}$$

where $P_{i-2}(\xi)$ is the Legendre polynomial of degree $i - 2$ and eq. (D.10) was used.

Since the exact solution is known, the exact value of the potential energy can be determined for any set of values of α, κ, c and ℓ. When κ and c are both constants then

$$\pi(u_{EX}) = -\frac{1}{2} \left[\kappa \left(\frac{\alpha^2}{2\alpha - 1} \ell^{2\alpha-1} - (\alpha + 1)\ell^{2\alpha} + \frac{(\alpha + 1)^2}{2\alpha + 1} \ell^{2\alpha+1} \right) \right.$$

$$\left. + c \left(\frac{1}{2\alpha + 1} \ell^{2\alpha+1} - \frac{1}{\alpha + 1} \ell^{2(\alpha+1)} + \frac{1}{2\alpha + 3} \ell^{2\alpha+3} \right) \right]. \tag{1.107}$$

The exact values of the potential energy for the data $\kappa = 1$, $c = 50$ and $\ell = 1$ and various values of α are shown in Table 1.2.

Table 1.2 Exact values of the potential energy for $\kappa = 1$, $c = 50$ and $\ell = 1$.

α	$\pi(u_{EX})$	α	$\pi(u_{EX})$
0.600	-2.3728354978	1.000	-1.0000000000
0.700	-1.7571858289	1.500	-0.5104166667
0.800	-1.4176885916	2.000	-0.3047619048
0.900	-1.1799028822	3.000	-0.1420634921

When α is a fractional number then derivatives higher than α will not be finite in $x = 0$. In the range $0.5 < \alpha < 1$ the first derivative in the point $x = 0$ is infinity. This range of α has considerable practical importance because the exact solutions of two- and three-dimensional problems often have analogous terms.

When α is an integer then all derivatives of u_{EX} are finite. Therefore u_{EX} can be approximated by Taylor series about any point of the domain $\bar{I} = [0, \ell]$. It is known that the error term of a Taylor series truncated at polynomial degree p is bounded by the $(p + 1)$th derivative of u_{EX}:

$$\max |u_{FE} - u_{EX}| \leq \frac{\ell^{p+1}}{(p + 1)!} \max_{x \in \bar{I}} \left| \frac{d^{p+1} u_{EX}}{dx^{p+1}} \right|. \tag{1.108}$$

In the special case when α is an integer and $p_{\min} \geq \alpha + 1$ then $u_{FE} = u_{EX}$.

Exercise 1.19 Show how eq. (1.106) is obtained from eq. (1.105). Provide details.

Example 1.10 Let us consider model problems in the form of eq. (1.103) with the following data: $\ell = 1$, $\kappa = 1$, $c = 50$ and exact solutions in the form of eq. (1.104) corresponding to $\alpha = 0.6$, 0.7, 0.8, 0.9. We will use a sequence of uniform finite element meshes with $M(\Delta) = 10$, 100, 1000 and $p_k = p = 2$ assigned to all elements. We are interested in the relationship between the estimated and true relative errors. The computed values of the potential energy and their estimated limit values computed by means of eq. (1.99) are listed in Table 1.3. These are comparable to the exact values of the potential energy listed in Table 1.2. The estimated limit values of the potential energy are denoted by $\pi_{M(\Delta) \to \infty}$.

With the information provided in Tables 1.2 and 1.3 it is possible to compare the estimated and exact values of the relative error. For example, using eq. (1.100) and

$$\|(u_{FE})_{M(\Delta)}\|_{E(I)}^2 = |\pi_{M(\Delta)}|$$

Table 1.3 Example: Computed and estimated values of the potential energy π. Uniform mesh refinement, $p_k = p = 2$ for all elements.

$M(\Delta)$	N	$\alpha = 0.6$	$\alpha = 0.7$	$\alpha = 0.8$	$\alpha = 0.9$
10	19	-2.17753673	-1.73038992	-1.41382648	-1.17955239
100	199	-2.25079984	-1.74673700	-1.41675042	-1.17984996
1000	1999	-2.29589857	-1.75303348	-1.41745363	-1.17989453
$\pi_{M(\Delta) \to \infty}$		-2.37254083	-1.75716094	-1.41768637	-1.17990276

the estimated relative error in energy norm for $M(\Delta) = 10$, $\alpha = 0.8$ is:

$$(e_r^*)_E = \sqrt{\frac{\pi(u_{FE}) - \pi_{M(\Delta)\to\infty}}{|\pi_{M(\Delta)\to\infty}|}} = \sqrt{\frac{-1.41382648 + 1.41768637}{1.41768637}} = 0.0522$$

or 5.22%. When using the exact value of the potential energy for reference then the relative error is the same as the estimated relative error to within three digits of accuracy:

$$(e_r)_E = \sqrt{\frac{\pi(u_{FE}) - \pi(u_{EX})}{\|u_{EX}\|_{E(I)}^2}} = \sqrt{\frac{-1.41382648 + 1.41768859}{1.41768859}} = 0.0522.$$

Exercise 1.20 Compare the estimated and exact values of the relative error in energy norm for the problem in Example 1.10 for $M(\Delta) = 100$, $\alpha = 0.7$.

Example 1.11 Let us consider once again model problems in the form of eq. (1.103) with the data: $\ell = 1$, $\kappa = 1$, $c = 50$ and exact solutions corresponding to $\alpha = 0.6$, 0.7, 0.8, 0.9, see eq. (1.104). Using a sequence of uniform finite element meshes with $M(\Delta) = 10$, 100, 1000, 10,000 and $p = 2$ assigned to each element, the results shown in Fig. 1.9 are obtained. The values of β were computed by linear regression using eq. (1.101). We observe that $\beta = \alpha - 1/2$. This is consistent with the asymptotic estimate given by eq. (1.91).

Example 1.12 Let us consider model problems in the form of eq. (1.103) with the data: $\ell = 1$, $\kappa = 1$, $c = 50$ and exact solutions corresponding to $\alpha = 0.6$, 0.7, 0.8, 0.9, see eq. (1.104). Using a uniform finite element mesh with $M(\Delta) = 10$ and $p = 2$, 3, 4, 5 assigned to each element, the results shown in Fig. 1.10 are obtained. The values of β were computed by linear regression using eq. (1.101). We observe that $\beta = 2(\alpha - 1/2)$, that is, the rate of convergence is twice that in Example 1.11. This is consistent with the theoretical results in [22, 84]: The rate of p-convergence is at least twice the rate of h-convergence when the singular point is a nodal point.

1.5.4 Error in the extracted QoI

In Example 1.9 it was demonstrated that the QoI can be extracted from the finite element solution efficiently and accurately even when the discretization was very poorly chosen. Let us consider a quantity of interest $\Phi(u)$ and the corresponding extraction function $w \in E(I)$. The extracted value of the QoI is

$$\Phi(u_{FE}) = F(w) - B(u_{FE}, w) \tag{1.109}$$

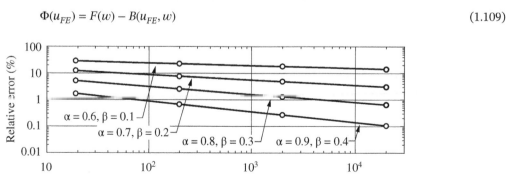

Figure 1.9 Relative error in energy norm. $M(\Delta) = 10$, 100, 1000, 10000, $p = 2$.

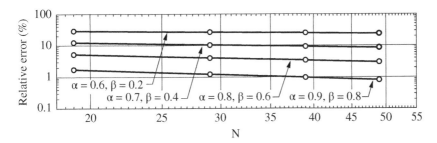

Figure 1.10 Relative error in energy norm. $M(\Delta) = 10$, $p = 2,\ 3,\ 4,\ 5$.

and the exact value of the QoI is

$$\Phi(u_{EX}) = F(w) - B(u_{EX}, w). \tag{1.110}$$

Subtracting eq. (1.109) from eq. (1.110) we get

$$\Phi(u_{EX}) - \Phi(u_{FE}) = -B(u_{EX} - u_{FE}, w). \tag{1.111}$$

We define a function $z_{EX} \in E^0(I)$ such that

$$B(z_{EX}, v) = B(w, v) \quad \text{for all } v \in E^0(I). \tag{1.112}$$

This operation projects $w \in E(I)$ onto the space $E^0(I)$. Letting $v = u_{EX} - u_{FE}$ we get:

$$B(z_{EX}, u_{EX} - u_{FE}) = B(w, u_{EX} - u_{FE}) \quad \text{for all } v \in E^0(I).$$

We will write this as

$$B(u_{EX} - u_{FE}, w) = B(u_{EX} - u_{FE}, z_{EX}). \tag{1.113}$$

Next we define $z_{FE} \in S^0(I)$ such that

$$B(z_{EX}, v) = B(z_{FE}, v) \quad \text{for all } v \in S^0(I). \tag{1.114}$$

This operation projects $z_{EX} \in E^0(I)$ onto the space $S^0(I)$. By Galerkin's orthogonality condition (see Theorem 1.3) we have

$$B(u_{EX} - u_{FE}, v) = 0 \quad \text{for all } v \in S^0(I).$$

Therefore, letting $v = z_{FE}$, we write eq. (1.113) as

$$B(u_{EX} - u_{FE}, w) = B(u_{EX} - u_{FE}, z_{EX} - z_{FE}) \tag{1.115}$$

and we can write eq. (1.111) as

$$\Phi(u_{EX}) - \Phi(u_{FE}) = -B(u_{EX} - u_{FE}, z_{EX} - z_{FE}). \tag{1.116}$$

Therefore the error in the extracted data is

$$\begin{aligned}
|\Phi(u_{EX}) - \Phi(u_{FE})| &= |B(u_{EX} - u_{FE}, z_{EX} - z_{FE})| \\
&\leq 2\|u_{EX} - u_{FE}\|_{E(I)}\|z_{EX} - z_{FE}\|_{E(I)}
\end{aligned} \tag{1.117}$$

where we used the Schwarz inequality, see Section A.3 in the appendix.

The function z_{FE} made it possible to write the error in the QoI in this form. It does not have to be computed.

Inequality (1.117) serves to explain why the error in the extracted data can converge to zero faster than the error in energy norm: If $\|z_{EX} - z_{FE}\|_{E(I)}$ is of comparable magnitude to $\|u_{EX} - u_{FE}\|_{E(I)}$

then the error in the extracted data is of comparable magnitude to the error in the strain energy, that is, the error in energy norm squared. But, as seen in Example 1.9, where w was much smoother than u_{EX}, it can be much smaller. In the exceptional case when the extraction function is Green's function, the error is zero.

1.6 The choice of discretization in 1D

In an ideal discretization the error (in energy norm) associated with each element would be the same. This ideal discretization can be approximated by automated adaptive methods in which the discretization is modified based on feedback information from previously obtained finite element solutions. Alternatively, based on a general understanding of the relationship between regularity and discretization, and understanding the strengths and limitations of the software tools available to them, analysts can formulate very efficient discretization schemes.

1.6.1 The exact solution lies in $H^k(I)$, $k - 1 > p$

When the solution is smooth then the most efficient finite element discretization scheme is uniform mesh and high polynomial degree. However, all implementations of finite element analysis software have limitations on how high the polynomial degree is allowed to be and therefore it may not be possible to increase the polynomial degree sufficiently to achieve the desired accuracy. In such cases the mesh has to be refined. Uniform refinement may not be optimal in all cases, however. Consider, for example, the following problem:

$$-\epsilon^2 u'' + cu = f(x), \quad u(0) = u'(\ell) = 0 \tag{1.118}$$

where $\epsilon \ll c$, and f is a smooth function. Intuitively, when ϵ^2 is small then the solution will be close to $u = f/c$ however, because of the boundary condition $u(0) = 0$, has to be satisfied, the function $u(x)$ will change sharply over some interval $0 < x < d(\epsilon) \ll \ell$.

Letting $c = 1$ and $f(x) = 1$ the exact solution of this problem is

$$u_{EX}(x) = 1 - \cosh x/\epsilon + \tanh(\ell/\epsilon) \sinh x/\epsilon \tag{1.119}$$

which is plotted for various values of ϵ on the interval $0 < x/\ell < 0.20$ in Fig. 1.11. It is seen that the gradient at $x = 0$ rapidly increases with respect to decreasing values of ϵ.

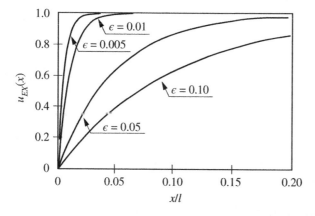

Figure 1.11 The solution $u_{EX}(x)$, given by eq. (1.119), in the neighborhood of $x = 0$ for various values of ϵ.

This is a simple example of boundary layer problems that arise in models of plates, shells and fluid flow. Despite the fact that u_{EX} is an analytic function, it may require unrealistically high polynomial degrees to obtain a close approximation to the solution when ϵ is small.

The optimal discretization scheme for problems with boundary layers is discussed in the context of the *hp*-version in [85]. The results of analysis indicate that the size of the element at the boundary is proportional to the product of the polynomial degree *p* and the parameter ϵ. Specifically, for the problem discussed here, the optimal mesh consists of two elements with the node points located at $x_1 = 0, x_2 = d, x_3 = \ell$, where $d = Cp\epsilon$ with $0 < C < 4/e$.

A practical approach to problems like this is to create an element at the boundary (in higher dimensions a layer of elements) the size of which is controlled by a parameter. The optimal value of that parameter is then selected adaptively.

1.6.2 The exact solution lies in $H^k(I)$, $k - 1 \le p$

In this section we consider a special case of the problem stated in eq. (1.103):

$$\int_0^\ell u'v'\,dx = F(v), \quad \text{for all } v \in E^0(I) \tag{1.120}$$

with the data $u(0) = u(\ell) = 0$, $\ell = 1$ and $F(v)$ defined such that the exact solution is

$$u_{EX} = x^\alpha(1 - x), \quad \alpha > 1/2, \quad 0 < x < 1 \tag{1.121}$$

that is,

$$F(v) = \int_0^\ell (\alpha x^{\alpha-1} - (\alpha + 1)x^\alpha)\,v'\,dx. \tag{1.122}$$

On integrating by parts, we get the following expression which is better suited for numerical evaluation:

$$F(v) = -\int_0^\ell u_{EX}v''dx. \tag{1.123}$$

We address the following questions: (a) How does the error in energy norm depend on the parameter α, the mesh Δ and the *p*-distribution **p**? and (b) How is this error distributed among the elements? Understanding these relationships is necessary for making sound choices of discretization based on a priori information concerning the regularity of the exact solution.

We compute the potential energy of the difference between the exact solution and its linear interpolant for the *k*th element:

$$\overline{\pi}_{EX}^{(k)} = \frac{1}{2} \int_{x_k}^{x_{k+1}} \left(u'_{EX} - \frac{u_{EX}(x_{k+1}) - u_{EX}(x_k)}{x_{k+1} - x_k} \right)^2 dx.$$

The exact solution for $\alpha = 0.75$ and its linear interpolant for $M(\Delta) = 5$, uniform mesh, are shown in Fig. 1.12.

To obtain the potential energy of the difference between the exact solution and its linear interpolant for the *k*th element, denoted by $\overline{\pi}_{FE}^{(k)}$, we need to solve:

$$\frac{2}{\ell_k} \begin{bmatrix} 1 & 0 & \cdots & 0 \\ 0 & 1 & \cdots & 0 \\ & & \ddots & \\ 0 & 0 & \cdots & 1 \end{bmatrix} \begin{Bmatrix} a_3^{(k)} \\ a_4^{(k)} \\ \vdots \\ a_{p_k+1}^{(k)} \end{Bmatrix} = \begin{Bmatrix} r_3^{(k)} \\ r_4^{(k)} \\ \vdots \\ r_{p_k+1}^{(k)} \end{Bmatrix}. \tag{1.124}$$

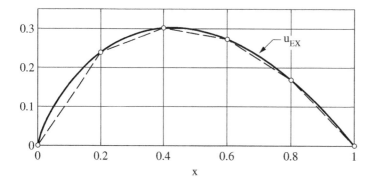

Figure 1.12 The exact solution for $\alpha = 0.75$ and its linear interpolant for $M(\Delta) = 5$, uniform mesh.

The solution is

$$a_i^{(k)} = \frac{\ell_k}{2} r_i^{(k)}, \quad i = 3, 4, \ldots, p_k + 1. \tag{1.125}$$

Using eq. (1.106) we get

$$r_i^{(k)} = -\sqrt{\frac{2i-3}{2}} \frac{2}{\ell_k} \int_{-1}^{1} \left(\kappa \, \tilde{u}_{EX}^{(k)} \right)_{x=Q_k(\xi)} \frac{dP_{i-2}}{d\xi} \, d\xi, \quad i = 3, 4, \ldots, p_k + 1 \tag{1.126}$$

where $\tilde{u}_{EX}^{(k)}$ is the difference between u_{EX} and its linear interpolant:

$$\tilde{u}_{EX}^{(k)} = (u_{EX})_{x=Q_k(\xi)} - \left(\frac{1-\xi}{2} u_{EX}(x_k) + \frac{1+\xi}{2} u_{EX}(x_{k+1}) \right) \tag{1.127}$$

and compute

$$\overline{\pi}_{FE}^{(k)} = -\frac{1}{2} \sum_{i=3}^{p_k+1} a_i^{(k)} r_i^{(k)}.$$

Referring to Theorem 1.5, the error in energy norm associated with the kth element is

$$\|e_k\|_{E(I_k)} = \sqrt{\overline{\pi}_{FE}^{(k)} - \overline{\pi}_{EX}^{(k)}} \tag{1.128}$$

and the relative error in energy norm associated with the kth element is:

$$(e_r^{(k)})_E = \frac{\|e_k\|_{E(I_k)}}{\sqrt{\left|\overline{\pi}_{EX}^{(k)}\right|}}. \tag{1.129}$$

The error of approximation over the entire domain is:

$$\|u_{EX} - u_{FE}\|_{E(I)} = \left(\sum_{k=1}^{M(\Delta)} \|e_k\|_{E(I_k)}^2 \right)^{1/2} = \sqrt{\pi_{FE} - \pi_{EX}} \tag{1.130}$$

By Theorem 1.2, the exact value of the potential energy is

$$\pi(u_{EX}) = -\frac{1}{2} \int_0^1 (u'_{EX})^2 dx = -\frac{1}{2} \left(\frac{\alpha^2}{2\alpha - 1} - (\alpha + 1) + \frac{(\alpha+1)^2}{2\alpha + 1} \right) \tag{1.131}$$

and the relative error in energy norm on the entire domain is:

$$(e_r)_E = \left(\frac{\pi_{FE} - \pi_{EX}}{|\pi(u_{EX})|} \right)^{1/2}. \tag{1.132}$$

Remark 1.13   In estimating the local error we used $u_{FE}(x_k) = u_{EX}(x_k)$. It can be shown that in the special case of this problem ($c = 0$) this relationship holds and therefore using the equal sign in eq. (1.128) is justified. In the general case ($c \neq 0$) however, $u_{FE}(x_k) \neq u_{EX}(x_k)$ and eq. (1.128) will be an estimate of the local error in the finite element solution. Therefore the equal sign in eq. (1.128) has to be replaced by the approximately equal ($\approx$) sign and the first equal sign in eq. (1.130) has to be replaced with the less or equal ($\leq$) sign.

Example 1.13   This example illustrates the distribution of the relative error among the elements for a fixed mesh and polynomial degree for selected fractional values of α. Uniform mesh on the domain $(0, 1)$ with $M(\Delta) = 5$ and $p_k = 2$ for $k = 1, 2, \dots, 5$ is used. The exact solution for $\alpha = 0.75$ is shown in Fig. 1.12. The percent relative error in energy norm associated with the kth element, given by eq. (1.129), is shown in Table 1.4 and the relative error for the entire domain is shown in the last column.

It is seen that for all values of α the maximum error is associated with the first element.

Example 1.14   This example illustrates the distribution of the relative error among the elements for a fixed mesh and polynomial degree for selected integer values of α. Uniform mesh on the domain $(0, 1)$ with $M(\Delta) = 5$ and $p_k = 2$ for $k = 1, 2, \dots, 5$ is used. The percent relative error in energy norm associated with the kth element, given by eq. (1.129), is shown in Table 1.5 and the relative error for the entire domain is shown in the last column.

The error of approximation for $\alpha = 1$ is zero. This follows directly from Theorem 1.4: The exact solution is a polynomial of degree 2. Therefore it lies in the finite element space and hence the finite element solution is the same as the exact solution.

Remark 1.14   In the foregoing discussion it was tacitly assumed that all data computed by numerical integration were accurate and the coefficient matrices of the linear equations were such that small changes in the right-hand-side vector produce small changes in the solution vector. This happens when the condition number of the coefficient matrix is reasonably small. In the finite element method the condition number depends on the choice of the shape functions, the mapping functions and the mesh. In one-dimensional setting the mapping is linear and the shape functions

Table 1.4   Example: Element-by-element and total relative errors in energy norm (percent) for selected fractional values of α.

	Element number					
α	1	2	3	4	5	$(e_r)_E$
1.25	79.49	7.50	2.80	1.63	1.12	4.80
1.15	99.52	4.06	1.63	0.97	0.67	3.92
1.05	29.56	1.24	0.53	0.32	0.22	1.77
0.95	18.89	1.16	0.52	0.32	0.22	2.41
0.85	42.94	3.26	1.52	0.94	0.67	9.84
0.75	60.39	5.14	2.47	1.56	1.11	22.37
0.65	76.07	6.86	3.39	2.16	1.56	42.91
0.55	91.80	8.44	4.28	2.76	2.00	76.22

Table 1.5 Example: Element-by-element and total relative errors in energy norm (percent) for selected integer values of α.

α	Element number 1	2	3	4	5	$(e_r)_E$
1	0	0	0	0	0	0
2	11.00	61.24	15.31	7.02	4.55	2.45
3	20.09	4.69	98.83	16.16	9.24	4.62
4	36.98	8.41	17.24	30.13	14.02	8.00

are energy-orthogonal, therefore round-off errors are not significant. This is not the case in two and three dimensions, however.

Errors in numerical integration can be particularly damaging. The reader should be mindful of this when applying the concepts and procedures discussed in this chapter to higher dimensions.

1.7 Eigenvalue problems

The following problem is a prototype of an important class of engineering problems which includes the undamped vibration of elastic structures:

$$-(\kappa u')' + cu = -\mu \frac{\partial^2 u}{\partial t^2}, \quad x \in (0, \ell), \quad t \in (0, \infty) \tag{1.133}$$

where the primes represent differentiation with respect to x. For example, we may think of an elastic bar of length ℓ, cross-section A, modulus of elasticity E, in which case $\kappa \equiv AE > 0$ given in units of Newton (N) or equivalent, the parameter $c \geq 0$ is the coefficient of distributed springs (N/mm^2) and the parameter $\mu > 0$ is mass per unit length (kg/m $= 10^{-6}$Ns^2/mm^2). The bar is vibrating in its longitudinal direction.

The boundary conditions are:

$$u(0, t) = 0, \quad u(\ell, t) = 0$$

and the initial conditions are

$$u(x, 0) = f(x), \quad \frac{\partial u}{\partial t}\Big|_{(x,0)} = g(x)$$

where $f(x)$ and $g(x)$ are given functions in $L^2(I)$. Here we consider homogeneous Dirichlet boundary conditions. However, the boundary conditions can be homogeneous Neumann or homogeneous Robin conditions, or any combination of those.

The generalized form is obtained by multiplying eq. (1.133) by a test function $v \in E^0(I)$ and integrating by parts.

$$\int_0^\ell (\kappa u'v' + cuv)\, dx = -\int_0^\ell \mu \frac{\partial^2 u}{\partial t^2} v\, dx. \tag{1.134}$$

We now introduce $u = U(x)T(t)$ where $U \in E^0(I)$, $T \in C^2(0, \infty)$. This is known as separation of variables. Therefore we get

$$T \int_0^\ell (\kappa U'v' + cUv)\, dx = -\frac{\partial^2 T}{\partial t^2} \int_0^\ell \mu Uv\, dx \tag{1.135}$$

which can be written as

$$\frac{\int_0^\ell (\kappa U' v' + cUv)\,dx}{\int_0^\ell \mu Uv\,dx} = -\frac{1}{T}\frac{\partial^2 T}{\partial t^2} = \omega^2. \tag{1.136}$$

Since the functions on the left are independent of t, the function T depends only on t, both expressions must equal some positive constant denoted by ω^2. That constant has to be positive because the expression on the left holds for all $v \in E^0(I)$ and if we select $v = U$ then the expression on the left is positive.

The function $T(t)$ satisfies the ordinary differential equation

$$\frac{\partial^2 T}{\partial t^2} + \omega^2 T = 0 \tag{1.137}$$

the solution of which is

$$T = a\cos(\omega t) + b\sin(\omega t) \tag{1.138}$$

where ω is the angular velocity (rad/s). Alternatively ω is written as $\omega = 2\pi f$ where f is the frequency (Hz).

To find ω and U we have to solve the problem

$$\int_0^\ell (\kappa U' v' + cUv)\,dx - \omega^2 \int_0^\ell \mu Uv\,dx = 0 \quad \text{for all } v \in E^0(I) \tag{1.139}$$

which will be abbreviated as

$$B(U, v) - \omega^2 D(U, v) = 0 \quad \text{for all } v \in E^0(I). \tag{1.140}$$

There are infinitely many solutions called eigenpairs (ω_i, U_i), $i = 1, 2, \ldots, \infty$. The set of eigenvalues is called the spectrum. If U_i is an eigenfunction and α is a real number then αU_i is also an eigenfunction. In the following we assume that the eigenfunctions have been normalized so that

$$D(U_i, U_i) \equiv \int_0^\ell \mu U_i^2\,dx = 1.$$

If the eigenvalues are distinct then the corresponding eigenfunctions are orthogonal: Let (ω_i, U_i) and (ω_j, U_j) be eigenpairs, $i \neq j$. Then from eq. (1.140) we have

$$B(U_i, U_j) - \omega_i^2 D(U_i, U_j) = 0$$
$$B(U_j, U_i) - \omega_j^2 D(U_j, U_i) = 0.$$

Subtracting the second equation from the first we see that if $\omega_i \neq \omega_j$ then U_i and U_j are orthogonal functions:

$$D(U_i, U_j) \equiv \int_0^\ell \mu U_i U_j\,dx = 0 \tag{1.141}$$

and hence $B(U_i, U_j) = 0$.

Importantly, it can be shown that any function $f \in E^0(I)$ can be written as a linear combination of the eigenfunctions:

$$\left\| f - \sum_{i=1}^{\infty} a_i U_i(x) \right\|_{L^2(I)} = 0 \tag{1.142}$$

where

$$a_i = \int_0^\ell \mu f U_i \, dx. \tag{1.143}$$

The Rayleigh[15] quotient is defined by

$$R(u) = \frac{B(u, u)}{D(u, u)}. \tag{1.144}$$

Eigenvalues are usually numbered in ascending order. Following that convention,

$$\omega_1^2 \equiv \omega_{\min}^2 = \min_{u \in E^0(I)} R(u) = R(U_1) \tag{1.145}$$

that is, the smallest eigenvalue is the minimum of the Rayleigh quotient and the corresponding eigenfunction is the minimizer of $R(u)$ on $E^0(I)$. This follows directly from eq. (1.140). The kth eigenvalue minimizes $R(u)$ on the space $E_k^0(I)$

$$\omega_k^2 = \min_{u \in E_k^0(I)} R(u) = R(U_k) \tag{1.146}$$

where

$$E_k^0(I) = \{u \mid u \in E^0(I), \ B(u, U_i) = 0, \ i = 1, 2, \ldots, k-1\}. \tag{1.147}$$

When the eigenvalues are computed numerically then the minimum of the Rayleigh quotient is sought on the finite-dimensional space $S^0(I)$. We see from the definition $R(u)$ that the error of approximation in the natural frequencies will depend on how well the eigenfunctions are approximated in energy norm, in the space $S^0(I)$.

The following example illustrates that in a sequence of numerically computed eigenvalues only the lower eigenvalues will be approximated well. It is possible, however, at least in principle, to obtain good approximation for any eigenvalue by suitably enlarging the space $S^0(I)$.

Example 1.15   Let us consider the eigenvalue problem

$$\kappa \frac{\partial^2 u}{\partial x^2} = \mu \frac{\partial^2 u}{\partial t^2}, \qquad u(0) = u(\ell) = 0, \ t \geq 0. \tag{1.148}$$

This equation models (among other things) the free vibration (natural frequencies and mode shapes) of a string of length ℓ stretched horizontally by the force $\kappa > 0$ (N) under the assumptions that the displacements are infinitesimal and confined to one plane, the plane of vibration, and the ends of the string are fixed. The mass per unit length is $\mu > 0$ (kg/m). We assume that κ and μ are constants. It is left to the reader to verify that the function u defined by

$$u = \sum_{i=1}^\infty \left(a_i \cos(\omega_i t) + b_i \sin(\omega_i t)\right) \sin(\lambda_i x) \tag{1.149}$$

where a_i, b_i are coefficients determined from the initial conditions and

$$\lambda_i = i\frac{\pi}{\ell}, \quad \omega_i = \lambda_i \sqrt{\frac{\kappa}{\mu}} \tag{1.150}$$

satisfies eq. (1.148).

If we approximate the eigenfunctions using uniform mesh, $p = 2$ and plot the ratio $(\omega_{FE}/\omega_{EX})_n$ against n/N, where n is the nth eigenvalue, then we get the curves shown in Fig. 1.13. The curves show that somewhat more than 20% of the numerically computed eigenvalues will be accurate.

---

15 John William Strutt, 3rd Baron Rayleigh 1842–1919.

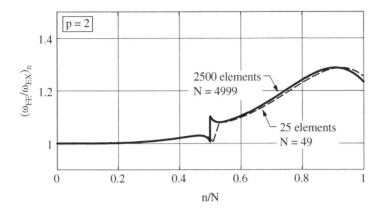

Figure 1.13 The ratio $(\omega_{FE}/\omega_{EX})_n$ corresponding to the h version, $p = 2$.

The higher eigenvalues cannot be well approximated in the space $S^0(I)$. The existence of the jump seen at $n/N = 0.5$ is a feature of numerically approximated eigenvalues by means of standard finite element spaces using the h-version [2]. The location of the jump depends on the polynomial degree of elements. There is no jump when $p = 1$.

If we approximate the eigenfunctions using a uniform mesh consisting of 5 elements, and increase the polynomial degrees uniformly then we get the curves shown in Fig. 1.14. The curves show that only about 40% of the numerically computed eigenvalues will be accurate. The error increases monotonically for the higher eigenvalues and the size of the error is virtually independent of p.

It is possible to reduce this error by enforcing the continuity of derivatives. Examples are available in [32]. There is a tradeoff, however: Enforcing continuity of derivatives on the basis functions reduces the number of degrees of freedom but entails a substantial programming burden because an adaptive scheme has to be devised for the general case to ensure that the proper degree of continuity is enforced. If, for example, μ would be a piecewise constant function then the continuity of the first and higher derivatives must not be enforced in those points where μ is discontinuous.

From the perspective of designing a finite element software, it is advantageous to design the software in such a way that it will work well for a broad class of problems. In the formulation presented in this chapter C^0 continuity is a requirement. Functions that lie in $C^k(I)$ where $k > 0$

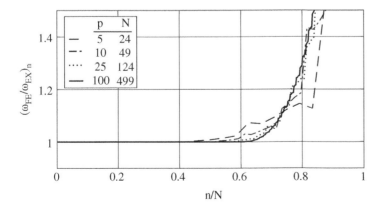

Figure 1.14 The ratio $(\omega_{FE}/\omega_{EX})_n$ corresponding to the p version. Uniform mesh, 5 elements.

Table 1.6 Example: p-Convergence of the 24th eigenvalue in Example 1.16.

p	5	10	15	20
ω_{24}	194.296	100.787	98.312	98.312

are also in $C^0(I)$. In other words, the space $C^k(I)$ is embedded in the space $C^0(I)$. Symbolically: $C^k(I) \subset C^0(I)$. The exact eigenfunctions in this example are in $C^\infty(I)$.

Example 1.16   Let us consider the problem in Example 1.15 modified so that μ is a piecewise constant function defined on a uniform mesh of 5 elements such that $\mu = 1$ on elements 1, 3 and 5, $\mu = 0.2$ on elements 2 and 4. In this case the exact eigenfunctions are not smooth and the exact eigenvalues are not known explicitly.

At $p = 5$ there are 24 degrees of freedom. Suppose that the 24th eigenvalue is of interest. If we increase p uniformly then this eigenvalue converges to 98.312. The results of computation are shown in Table 1.6.

Any eigenvalue can be approximated to an arbitrary degree of precision on a suitably defined mesh and uniform increase in the degrees of freedom. When κ and/or μ are discontinuous functions then the points of discontinuity must be node points.

Observe that the numerically computed eigenvalues converge monotonically from above. This follows directly from the fact that the eigenfunctions are minimizers of the Rayleigh quotient.

Exercise 1.21   Prove eq. (1.143).

Exercise 1.22   Find the eigenvalues for the problem of Example 1.15 using the generalized formulation and the basis functions $\varphi_n(x) = \sin(n\pi x/\ell)$, $(n = 1, 2, \dots, N)$. Assume that κ and μ are constants and $\mu/\kappa = 1$. Let $\ell = 10$. Explain what makes this choice of basis functions very special. Hint: Owing to the orthogonality of the basis functions, only hand calculations are involved.

1.8   Other finite element methods

Up to this point we have been concerned with the finite element method based on the generalized formulation, called the principle of virtual work. There are many other finite element methods. All finite element methods share the following attributes:

1. Formulation. A bilinear form $B(u, v)$ is defined on the normed linear spaces X, Y (i.e. $u \in X$, $v \in Y$) and the functional $F(v)$ is defined on Y. The exact solution u_{EX} lies in X and satisfies:

$$B(u_{EX}, v) = F(v) \quad \text{for all } v \in Y \tag{1.151}$$

The normed linear spaces, X, Y, the linear functional F and the bilinear form B satisfy the respective properties listed in sections A.1.1 and A.1.2.

2. Finite element spaces. The finite-dimensional subspaces $S_i \subset X$, $V_i \subset Y$ $(i = 1, 2, \dots)$ are defined and it is assumed that there are $\hat{u}_i \in S_i$ such that the sequence of functions $\hat{u}_i$ $(i = 1, 2, \dots)$ converges in the space X to u_{EX}, that is:

$$\|u_{EX} - \hat{u}_i\|_X \leq \epsilon_i \quad \epsilon_i \to 0 \text{ as } i \to \infty. \tag{1.152}$$

The functions $\hat{u}_i$ are not the finite element solutions in general.

3. The finite element solution. The finite element solution $u_{i|FE} \in S_i$ satisfies:

$$B(u_{i|FE}, v) = F(v) \quad \text{for all } v \in V_i. \tag{1.153}$$

4. The stability criterion. The finite element method is said to be stable if

$$\|u_{i|FE} - \hat{u}_i\|_X \le C\|U - \hat{u}_i\|_X \quad i = 1, 2, \ldots \tag{1.154}$$

for all possible $U \in X$. The necessary and sufficient condition for a finite element method to be stable is that for every $u \in S_i$ there is a $v \in V_i$ so that

$$|B(u, v)| \ge C \|u\|_X \|v\|_Y \tag{1.155}$$

where $C > 0$ is a constant, independent of i, or for every $v \in V_i$ there is a $u \in S_i$ so that this inequality holds. This inequality is known as the Babuška-Brezzi condition, usually abbreviated to "the BB condition". This condition was formulated by Babuška in 1971 [9] and independently by Brezzi in 1974 [29].

If the Babuška-Brezzi condition is not satisfied then there will be at least some $u_{EX} \in X$ for which $\|u_{EX} - u_{i|FE}\|_X \nrightarrow 0$ as $i \to \infty$ even though there may be $u_{EX} \in X$ for which $\|u_{EX} - u_{i|FE}\|_X \to 0$ as $i \to \infty$. Examples are presented in [6]. In general it is difficult, or may even be impossible, to separate those u_{EX} for which the method works well from those for which it does not. The Babuška-Brezzi condition guarantees that the condition number of the stiffness matrix will not become too large as i increases.

Remark 1.15 Any implementation of the finite element method must be shown to satisfy the Babuška-Brezzi condition otherwise there will be some input data for which the method will fail even though it may work well for other input data. The formulation based on the principle of virtual work satisfies the Babuška-Brezzi condition.

Exercise 1.23 Show that the finite element method based on the principle of virtual work satisfies the Babuška-Brezzi condition.

1.8.1 The mixed method

Consider writing eq. (1.5) in the following form:

$$\kappa u' - F = 0 \tag{1.156}$$

$$-F' + cu = f \tag{1.157}$$

and assume that the boundary conditions are $u(0) = u(\ell) = 0$.

In the following we will use the one-dimensional equivalent of the notation introduced in sections A.2.2 and A.2.3. Multiply eq. (1.156) by $G \in L^2(I)$ and eq. (1.157) by $v \in H^1(I)$, integrate by parts and sum the resulting equations to obtain:

$$\int_0^\ell \left(\kappa \frac{du}{dx} G - FG \right) dx + \int_0^\ell \left(F \frac{dv}{dx} + cuv \right) dx = \int_0^\ell Tv dx. \tag{1.158}$$

We define the bilinear form:

$$B(u, F; v, G) \stackrel{\text{def}}{=} \int_0^\ell \left(\kappa u' G - FG \right) dx + \int_0^\ell \left(Fv' + cuv \right) dx \tag{1.159}$$

and the linear form

$$F(v) \stackrel{\text{def}}{=} \int_0^\ell fv \; dx. \tag{1.160}$$

The problem is now stated as follows: Find $u_{EX} \in H_0^1(I)$, $F_{EX} \in L^2(I)$ such that

$$B\left(u_{EX}, F_{EX}; v, G\right) = F(v) \quad \text{for all } v \in H_0^1(I), \; G \in L^2(I). \tag{1.161}$$

The finite element problem is formulated as follows: Find $u_{FE} \in S^0(I)$ where $S^0(I)$ is a subspace of $H_0^1(I)$ and $F_{FE} \in V(I)$ where $V(I)$ is a subspace of $L^2(I)$ such that

$$B\left(u_{FE}, F_{FE}; v, G\right) = F(v) \quad \text{for all } v \in S^0(I), \; G \in V(I). \tag{1.162}$$

We now ask: In what sense will $\left(u_{FE}, F_{FE}\right)$ be close to $\left(u_{EX}, F_{EX}\right)$? The answer is that there is a constant C, independent of the finite element mesh and $\left(u_{EX}, F_{EX}\right)$, such that

$$\begin{aligned}
\|u_{EX} - u_{FE}\|_{H^1(I)} &+ \|F_{EX} - F_{FE}\|_{L^2(I)} \\
&\leq C \left[\min \|u_{EX} - u\|_{H^1(I)} + \min \|F_{EX} - F\|_{L^2(I)}\right]
\end{aligned} \tag{1.163}$$

provided, however, that $S^0(I)$ and $V(I)$ were properly selected.

For example, let S be the space defined in eq. (1.61) with $p_k = 1$, $k = 1, 2 \ldots M(\Delta)$. The space $S^0(I)$ has the dimension $M(\Delta) - 1$. For $V(I)$ consider three choices:

1. $V_1(I)$ is the set of functions which are constant on each finite element. $V_1(I)$ has the dimension $M(\Delta)$.
2. $V_2(I)$ is the space S defined in (3.11) with $p_k = 1$, $k = 1, 2, \ldots, M(\Delta)$ (dimension $M(\Delta) + 1$).
3. $V_3(I)$ is the set of functions which are linear on every element and discontinuous at the nodes (dimension $2M(\Delta)$).

For these choices of $V(I)$ the mixed formulation leads to systems of linear equations with $2M(\Delta) - 1$, $2M(\Delta)$ and $3M(\Delta) - 1$ unknowns, respectively. In the cases $V = V_1$ and $V = V_3$, a constant C exists such that the inequality (1.163) holds for all u_{EX}, and F_{EX}. In the case of $V = V_2$, however, such a constant does not exist. This means that no matter how large C is, there exist some $u_{EX} \in H_0^1(I)$ and $F_{EX} \in L^2(I)$ and mesh Δ so that the inequality (1.163) is not satisfied. On the other hand, there will be $u_{EX} \in H_0^1(I)$ and $F_{EX} \in L^2(I)$ for which the inequality is satisfied and therefore the finite element solutions will converge to the underlying exact solution.

1.8.2 Nitsche's method

Nitsche's method[16] allows the treatment of essential boundary conditions as natural boundary conditions. This has certain advantages in two and three dimensions. An outline of the algorithmic aspects of the method is presented in the following. For additional details we refer to [51].

Consider the problem:

$$-u'' + cu = f(x), \qquad x \in (0, \; \ell) \tag{1.164}$$

with the boundary conditions $u'(0) = 0$ and $u(\ell) = \hat{u}_\ell$. However, at $x = \ell$ we substitute the natural boundary condition:

$$u'(\ell) = \frac{1}{\epsilon}(\hat{u}_\ell - u(\ell)) \tag{1.165}$$

---

16 Joachim Nitsche 1926–1996.

where ϵ is a small positive number, $1/\epsilon$ is called penalty parameter. The role of the penalty parameter becomes clearly visible if we consider the potential energy

$$\Pi(u) = \frac{1}{2} \int_0^\ell \left((u')^2 + cu^2\right) dx + \frac{1}{2\epsilon}(u(\ell) - \hat{u}_\ell)^2 - \int_0^\ell f(x)u \ dx. \tag{1.166}$$

Letting $\epsilon \to 0$, the minimizer of the potential energy converges to the solution of the Dirichlet problem; however, the numerical problem becomes ill-conditioned. Nitsche's method stabilizes the numerical problem making it possible to solve it for the full range of boundary conditions, including $\epsilon = 0$.

Stabilization

On multiplying eq. (1.164) by v and integrating by parts we get

$$-u'(\ell)v(\ell) + \int_0^\ell (u'v' + cuv) \ dx = \int_0^\ell f(x)v \ dx. \tag{1.167}$$

We introduce the stability parameter γ and multiply eq. (1.165) by $v(\ell)\epsilon/(\epsilon + \gamma\ell)$ to get

$$\frac{1}{\epsilon + \gamma\ell} \left(\epsilon u'(\ell)v(\ell) + u(\ell)v(\ell)\right) = \frac{1}{\epsilon + \gamma\ell}\hat{u}_\ell v(\ell). \tag{1.168}$$

Adding eq. (1.167) and eq. (1.168) we get

$$\int_0^\ell (u'v' + cuv) \ dx - \frac{\gamma\ell}{\epsilon + \gamma\ell}u'(\ell)v(\ell) + \frac{1}{\epsilon + \gamma\ell}u(\ell)v(\ell)$$

$$= \int_0^\ell f(x)v \ dx + \frac{1}{\epsilon + \gamma\ell}\hat{u}_\ell v(\ell) \tag{1.169}$$

and, multiplying eq. (1.165) by $v'(\ell)\epsilon\gamma\ell/(\epsilon + \gamma\ell)$, we have

$$\frac{\epsilon\gamma\ell}{\epsilon + \gamma\ell}u'(\ell)v'(\ell) + \frac{\gamma\ell}{\epsilon + \gamma\ell}u(\ell)v'(\ell) = \frac{\gamma\ell}{\epsilon + \gamma\ell}\hat{u}_\ell v'(\ell). \tag{1.170}$$

Subtracting eq. (1.170) from eq. (1.169) we obtain the generalized formulation:

$$\int_0^\ell (u'v' + cuv) \ dx - \frac{\gamma\ell}{\epsilon + \gamma\ell} \left(u'(\ell)v(\ell) + u(\ell)v'(\ell)\right)$$

$$+ \frac{1}{\epsilon + \gamma\ell}u(\ell)v(\ell) - \frac{\epsilon\gamma\ell}{\epsilon + \gamma\ell}u'(\ell)v'(\ell)$$

$$= \int_0^\ell f(x)v \ dx + \frac{1}{\epsilon + \gamma\ell}\hat{u}_\ell v(\ell) - \frac{\gamma\ell}{\epsilon + \gamma\ell}\hat{u}_\ell v'(\ell). \tag{1.171}$$

Letting $\epsilon = 0$ in eq. (1.171) we get the stabilized method proposed by Nitsche [67]:

$$\int_0^\ell (u'v' + cuv) \ dx - \left(u'(\ell)v(\ell) + u(\ell)v'(\ell)\right) + \frac{1}{\gamma\ell}u(\ell)v(\ell)$$

$$= \int_0^\ell f(x)v \ dx + \frac{1}{\gamma\ell}\hat{u}_\ell v(\ell) - \hat{u}_\ell v'(\ell). \tag{1.172}$$

Numerical example

Letting $c = 1$, $f(x) = 1$, $\ell = 10$ and $\hat{u}_\ell = 0.25$ we construct the numerical problem using one element and the hierarchic shape functions defined in Section 1.3.1. By definition:

$$u = \sum_{j=1}^{p+1} a_j N_j(\xi), \qquad v = \sum_{i=1}^{p+1} b_i N_i(\xi) \tag{1.173}$$

Table 1.7 The computed values of $u(\ell)$.

γ	10^{-3}	10^{-6}	10^{-9}	10^{-12}	10^{-15}
$u(\ell)$	0.2540348	0.2500004	$0.25(0)_6 4$	$0.25(0)_9 4$	$0.25(0)_{12} 4$

where p is the polynomial degree. Therefore $u(\ell) = a_2$ and $v(\ell) = b_2$ and, using the Legendre shape functions, for $p = 3$ the unconstrained coefficient matrix, without the modifications of Nitsche, is

$$[M] = \frac{2}{\ell}\begin{bmatrix} 1/2 & -1/2 & 0 & 0 \\ & 1/2 & 0 & 0 \\ (\text{sym.}) & & 1 & 0 \\ & & & 1 \end{bmatrix} + \frac{c\ell}{2}\begin{bmatrix} 2/3 & 1/3 & -1/\sqrt{6} & 1/3\sqrt{10} \\ & 2/3 & -1/\sqrt{6} & -1/3\sqrt{10} \\ (\text{sym.}) & & 2/5 & 0 \\ & & & 2/21 \end{bmatrix}$$

Referring to eq. (1.172), the coefficient matrix is modified by the application of Nitsche's method. Those modifications in the present case are:

$$[N] = \begin{bmatrix} 0 & 1/\ell & 0 & 0 \\ & -1/\ell + 1/(\gamma\ell) & -\sqrt{12}/\ell & -\sqrt{18}/\ell \\ (\text{sym.}) & & 0 & 0 \\ & & & 0 \end{bmatrix}.$$

The unconstrained right hand side vector without the modifications of Nitsche is:

$$\{r\} = \{\ell/2 \quad \ell/2 - (\ell/2)\sqrt{2/3} \quad 0\}^T$$

and with the modifications of Nitsche it is:

$$\{r_N\} = \{\hat{u}_\ell/\ell \quad \hat{u}_\ell/(\gamma\ell) - \hat{u}_\ell/\ell - \hat{u}_\ell\sqrt{12}/\ell - \hat{u}_\ell\sqrt{18}/\ell\}^T.$$

The numerical results shown in Table 1.7 indicate that the stabilized formulation is remarkably robust. The notation $(0)_n$ indicates that there are n zeros.

2

Boundary value problems

The strong forms of boundary value problems are formulated from first principles with reference the problems of heat conduction in solid bodies and elasticity. Such mathematical problems appear in various mathematical models used in numerical simulation of physical systems in structural, mechanical and aerospace engineering. The generalized (weak) formulations are derived and examples are presented. Simplifications through dimensional reduction and the use of symmetry, antisymmetry and periodicity are illustrated by examples.

It is assumed that the reader is familiar with, and has access to, at least one finite element software product. The numerical solutions presented in this book were obtained with StressCheck[1] unless otherwise noted.

2.1 Notation

The Euclidean space in n dimensions is denoted by $\mathbb{R}^n$. The Cartesian[2] coordinate axes in $\mathbb{R}^3$ are labeled x, y, z (in cylindrical systems r, θ, z) and a vector in $\mathbb{R}^n$ is denoted by $\mathbf{u}$. For example, $\mathbf{u} \equiv \{u_x\ u_y\ u_z\}$ represents a vector in $\mathbb{R}^3$.

The index notation will be introduced gradually, in parallel with the familiar Cartesian notation, so that readers who are not yet acquainted with this notation can become familiar with it. The basic rules of index notation are as follows.

1. The Cartesian coordinate axes are labeled $x = x_1, y = x_2, z = x_3$.
2. In conventional notation the position vector in $\mathbb{R}^3$ is $\mathbf{x} \equiv \{x\ y\ z\}^T$. In index notation it is simply x_i. A general vector $\mathbf{a} \equiv \{a_x\ a_y\ a_z\}$ and its transpose is written simply as a_i.
3. A free index in $\mathbb{R}^n$ is understood to range from 1 to n.
4. Two free indices represent a matrix. The size of the matrix depends on the range of indices. Thus, in three dimensions ($\mathbb{R}^3$):

$$a_{ij} \equiv \begin{bmatrix} a_{11} & a_{12} & a_{13} \\ a_{21} & a_{22} & a_{23} \\ a_{31} & a_{32} & a_{33} \end{bmatrix} \equiv \begin{bmatrix} a_{xx} & a_{xy} & a_{xz} \\ a_{yx} & a_{yy} & a_{yz} \\ a_{zx} & a_{zy} & a_{zz} \end{bmatrix}.$$

The identity matrix is represented by the Kronecker[3] delta δ_{ij}, defined as follows:

$$\delta_{ij} = \begin{cases} 1 & \text{if } i = j \\ 0 & \text{if } i \neq j. \end{cases} \tag{2.1}$$

---

1 StressCheck is a trademark of Engineering Software Research and Development, Inc.
2 René Descartes (in Latin: Renatus Cartesius) 1596–1650.
3 Leopold Kronecker 1823–1891.

Finite Element Analysis: Method, Verification and Validation, Second Edition. Barna Szabó and Ivo Babuška.
© 2021 John Wiley & Sons, Inc. Published 2021 by John Wiley & Sons, Inc.
Companion Website: www.wiley.com/go/szabo/finite_element_analysis

5. Repeated indices imply summation. For example, the scalar product of two vectors a_i and b_j is $a_i b_i \equiv a_1 b_1 + a_2 b_2 + a_3 b_3$. The product of two matrices a_{ij} and b_{ij} is written as $c_{ij} = a_{ik} b_{kj}$.

Definition 2.1 Repeated indices are also called dummy indices. This is because summation is performed therefore the index designation is immaterial. For example, $a_i b_i \equiv a_k b_k$.

6. In order to represent the cross product in index notation, it is necessary to introduce the permutation symbol e_{ijk}. The components of the permutation symbol are defined as follows:

$e_{ijk} = 0$ if the values of i, j, k do not form a permutation of 1, 2, 3

$e_{ijk} = 1$ if the values of i, j, k form an even permutation of 1, 2, 3

$e_{ijk} = -1$ if the values of i, j, k form an odd permutation of 1, 2, 3.

The cross product of vectors a_j and b_k is written as

$$c_i = e_{ijk} a_j b_k.$$

Definition 2.2 The permutations (1, 2, 3), (2, 3, 1) and (3, 1, 2) are even permutations. The permutations (1, 3, 2), (2, 1, 3) and (3, 2, 1) are odd permutations.

7. Indices following a comma represent differentiation with respect to the variables identified by the indices. For example, if $u(x_i)$ is a scalar function then

$$u_{,2} \equiv \frac{\partial u}{\partial x_2}, \qquad u_{,23} \equiv \frac{\partial^2 u}{\partial x_2 \partial x_3}.$$

The gradient of u is simply $u_{,i}$.
If $u_i = u_i(x_k)$ is a vector function in $\mathbb{R}^3$ then

$$u_{i,i} \equiv \frac{\partial u_1}{\partial x_1} + \frac{\partial u_2}{\partial x_2} + \frac{\partial u_3}{\partial x_3}$$

is the divergence of u_i.

8. The transformation rules for Cartesian vectors and tensors are presented in Appendix K.

Example 2.1 The divergence theorem in index notation is:

$$\int_\Omega u_{i,i} \, dV = \int_{\partial\Omega} u_i n_i \, dS \tag{2.2}$$

where u_i and $u_{i,i}$ are continuous on the domain Ω and its boundary $\partial\Omega$, n_i is the outward unit normal vector to the boundary, dV is the differential volume and dS is the differential surface. We will make use of the divergence theorem in the derivation of generalized formulations.

Exercise 2.1 Write out each term of $c_i = e_{ijk} a_j b_k$ where $a_i = \{a_1 \ a_2 \ a_3\}^T$ and $b_i = \{b_1 \ b_2 \ b_3\}^T$.

Exercise 2.2 Outline a derivation of the divergence theorem in two dimensions. Hint: Review the derivation of Green's theorem and cast it in the form of eq. (2.2).

2.2   The scalar elliptic boundary value problem

The three-dimensional analogue of the model problem introduced in Section 1.1 is the scalar elliptic boundary value problem

$$-\text{div}\left([\kappa]\text{grad } u\right) + cu = f(x, y, z), \qquad (x, y, z) \in \Omega \tag{2.3}$$

where

$$[\kappa] = \begin{bmatrix} \kappa_x & \kappa_{xy} & \kappa_{xz} \\ \kappa_{yx} & \kappa_y & \kappa_{yz} \\ \kappa_{zx} & \kappa_{zy} & \kappa_z \end{bmatrix} \tag{2.4}$$

is a positive-definite matrix[4] and $c = c(x, y, z) \geq 0$. In index notation eq. (2.3) reads:

$$-(\kappa_{ij}u_{,j})_{,i} + cu = f. \tag{2.5}$$

We will be concerned with the following linear boundary conditions:

1. Dirichlet boundary condition: $u = \hat{u}$ is prescribed on boundary region $\partial\Omega_u$. When $\hat{u} = 0$ on $\partial\Omega_u$ then the Dirichlet boundary condition is said to be homogeneous.
2. Neumann boundary condition: The flux vector is defined by

$$\mathbf{q} \stackrel{\text{def}}{=} -[\kappa]\text{grad } u, \quad \text{equivalently;} \quad q_i \stackrel{\text{def}}{=} \kappa_{ij}u_{,j}. \tag{2.6}$$

   The normal flux is defined by $q_n \stackrel{\text{def}}{=} \mathbf{q} \cdot \mathbf{n} \equiv q_i n_i$ where $\mathbf{n} \equiv n_i$ is the unit outward normal to the boundary. When $q_n = \hat{q}_n$ is prescribed on boundary region $\partial\Omega_q$ then the boundary condition is called a Neumann boundary condition. When $\hat{q}_n = 0$ on $\partial\Omega_q$ then the Neumann boundary condition is said to be homogeneous.
3. Robin boundary condition: $q_n = h_R(u - u_R)$ is given on boundary segment $\partial\Omega_R$. In this expression $h_R > 0$ and u_R are given functions. When $u_R = 0$ on $\partial\Omega_R$ then the Robin boundary condition is said to be homogeneous.
4. Boundary conditions of convenience: In many instances the solution domain can be simplified through taking advantage of symmetry, antisymmetry and/or periodicity. These boundary conditions are called boundary conditions of convenience.

The boundary segments $\partial\Omega_u$, $\partial\Omega_q$, $\partial\Omega_R$ and $\partial\Omega_p$ are non-overlapping and collectively cover the entire boundary $\partial\Omega$. Any of the boundary segments may be empty.

Definition 2.3   The Dirichlet boundary condition is also called essential boundary condition. The Neumann and Robin conditions area called natural boundary conditions.

2.2.1   Generalized formulation

To obtain the generalized formulation for the scalar elliptic boundary value problem we multiply eq. (2.5) by a test function v and integrate over the domain Ω:

$$-\int_\Omega (\kappa_{ij}u_{,j})_{,i} v \, dV + \int_\Omega cuv \, dV = \int_\Omega fv \, dV. \tag{2.7}$$

---

4 A symmetric matrix $[\kappa]$ of real numbers is positive-definite if $\mathbf{x}^T[\kappa]\mathbf{x} > 0$ for any $\mathbf{x} \neq \mathbf{0}$.

This equation must hold for arbitrary v, provided that the indicated operations are defined. The first integral can be written as:

$$\int_\Omega (\kappa_{ij} u_j)_{,i} v \, dV = \int_\Omega (\kappa_{ij} u_j v)_{,i} \, dV - \int_\Omega \kappa_{ij} u_j v_{,i} \, dV.$$

Applying the divergence theorem (eq. (2.2)) we have:

$$\int_\Omega (\kappa_{ij} u_j v)_{,i} \, dV = \int_{\partial\Omega} \kappa_{ij} u_j n_i v \, dS$$

where n_i is the unit normal vector to the boundary surface. Therefore eq. (2.7) can be written in the following form:

$$-\int_{\partial\Omega} \kappa_{ij} u_j n_i v \, dS + \int_\Omega \kappa_{ij} u_j v_{,i} \, dV + \int_\Omega cuv \, dV = \int_\Omega fv \, dV. \tag{2.8}$$

It is customary to write

$$q_i = -\kappa_{ij} u_j \quad \text{and} \quad q_n = q_i n_i.$$

With this notation we have:

$$\int_\Omega \kappa_{ij} u_j v_{,i} \, dV + \int_\Omega cuv \, dV = \int_\Omega fv \, dV - \int_{\partial\Omega} q_n v \, dS. \tag{2.9}$$

This is the generalization of eq. (1.18) to two and three dimensions. As we have seen in Section 1.2, the specific statement of a generalized formulation depends on the boundary conditions. In the general case $u = \hat{u}$ is prescribed on $\partial\Omega_u$ (Dirichlet boundary condition); $q_n = \hat{q}_n$ is prescribed on $\partial\Omega_q$ (Neumann boundary condition) and $q_n = h_R(u - u_R)$ is prescribed on Ω_R (Robin boundary condition), see Section 2.2. We now define the bilinear form as follows:

$$B(u, v) = \int_\Omega \kappa_{ij} u_j v_{,i} \, dV + \int_\Omega cuv \, dV + \int_{\partial\Omega_R} h_R uv \, dS \tag{2.10}$$

and the linear functional:

$$F(v) = \int_\Omega fv \, dV - \int_{\partial\Omega_q} q_n v \, dS + \int_{\partial\Omega_R} h_R u_R v \, dS. \tag{2.11}$$

When $\partial\Omega_R$ is empty then the last terms in equations (2.10) and (2.11) are omitted. When Neumann condition is prescribed on the entire boundary and $c = 0$ then the data must satisfy the following condition:

$$\int_\Omega f \, dV = \int_{\partial\Omega} q_n \, dS. \tag{2.12}$$

The space $E(\Omega)$ is defined by

$$E(\Omega) \stackrel{\text{def}}{=} \{u \mid B(u, u) < \infty\}$$

and the energy norm

$$\|u\|_E \stackrel{\text{def}}{=} \sqrt{\frac{1}{2} B(u, u)}$$

is associated with $E(\Omega)$. The space of admissible functions is defined by:

$$\tilde{E}(\Omega) \stackrel{\text{def}}{=} \{u \mid u \in E(\Omega), \ u = \hat{u} \text{ on } \partial\Omega_u\}.$$

Here we assume that corresponding to any $u = \hat{u}$ specified on $\partial\Omega_u$ there is a $u^\star \in E(\Omega)$ such that $u^\star = \hat{u}$ on $\partial\Omega_u$. This imposes certain restrictions on $\hat{u}$ and ensures that $\tilde{E}(\Omega)$ is not empty. The space of test functions is defined by:

$$E^0(\Omega) \overset{\text{def}}{=} \{u \mid u \in E(\Omega),\ u = 0 \text{ on } \partial\Omega_u\}.$$

The generalized formulation is now stated as follows: "Find $u \in \tilde{E}(\Omega)$ such that $B(u, v) = F(v)$ for all $v \in E^0(\Omega)$". A function u that satisfies this condition is called a generalized solution.

The generalized formulation is often stated as a minimization problem. The potential energy defined by

$$\pi(u) \overset{\text{def}}{=} \frac{1}{2}B(u, u) - F(u) \tag{2.13}$$

is formally identical to eq. (1.36) and Theorem 1.2 is applicable: The exact solution of the generalized formulation minimizes the potential energy on the space $\tilde{E}(\Omega)$. Alternatively, the potential energy is defined by

$$\Pi(u) \overset{\text{def}}{=} \frac{1}{2}\int_\Omega \kappa_{ij}u_{,i}u_{,j}\ dV + \frac{1}{2}\int_\Omega cu^2\ dV + \frac{1}{2}\int_{\partial\Omega_R} h_R(u - u_R)^2\ dS$$
$$- \int_\Omega fu\ dV + \int_{\partial\Omega_q} q_n u\ dS \tag{2.14}$$

so that when $u = u_R$ on $\partial\Omega_R$ and $\Omega_R = \Omega$ then $\Pi(u) = 0$. Note that $\pi(u)$ differs from $\Pi(u)$ only by a constant. Therefore the minimizer of $\pi(u)$ is the same as the minimizer of $\Pi(u)$.

Exercise 2.3   Consider the case $\Omega_R = \Omega$, that is Robin boundary conditions are prescribed on the entire boundary. Compare the definitions of the potential energy given by equations (2.13) and (2.14).

Exercise 2.4   Following the proof of Theorem 1.2, show that the generalized formulation minimizes $\Pi(u)$ given by eq. (2.14).

2.2.2   Continuity

In two and three dimensions $u \in E(\Omega)$ is not necessarily continuous or bounded. For example, the function $u = \log|\log r|$ where $r = (x^2 + y^2)^{1/2}$ is discontinuous and unbounded in the point $r = 0$, yet it lies in $E(\Omega)$. This has important implications: Concentrated fluxes are inadmissible data. Similarly, point constraints are inadmissible except for the enforcement of the uniqueness of the solution when $c = 0$ and Neumann conditions are specified on the entire boundary and eq. (2.12) is satisfied.

Exercise 2.5   Let $r = (x^2 + y^2)^{1/2}$ and $\Omega = \{r \mid r \le \rho_0 < 1\}$. Show that $u_1 = \log r$ does not lie in $E(\Omega)$ but $u_2 = \log|\log r|$ does. Hint: $u_2 = \log(-\log r)$ when $r < 1$.

2.3   Heat conduction

Steady state potential flow problems are among the physical phenomena that can be modeled as scalar elliptic boundary value problems. In this section the formulation of a mathematical problem that models heat flow by conduction in solid bodies is described.

Mathematical models of heat conduction are based on two fundamental relationships: the conservation law and Fourier's law of heat conduction described in the following.

1. **The conservation law** states that the quantity of heat entering any volume element of the conducting medium equals the quantity of heat exiting the volume element plus the quantity of heat retained in the volume element. The heat retained causes a change in temperature in the volume element which is proportional to the specific heat of the conducting medium c (in J/(kg K) units) multiplied by the density ρ (in kg/m^3 units). The temperature will be denoted by $u(x, y, z, t)$ where t is time.

   The heat flow rate across a unit area is represented by a vector quantity called heat flux. The heat flux is in W/m^2 units, or equivalent, and will be denoted by $\mathbf{q} = \mathbf{q}(x, y, z, t) = \{q_x(x, y, z, t)\ q_y(x, y, z, t)\ q_z(x, y, z, t)\}^T$. In addition to heat flux entering and leaving the volume element, heat may be generated within the volume element, for example from chemical reactions. The heat generated per unit volume and unit time will be denoted by Q (in W/m^3 units).

   Applying the conservation law to the volume element shown in Fig. 2.1, we have:

   $$\Delta t[q_x \Delta y \Delta z - (q_x + \Delta q_x)\Delta y \Delta z + q_y \Delta x \Delta z - (q_y + \Delta q_y)\Delta x \Delta z +$$
   $$q_z \Delta x \Delta y - (q_z + \Delta q_z)\Delta x \Delta y + Q \Delta x \Delta y \Delta z] = c\rho \Delta u \Delta x \Delta y \Delta z. \quad (2.15)$$

   Assuming that u and $\mathbf{q}$ are continuous and differentiable and neglecting terms that go to zero faster than $\Delta x, \Delta y, \Delta z, \Delta t$, we have:

   $$\Delta q_x = \frac{\partial q_x}{\partial x}\Delta x, \quad \Delta q_y = \frac{\partial q_y}{\partial y}\Delta y, \quad \Delta q_z = \frac{\partial q_z}{\partial z}\Delta z, \quad \Delta u = \frac{\partial u}{\partial t}\Delta t.$$

   On factoring $\Delta x \Delta y \Delta z \Delta t$ the conservation law is obtained:

   $$-\frac{\partial q_x}{\partial x} - \frac{\partial q_y}{\partial y} - \frac{\partial q_z}{\partial z} + Q = c\rho \frac{\partial u}{\partial t}. \quad (2.16)$$

   In index notation:

   $$-q_{i,i} + Q = c\rho \frac{\partial u}{\partial t}. \quad (2.17)$$

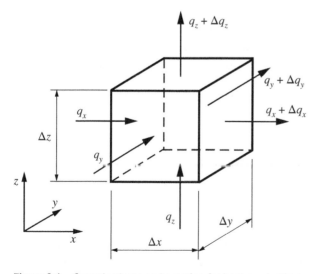

Figure 2.1 Control volume and notation for heat conduction.

2. **Fourier's law** states that the heat flux vector is related to the temperature gradient in the following way:

$$q_x = -\left(k_{xx}\frac{\partial u}{\partial x} + k_{xy}\frac{\partial u}{\partial y} + k_{xz}\frac{\partial u}{\partial z}\right) \tag{2.18}$$

$$q_y = -\left(k_{yx}\frac{\partial u}{\partial x} + k_{yy}\frac{\partial u}{\partial y} + k_{yz}\frac{\partial u}{\partial z}\right) \tag{2.19}$$

$$q_z = -\left(k_{zx}\frac{\partial u}{\partial x} + k_{zy}\frac{\partial u}{\partial y} + k_{zz}\frac{\partial u}{\partial z}\right) \tag{2.20}$$

where the coefficients k_{xx}, k_{xy}, $\ldots$, k_{zz} are called coefficients of thermal conduction (measured in W/(mK) units). It is customary to write $k_x = k_{xx}, k_y = k_{yy}, k_z = k_{zz}$. The coefficients of thermal conduction will be assumed to be independent of the temperature u, unless otherwise stated. Denoting the matrix of coefficients by $[K]$, Fourier's law of heat conduction can be written as:

$$\mathbf{q} = -[K]\operatorname{grad} u. \tag{2.21}$$

The matrix of coefficients $[K]$ is symmetric and positive-definite. The negative sign indicates that the direction of heat flow is opposite to the direction of the temperature gradient, that is, the direction of heat flow is from high to low temperature. In index notation eq. (2.21) is written as:

$$q_i = -k_{ij}u_j. \tag{2.22}$$

For isotropic materials $k_{ij} = k\delta_{ij}$.

2.3.1   The differential equation

Combining equations (2.16) through (2.20), we have:

$$\frac{\partial}{\partial x}\left(k_x\frac{\partial u}{\partial x} + k_{xy}\frac{\partial u}{\partial y} + k_{xz}\frac{\partial u}{\partial z}\right) + \frac{\partial}{\partial y}\left(k_{yx}\frac{\partial u}{\partial x} + k_y\frac{\partial u}{\partial y} + k_{yz}\frac{\partial u}{\partial z}\right) +$$

$$\frac{\partial}{\partial z}\left(k_{zx}\frac{\partial u}{\partial x} + k_{zy}\frac{\partial u}{\partial y} + k_z\frac{\partial u}{\partial z}\right) + Q = c\rho\frac{\partial u}{\partial t} \tag{2.23}$$

which can be written in the following compact form:

$$\operatorname{div}\left([K]\operatorname{grad} u\right) + Q = c\rho\frac{\partial u}{\partial t} \tag{2.24}$$

or in index notation:

$$(k_{ij}u_j)_{,i} + Q = c\rho\frac{\partial u}{\partial t}. \tag{2.25}$$

In many practical problems u is independent of time. Such problems are called stationary or steady state problems. The solution of a stationary problem can be viewed as the solution of some time-dependent problem, with time-independent boundary conditions, at $t = \infty$.

In formulating eq. (2.23) we assumed that k_{ij} are differentiable functions. In many practical problems the solution domain is comprised of subdomains Ω_i that have different material properties. In such cases eq. (2.23) is valid on each subdomain. On the boundaries of adjoining subdomains continuous temperature and flux are prescribed.

To complete the definition of a mathematical model, initial and boundary conditions have to be specified. This is discussed in the following section.

2.3.2 Boundary and initial conditions

The solution domain will be denoted by Ω and its boundary by $\partial\Omega$. We will consider three kinds of boundary conditions:

1. Prescribed temperature (Dirichlet condition): The temperature $u = \hat{u}$ is prescribed on boundary region $\partial\Omega_u$.
2. Prescribed flux (Neumann condition): The flux vector component normal to the boundary, denoted by q_n, is prescribed on the boundary region $\partial\Omega_q$. By definition;

$$q_n \overset{\text{def}}{=} \mathbf{q} \cdot \mathbf{n} \equiv -([K]\,\text{grad}\,u) \cdot \mathbf{n} \equiv -k_{ij}u_{,j}n_i \tag{2.26}$$

where $\mathbf{n} \equiv n_i$ is the (outward) unit normal to the boundary. The prescribed flux on $\partial\Omega_q$ will be denoted by $\hat{q}_n$.
3. Convection (Robin condition): On boundary region $\partial\Omega_c$ the flux vector component q_n is proportional to the difference between the temperature of the boundary and the temperature of a convective medium:

$$q_n = h_c(u - u_c), \qquad (x, y, z) \in \partial\Omega_c \tag{2.27}$$

where h_c is the coefficient of convective heat transfer in $\text{W}/(\text{m}^2\text{K})$ units and u_c is the (known) temperature of the convective medium.

The sets $\partial\Omega_u$, $\partial\Omega_q$ and $\partial\Omega_c$ are non-overlapping and collectively cover the entire boundary. Any of the sets may be empty.

The boundary conditions may be time-dependent. For time-dependent problems an initial condition has to be prescribed on Ω: $u(x, y, z, 0) = U(x, y, z)$.

It is possible to show that eq. (2.23), subject to the enumerated boundary conditions, has a unique solution. Stationary problems also have unique solutions, subject to the condition that when flux is prescribed over the entire boundary $\partial\Omega$ then the following condition must be satisfied:

$$\int_\Omega Q\, dV = \int_{\partial\Omega} q_n\, dS. \tag{2.28}$$

This is easily seen by integrating

$$(k_{ij}u_{,j})_{,i} + Q = 0 \tag{2.29}$$

on Ω and using the divergence theorem, eq. (2.2) and the definition (2.26).

Note that if u_i is a solution of eq. (2.29) then $u_i + C$ is also a solution, where C is an arbitrary constant. Therefore the solution is unique up to an arbitrary constant.

In addition to the three types of boundary conditions discussed in this section, radiation may have to be considered. When two bodies exchange heat by radiation then the flux is proportional to the difference of the fourth power of their absolute temperatures, therefore radiation is a non-linear boundary condition. The boundary region subject to radiation, denoted by $\partial\Omega_r$, may overlap $\partial\Omega_c$. Radiation is discussed in Section 9.1.1.

In the following it will be assumed that the coefficients of thermal conduction, the flux prescribed on Ω_q and the coefficient h_c prescribed on Ω_c are independent of the temperature. This assumption can be justified on the basis of empirical data in a narrow range of temperatures only.

Exercise 2.6 Discuss the physical interpretation of eq. (2.28).

Exercise 2.7 Show that in cylindrical coordinates r, θ, z the conservation law is of the form:

$$-\frac{1}{r}\frac{\partial(rq_r)}{\partial r} - \frac{1}{r}\frac{\partial q_\theta}{\partial \theta} - \frac{\partial q_z}{\partial z} + Q = c\rho\frac{\partial u}{\partial t}. \tag{2.30}$$

Use two methods: (a) apply the conservation law to an infinitesimal volume element in cylindrical coordinates and (b) transform eq. (2.16) to cylindrical coordinates.

Exercise 2.8 Show that there are three mutually perpendicular directions (called principal directions) such that the heat flux is proportional to the (negative) gradient vector. Hint: Consider steady state heat conduction and let:

$$[K]\,\text{grad}\,u = \lambda\,\text{grad}u$$

then show that the principal directions are defined by the normalized eigenvectors.

Remark 2.1 The result of Exercise 2.8 implies that the general form of matrix [K] can be obtained by rotation from orthotropic material axes.

Exercise 2.9 List all of the physical assumptions incorporated into the mathematical model represented by eq. (2.23) and the boundary conditions described in this section.

2.3.3 Boundary conditions of convenience

A scalar function is said to be symmetric with respect to a plane of symmetry if in symmetrically located points the function has equal values. On a plane of symmetry $q_n = 0$. A function is said to be antisymmetric with respect to a plane of symmetry if in symmetrically located points the function has equal absolute values but opposite sign. On a plane of antisymmetry $u = 0$.

In many instances the domain has one or more planes of symmetry, antisymmetry and/or it may be periodic. For example, the domain shown in Fig. 2.2 has one plane of symmetry; the plane $x = 0$, and it is periodic; the shaded subdomain is replicated five times. If the material properties, source function and boundary conditions are also symmetric, antisymmetric and/or periodic then it is often advantageous and convenient to formulate the problem on a subdomain and extend the solution to the entire domain by symmetry, antisymmetry or periodicity.

Figure 2.2 Notation for Example 2.2.

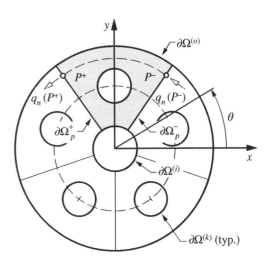

When Ω, $[K]$, Q and the boundary conditions are periodic then a periodic sector of Ω has a periodic boundary segment pair denoted by $\partial\Omega_p^+$ and $\partial\Omega_p^-$. On corresponding points of a periodic boundary segment pair, $P^+ \in \partial\Omega_p^+$ and $P^- \in \partial\Omega_p^-$, the boundary conditions are: $u(P^+) = u(P^-)$ and $q_n^+ = -q_n^-$.

Symmetric, antisymmetric and periodic boundary conditions are illustrated by the following example.

Example 2.2 Fig, 2.2 is the plan view of a plate-like body of constant thickness t. We assume that the surfaces $z = \pm t/2$ are perfectly insulated, the source function is zero and the material is isotropic. We further assume that on the cylindrical boundary represented by the inner circle ($\partial\Omega^{(i)}$) constant temperature u_0 is prescribed and on the boundary represented by the outer circle ($\partial\Omega^{(o)}$) flux $q_n(\theta)$ is prescribed. On the boundaries of the five circular cut-outs ($\partial\Omega^{(k)}$, $k = 1, 2, \ldots, 5$) $q_n = 0$.

Some examples of symmetric, antisymmetric and periodic functions are given here with reference to Fig. 2.2. Let a, b and $c > 0$ be given real numbers. The restriction $c > 0$ is necessary because the temperature u is in K (Kelvin[5]) units.

1. The functions $q_n = a + b\cos n\theta$ and $u_0 = c$ are symmetric with respect to the y axis for $n = 2, 4, 6, \ldots$ This problem can be solved on the half domain with $q_n = 0$ on the boundary coincident with the y axis.
2. The functions $q_n = b\cos n\theta$ and $u_0 = 0$ are antisymmetric with respect to the y axis for $n = 1, 3, 5, \ldots$ This problem can be solved on the half domain with $u = 0$ on the boundary coincident with the y axis.
3. The functions $q_n = a + b\cos(5n(\theta - \pi/2))$ and $u_0 = c$ are periodic for $n = 1, 2, 3, \ldots$ This problem can be solved on a periodic subdomain, for example the shaded sector shown in Fig. 2.2.

Exercise 2.10 Suppose that the problem described in Example 2.2 is modified such that on the boundary $\partial\Omega^{(i)}$ flux $q_n^{(i)}(\theta)$ is prescribed, which is a periodic function. Assuming that on $\partial\Omega^{(o)}$ $q_n = a + b\cos(5(\theta - \pi/2))$, what restriction must be imposed on $q_n^{(i)}(\theta)$? Hint: Refer to eq. (2.28.)

Numerical treatment of periodic functions

Direct implementation of periodic boundary conditions for arbitrary forcing functions leads to minimization of the potential energy subject to linear constraints. The constraint conditions are coupled on periodic boundary segment pairs. However, in the special cases of even and odd loading functions, symmetric periodic subdomains and symmetric material properties on periodic subdomains, the periodic boundary conditions are not coupled. In fact, the boundary conditions are the same as the conditions of symmetry and antisymmetry respectively, which makes implementation much simpler.

A function $f(x)$ defined on the interval $-\ell/2 < x < \ell/2$ is an even function if $f(x) = f(-x)$. It is an odd function if $f(x) = -f(-x)$. Examples of even functions on $-\infty < x < \infty$ are $\cos x$, $\cosh x$. Examples of odd functions are $\sin x$, $\sinh x$. Any function can written as the sum of an even and an odd function:

$$f(x) = f_e(x) + f_o(x) \tag{2.31}$$

where

$$f_e(x) = \frac{1}{2}(f(x) + f(-x)) \tag{2.32}$$

---

5 William Thomson, Lord Kelvin 1824–1907.

is an even function and

$$f_o(x) = \frac{1}{2}(f(x) - f(-x)) \tag{2.33}$$

is an odd function.

Example 2.3 Consider the domain shown in Fig. 2.3. We assume that the source function is zero and the loading function on the boundary $\partial\Omega^{(o)}$ is the sum of a symmetric (even) and an antisymmetric (odd) function:

$$q_n = \cos(5(\theta - \pi/2)) + \sin(5(\theta - \pi/2)), \quad 0 \leq \theta < 2\pi.$$

Let us assume that on all other boundary segments $q_n = 0$ and note that q_n satisfies eq. (2.28). The temperature $u(x, y)$ is determined by these boundary conditions up to an arbitrary constant. We will fix this constant to be 300 K.

The radius of circle $\partial\Omega^{(o)}$ is $r_o = 100$ mm. The radius of circle $\partial\Omega^{(i)}$ is $r_o = 22.5$ mm. The radius of the cut-out circles $\partial\Omega^{(k)}$ ($k = 1, 2, \ldots, 5$) is $r_k = 16.875$ mm. The centers of the five cut-out circles are evenly distributed on a circle of radius $r_m = 61.25$ mm. The thickness is $t = 5$ mm. The material is homogeneous and isotropic, the coefficient of thermal conduction is 0.161 W/(mm²K).

The solution u on a periodic segment is shown in Fig. 2.3(a). It is the sum of its symmetric part shown in Fig. 2.3(b) and its antisymmetric part shown in Fig. 2.3(c). For the symmetric part the boundary condition on the periodic boundary segment pair $\partial\Omega_p^+$, $\partial\Omega_p^-$ is $q_n = 0$, for the antisymmetric part it is $u = 0$.

2.3.4 Dimensional reduction

In many important practical applications reduction of the number of dimensions is possible without significantly affecting the quantities of interest. In other words, a mathematical model in one or two dimensions may be an acceptable substitute for the fully three-dimensional model.

Planar problems
Consider a plate-like body shown in Fig. 2.4. The thickness, denoted by t_z, will be assumed to be constant. The mid-surface is the solution domain which will be denoted by Ω. The solution domain lies in the xy plane. The boundary points of Ω (shown as a dotted line) is denoted by Γ. The unit outward normal to the boundary is denoted by **n**.

The heat conduction problem in two dimensions is a special case of the three-dimensional problem, applicable when the boundary conditions on the surface parallel to the z axis and Q are

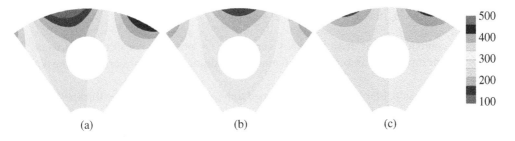

Figure 2.3 Example 2.3: The solution u of (a) the periodic problem, (b) the symmetric part of the periodic problem, (c) the antisymmetric part of the periodic problem.

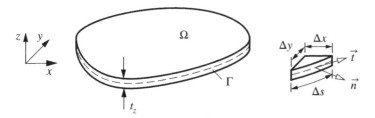

Figure 2.4 Notation for two-dimensional domains.

independent of z

$$\frac{\partial}{\partial x}\left(k_x\frac{\partial u}{\partial x}+k_{xy}\frac{\partial u}{\partial y}\right)+\frac{\partial}{\partial y}\left(k_{yx}\frac{\partial u}{\partial x}+k_y\frac{\partial u}{\partial y}\right)+\overline{Q}=c\rho\frac{\partial u}{\partial t} \qquad (2.34)$$

where the meaning of $\overline{Q}$ depends on the boundary conditions prescribed on the top and bottom surfaces ($z=\pm t_z/2$) as described in the following.

Eq. (2.34) represents one of two cases:

1. The thickness is large in comparison with the other dimensions and both the material properties and boundary conditions are independent of z, i.e., $u(x,y,z)=u(x,y)$. This is equivalent to the case of finite thickness with the top and bottom surfaces ($z=\pm t_z/2$) perfectly insulated and $\overline{Q}=Q$.

2. The thickness is small in relation to the other dimensions. In this case the two-dimensional solution is an approximation of the three-dimensional solution. It can be interpreted as the first term in the expansion of the three-dimensional solution with respect to the z coordinate. The definition of $\overline{Q}$ depends on the boundary conditions on the top and bottom surfaces as explained here.

   a) Prescribed flux: Let us denote the heat flux prescribed on the top (resp. bottom) surface by $\hat{q}_n^+$ (resp. $\hat{q}_n^-$). Note that positive $\hat{q}_n$ is heat flux exiting the body. The amount of heat exiting the body over a small area ΔA per unit time is $(\hat{q}_n^+ + \hat{q}_n^-)\Delta A$. Dividing by $\Delta A t_z$, the heat generated per unit volume becomes

   $$\overline{Q}=Q-(\hat{q}_n^+ + \hat{q}_n^-)\frac{1}{t_z}. \qquad (2.35)$$

   b) Convective heat transfer: Let us denote the coefficient of convective heat transfer at $z=t_z/2$ (resp. $z=-t_z/2$) by h_c^+ (resp. h_c^-) and the corresponding temperature of the convective medium by u_c^+ (resp. u_c^-) then the amount of heat exiting the body over a small area ΔA per unit time is $[h_c^+(u-u_c^+)+h_c^-(u-u_c^-)]\Delta A$. Therefore the heat generated per unit volume is changed to

   $$\overline{Q}=Q-[h_c^+(u-u_c^+)+h_c^-(u-u_c^-)]\frac{1}{t_z}. \qquad (2.36)$$

Of course, combinations of these boundary conditions are possible. For example flux may be prescribed on the top surface and convective boundary conditions may be prescribed on the bottom surface.

In the two-dimensional formulation u is assumed to be constant through the thickness. Therefore prescribing a temperature has meaning only if the temperature is the same on the top and bottom surfaces. Furthermore, the temperature is a continuous function therefore specification of the temperature on the top and bottom surfaces determines the temperature on the side surfaces as well. In this case the solution is the prescribed temperature.

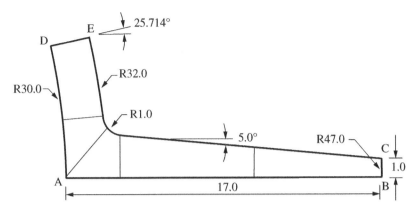

Figure 2.5 Example 2.4: The solution domain and finite element mesh (mm).

Figure 2.6 Control volume and notation for heat conduction in 2D.

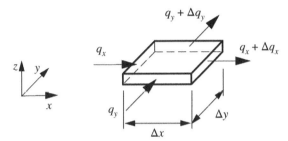

Example 2.4 Consider the mathematical problem associated with estimating heat loss per unit length in a stainless steel pipe with cooling fins. The inner radius of the pipe is 30.0 mm, the outer radius is 32.0 mm. There are seven equally spaced cooling fins. One-fourteenth of the cross-section is shown in Fig. 2.5.

The coefficient of thermal conductivity is 0.0236 W/(mmK) and the coefficient of convective heat transfer is 1.8×10^{-4} W/(mm²K). The temperature of the internal surface is 800 K. The external surface is cooled by convection. The temperature of the convective medium is 300 K.

The solution is a periodic function which is even with respect to segment AB. Therefore the normal flux is zero on boundary segments AB and DE, i.e., the boundary conditions are symmetry boundary condition. The estimated heat loss converges to 1816 W/m for the sector shown in Fig. 2.5. The heat loss for the entire pipe is 14 times this number.

Exercise 2.11 Investigate how the length of the fin influences heat loss. Solve the problem of Example 2.4 letting dimension AB vary between 7 and 17 mm.

Exercise 2.12 Using the control volume shown in Fig. 2.6, derive eq. (2.34) from first principles.

Exercise 2.13 Assume that $t_z = t_z(x, y) > 0$ is a continuous and differentiable function and the maximum value of t_z is small in comparison with the other dimensions. Assume further that convective heat transfer occurs on the surfaces $z = \pm t_z/2$ with $h_c^+ = h_c^- = h_c$ and $u_c^+ = u_c^- = u_c$. Using a control volume similar to that shown in Fig. 2.6, but accounting for variable thickness, show that in this case the conservation law for heat conduction in two dimensions is:

$$-\frac{\partial}{\partial x}(t_z q_x) - \frac{\partial}{\partial y}(t_z q_y) - 2h_c(u - u_c) + Qt_z = c\rho t_z \frac{\partial u}{\partial t}. \tag{2.37}$$

Remark 2.2 In Exercise 2.13 the thickness t_z was assumed to be continuous and differentiable on Ω. If t_z is continuous and differentiable over two or more subdomains of Ω, but discontinuous on the boundaries of the subdomains, then eq. (2.37) is applicable on each subdomain subject to the requirement that $q_n t_z$ is continuous on the boundaries of the subdomains.

Example 2.5 Let the domain

$$\Omega = \{x, y \mid -b < x < b, \; -c < x < c\}$$

represent the mid-surface of a rectangular plate of constant thickness t_z. Assume that the surfaces $(z = \pm t_z/2)$ are perfectly insulated. On the side surfaces $(x = \pm b, y = \pm c, -t/2 < z < t/2)$ the temperature is constant ($u = \hat{u}_0$). Let $k_x = k_y = k$, $k_{xy} = 0$ and $Q = Q_0$ where k and Q_0 are constants. The goal is to determine the stationary temperature distribution in the plate. The mathematical problem is to solve:

$$k\left(\frac{\partial^2 u}{\partial x^2} + \frac{\partial^2 u}{\partial y^2}\right) + Q_0 = 0 \tag{2.38}$$

on Ω with the boundary condition $u = \hat{u}_0$.

The solution of this problem can be determined by classical methods[6]:

$$u = \hat{u}_0 + 16\frac{Q_0}{k}\frac{b^2}{\pi^3}\sum_{n=1,3,5,\ldots}^{\infty}\frac{(-1)^{\frac{n-1}{2}}}{n^3}\left(1 - \frac{\cosh n\pi y/2b}{\cosh n\pi c/2b}\right)\cos n\pi x/2b. \tag{2.39}$$

This infinite series converges absolutely. It is seen that the classical solution of this seemingly simple problem is rather complicated and in fact the exact solution can be computed only approximately. However, the truncation error can be made arbitrarily small by computing a sufficiently large number of terms of the infinite series.

The solution would be far more complicated if (for example) Ω were a general polygonal domain.

Exercise 2.14 Write down the generalized formulation of the problem in Example 2.5 for (a) the entire domain and (b) taking advantage of symmetry, for one quarter of the domain. Specify $B(u, v)$, $F(v)$ and the trial and test spaces for both cases.

Exercise 2.15 Let Ω be a regular pentagon (i.e. a pentagon with equal sides). Taking advantage of the periodicity of the solution, write down the generalized formulation corresponding to the problem given by eq. (2.38) on the smallest domain possible. Assume that $u = 0$ on the boundaries. Specify the domain, $B(u, v)$, $F(v)$ and the trial and test spaces.

Exercise 2.16 Assume that the coefficients of thermal conduction k_x, k_{xy}, k_y are given. Show that in a Cartesian coordinate system $x'y'$, rotated counterclockwise by the angle α relative to the xy system, the coefficients will be:

$$\begin{bmatrix} k_{x'} & k_{x'y'} \\ k_{y'x'} & k_{y'} \end{bmatrix} = \begin{bmatrix} \cos\alpha & \sin\alpha \\ -\sin\alpha & \cos\alpha \end{bmatrix} \begin{bmatrix} k_x & k_{xy} \\ k_{yx} & k_y \end{bmatrix} \begin{bmatrix} \cos\alpha & -\sin\alpha \\ \sin\alpha & \cos\alpha \end{bmatrix}.$$

---

6 See, for example, Timoshenko, S. and Goodier, J. N., *Theory of Elasticity*, McGraw-Hill, New York, 2nd ed. 1951, pp. 275-276.

Hint: A scalar $\{a\}^T[K]\{a\}$, where $\{a\}$ is an arbitrary vector, is invariant under coordinate transformation by rotation.

Axisymmetric models

Axial symmetry exists when the solution domain can be generated by sweeping a plane figure around an axis, known as the axis of symmetry, and the material properties and boundary conditions are axially symmetric. For example, pipes, cylindrical and spherical pressure vessels are often idealized in this way. In such cases the problem can be formulated in cylindrical coordinates and, since the solution is independent of the circumferential variable, the number of dimensions is reduced to two. In the following the z axis will be the axis of symmetry and the radial (resp. circumferential) coordinates will be denoted by r (resp. θ). Referring to the result of Exercise 2.7, and letting $q_\theta = 0$, the conservation law is

$$-\frac{1}{r}\frac{\partial(rq_r)}{\partial r} - \frac{\partial q_z}{\partial z} + Q = c\rho\frac{\partial u}{\partial t}.$$

Substituting the axisymmetric form of Fourier's law:

$$q_r = -k_r\frac{\partial u}{\partial r}, \qquad q_z = -k_z\frac{\partial u}{\partial z}$$

we have the formulation of the axisymmetric heat conduction problem in cylindrical coordinates:

$$\frac{1}{r}\frac{\partial}{\partial r}\left(rk_r\frac{\partial u}{\partial r}\right) + \frac{\partial}{\partial z}\left(k_z\frac{\partial u}{\partial z}\right) + Q = c\rho\frac{\partial u}{\partial t}. \tag{2.40}$$

One or more segments of the boundary may lie on the z axis. Implied in the formulation is that the boundary condition is the zero flux condition on those segments. Therefore it would not be meaningful to prescribe essential boundary conditions on those boundary segments. To show this, consider an axisymmetric problem of heat conduction, the solution of which is independent of z. For simplicity we assume that $k_r = 1$. In this case the problem is essentially one-dimensional:

$$\frac{1}{r}\frac{d}{dr}\left(r\frac{du}{dr}\right) = 0 \qquad r_i < r < r_o.$$

Assuming that the boundary conditions $u(r_i) = \hat{u}_i$, $u(r_o) = \hat{u}_o$ are prescribed. the exact solution of this problem is:

$$u(r) = \frac{\hat{u}_o - \hat{u}_i}{\ln r_o - \ln r_i}\ln r + \frac{\hat{u}_i\ln r_o - \hat{u}_o\ln r_i}{\ln r_o - \ln r_i}. \tag{2.41}$$

Consider now the solution in an arbitrary fixed point $r = \varrho$ where $r_i < \varrho < r_o$ and let $r_i \to 0$:

$$\lim_{r_i \to 0} u(\varrho) = \lim_{r_i \to 0}\left(\frac{\hat{u}_o - \hat{u}_i}{\ln r_o/\ln r_i - 1}\frac{\ln \varrho}{\ln r_i} + \frac{\hat{u}_i\ln r_o/\ln r_i - \hat{u}_o}{\ln r_o/\ln r_i - 1}\right) = \hat{u}_o.$$

Therefore the solution is independent of $\hat{u}_i$ when $r_i = 0$. It is left to the reader in Exercise 2.17 to show that $du/dr \to 0$ as $r_i \to 0$, hence the boundary condition at $r = 0$ is the zero flux boundary condition.

Exercise 2.17   Refer to the solution given by eq. (2.41). Show that for any $\varrho > 0$

$$\left(\frac{du}{dr}\right)_{r=\varrho} \to 0 \quad \text{as} \quad r_i \to 0$$

independently of $\hat{u}_i$ and $\hat{u}_o$.

Exercise 2.18 Derive eq. (2.40) by considering a control volume in cylindrical coordinates and using the assumption that the temperature is independent of the circumferential variable.

Exercise 2.19 Consider the generalized formulation of steady state heat conduction in cylindrical coordinates in the special case when the solution depends only on the radial variable r:

$$\int_{r_i}^{r_o} k(r) \frac{du}{dr} \frac{dv}{dr} r \, dr = \left(rk \frac{du}{dr} v \right)_{r=r_o} - \left(rk \frac{du}{dr} v \right)_{r=r_i}.$$

(a) Derive this formulation from eq. (2.40) and (b) apply this formulation to a long pipe of internal radius r_i, external radius r_o using the boundary conditions $u(r_i) = \hat{u}_i$ and

$$q_n = -k \frac{du}{dr} = h_c(u - u_c) \quad \text{at} \quad r = r_o.$$

Exercise 2.20 Consider water flowing in a stainless steel pipe. The temperature of the water is 80 °C. The outer surface of the pipe is cooled by air flow. The temperature of the air is 20 °C. The outer diameter of the pipe is 0.20 m and its wall thickness is 0.01 m. (a) Assuming that convective heat transfer occurs on both the inner and outer surfaces of the pipe and u is a function of r only, formulate the mathematical model for stationary heat transfer. (b) Assume that the coefficient of thermal conduction of stainless steel is 20 W/mK. Using $h_c^{(w)} = 750$ W/m^{2}K for the water and $h_c^{(a)} = 10$ W/m^{2}K for the air, determine the temperature of the external surface of the pipe and the rate of heat loss per unit length.

Heat conduction in a bar

One-dimensional models of heat conduction are discussed in this section. It is assumed that (a) the solution domain is a bar, one end of which is located in $x = 0$, the other end in $x = \ell$; (b) the dimensions of the cross-section are small in comparison with ℓ and the cross-sectional area $A = A(x) > 0$ is a continuous and differentiable function.

If convective heat transfer occurs along the bar, as described in Section 2.3.4, the conservation law is:

$$-\frac{\partial(Aq)}{\partial x} - c_b(u - u_a) + QA = c\rho A \frac{\partial u}{\partial t}$$

where $c_b = c_b(x)$ is the coefficient of convective heat transfer of the bar (in W/(mK) units) obtained from h_c by integration:

$$c_b = \oint h_c \, ds$$

the contour integral taken along the perimeter of the cross-section. Therefore the differential equation of heat conduction in a bar is:

$$\frac{\partial}{\partial x} \left(Ak \frac{\partial u}{\partial x} \right) - c_h(u - u_n) + QA = c\rho A \frac{\partial u}{\partial t}. \tag{2.12}$$

One of the boundary conditions described in Section 2.3 is prescribed at $x = 0$ and $x = \ell$.

Example 2.6 Consider stationary heat flow in a partially insulated bar of length ℓ and constant cross-section A. The coefficients k and c_b are constant and $Q = 0$. Therefore eq. (2.42) can be cast in the following form:

$$u'' - \lambda^2(u - u_a) = 0, \qquad \lambda^2 = \frac{c_b}{Ak}.$$

If the temperature u_a is a linear function of x, i.e., $u_a(x) = a + bx$ and the boundary conditions are $u(0) = \hat{u}_0$, $q(\ell) = \hat{q}_\ell$ then the solution of this problem is:

$$u = C_1 \cosh \lambda x + C_2 \sinh \lambda x + a + bx$$

where:

$$C_1 = \hat{u}_0 - a, \quad C_2 = -\frac{1}{\lambda \cosh \lambda \ell} \left(\frac{\hat{q}_\ell}{k} + (\hat{u}_0 - a)\lambda \sinh \lambda \ell + b \right).$$

Exercise 2.21 Solve the problem described in Example 2.6 using the following boundary conditions: $q(0) = \hat{q}_0$, $q(\ell) = h_\ell(u(\ell) - U_a)$ where $\hat{q}_0$, h_ℓ, U_a are given data.

Exercise 2.22 A perfectly insulated bar of constant cross-section, length ℓ, thermal conduction k, density ρ and specific heat c is subject to the initial condition $u(x, 0) = U_0$ (constant) and the boundary conditions $u(0, t) = u(\ell, t) = 0$. Assuming that $Q = 0$, verify that the solution of this problem is:

$$u = 2U_0 \sum_{n=1}^{\infty} \frac{1 - \cos(n\pi)}{n\pi} \exp\left(-\frac{n^2\pi^2 k}{\ell^2 c\rho}t\right) \sin\left(n\pi\frac{x}{\ell}\right).$$

It is sufficient to show that the differential equation, the boundary conditions and the initial condition are satisfied by $u = u(x, t)$.

2.4 Equations of linear elasticity – strong form

Mathematical problems of linear elasticity belong in the category of vector elliptic boundary value problems. The unknown functions are the components of the displacement vector. In Cartesian coordinates the displacement vector is:

$$
\begin{aligned}
\mathbf{u} &\overset{\text{def}}{=} u_x(x, y, z)\, \mathbf{e}_x + u_y(x, y, z)\, \mathbf{e}_y + u_z(x, y, z)\, \mathbf{e}_z \\
&\equiv \{u_x(x, y, z)\ u_y(x, y, z)\ u_z(x, y, z)\}^T \\
&\equiv u_i(x_j)
\end{aligned}
\tag{2.43}
$$

where $\mathbf{e}_x$, $\mathbf{e}_y$, $\mathbf{e}_z$ are the Cartesian basis vectors.

The formulation of problems of linear elasticity is based on three fundamental relationships: the strain-displacement equations, the stress-strain relationships and the equilibrium equations.

1. **Strain-displacement relationships.** We introduce the infinitesimal strain-displacement relationships here. A detailed derivation of these relationships is presented in Section 9.2.1. By definition, the infinitesimal normal strain components are:

$$\epsilon_x \equiv \epsilon_{xx} \overset{\text{def}}{=} \frac{\partial u_x}{\partial x} \quad \epsilon_y \equiv \epsilon_{yy} \overset{\text{def}}{=} \frac{\partial u_y}{\partial y} \quad \epsilon_z \equiv \epsilon_{zz} \overset{\text{def}}{=} \frac{\partial u_z}{\partial z} \tag{2.44}$$

and the shear strain components are:

$$\epsilon_{xy} = \epsilon_{yx} \equiv \frac{\gamma_{xy}}{2} \overset{\text{def}}{=} \frac{1}{2}\left(\frac{\partial u_x}{\partial y} + \frac{\partial u_y}{\partial x}\right)$$

$$\epsilon_{yz} = \epsilon_{zy} \equiv \frac{\gamma_{yz}}{2} \overset{\text{def}}{=} \frac{1}{2}\left(\frac{\partial u_y}{\partial z} + \frac{\partial u_z}{\partial y}\right)$$

$$\epsilon_{zx} = \epsilon_{xz} \equiv \frac{\gamma_{zx}}{2} \overset{\text{def}}{=} \frac{1}{2}\left(\frac{\partial u_z}{\partial x} + \frac{\partial u_x}{\partial z}\right) \tag{2.45}$$

where γ_{xy}, γ_{yz}, γ_{zx} are called the engineering shear strain components. In index notation, the infinitesimal strain at a point is characterized by the strain tensor

$$\epsilon_{ij} \stackrel{\text{def}}{=} \frac{1}{2}\left(u_{i,j} + u_{j,i}\right). \tag{2.46}$$

2. **Stress-strain relationships** Mechanical stress is defined as force per unit area ($\text{N/m}^2 \equiv \text{Pa}$). Since one Pascal (Pa) is a very small stress, the usual unit of mechanical stress is the Megapascal (MPa) which can be understood to mean either 10^6 N/m^2 or 1 N/mm^2.

The usual notation for stress components is illustrated on an infinitesimal volume element shown in Fig. 2.7. The indexing rules are as follows: Faces to which the positive x, y, z axes are normal are called positive faces, the opposite faces are called negative faces. The normal stress components are denoted by σ, the shear stresses components by τ. The normal stress components are assigned one subscript only, since the orientation of the face and the direction of the stress component are the same. For example, σ_x is the stress component acting on the faces to which the x-axis is normal and the stress component is acting in the positive (resp. negative) coordinate direction on the positive (resp. negative) face. For the shear stresses, the first index refers to the coordinate direction of the normal to the face on which the shear stress is acting. The second index refers to the direction in which the shear stress component is acting.

On a positive (resp. negative) face the positive stress components are oriented in the positive (resp. negative) coordinate directions. The reason for this is that if we subdivide a solid body into infinitesimal hexahedral volume elements, similar to the element shown in Fig. 2.7, then each negative face will be coincident with a positive face. By the action-reaction principle, the forces acting on those faces must have equal absolute value and opposite sign. In index notation $\sigma_{11} \equiv \sigma_x$, $\sigma_{12} \equiv \sigma_{xy} \equiv \tau_{xy}$, etc.

The mechanical properties of isotropic elastic materials are characterized by the modulus of elasticity $E > 0$, Poisson's ratio[7] $\nu < 1/2$ and the coefficient of thermal expansion $\alpha > 0$. The

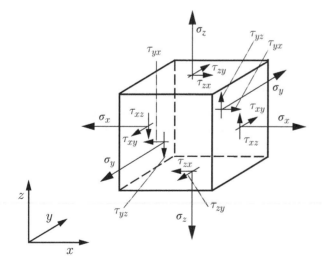

Figure 2.7 Notation for stress components.

---

7 Simeon Denis Poisson 1781–1840.

stress-strain relationships, known as Hooke's law[8], are:

$$\epsilon_x = \frac{1}{E}\left(\sigma_x - \nu\sigma_y - \nu\sigma_z\right) + \alpha T_\Delta \tag{2.47}$$

$$\epsilon_y = \frac{1}{E}\left(-\nu\sigma_x + \sigma_y - \nu\sigma_z\right) + \alpha T_\Delta \tag{2.48}$$

$$\epsilon_z = \frac{1}{E}\left(-\nu\sigma_x - \nu\sigma_y + \sigma_z\right) + \alpha T_\Delta \tag{2.49}$$

$$\gamma_{xy} \equiv 2\epsilon_{xy} = \frac{2(1+\nu)}{E}\tau_{xy} \tag{2.50}$$

$$\gamma_{yz} \equiv 2\epsilon_{yz} = \frac{2(1+\nu)}{E}\tau_{yz} \tag{2.51}$$

$$\gamma_{zx} \equiv 2\epsilon_{zx} = \frac{2(1+\nu)}{E}\tau_{zx} \tag{2.52}$$

where $T_\Delta = T_\Delta(x, y, z)$ is the temperature change with respect to a reference temperature at which the strain is zero. The strain components represent the total strain; αT_Δ is the thermal strain. Mechanical strain is defined as the total strain minus the thermal strain.

In index notation Hooke's law can be written as

$$\epsilon_{ij} = \frac{1+\nu}{E}\sigma_{ij} - \frac{\nu}{E}\sigma_{kk}\delta_{ij} + \alpha T_\Delta \delta_{ij}. \tag{2.53}$$

The inverse is:

$$\sigma_{ij} = \lambda\epsilon_{kk}\delta_{ij} + 2G\epsilon_{ij} - (3\lambda + 2G)\alpha T_\Delta \delta_{ij} \tag{2.54}$$

where λ and G, called the Lamé constants[9], are defined by

$$\lambda \overset{\text{def}}{=} \frac{E\nu}{(1+\nu)(1-2\nu)}, \qquad G \overset{\text{def}}{=} \frac{E}{2(1+\nu)}. \tag{2.55}$$

G is also called shear modulus or the modulus of rigidity. Since λ and G are positive, the admissible range of Poisson's ratio is $-1 < \nu < 1/2$. Typically it is $0 \le \nu < 1/2$.

The generalized Hooke's law states that the components of the stress tensor are linear functions of the mechanical strain tensor:

$$\sigma_{ij} = C_{ijkl}(\epsilon_{kl} - \alpha_{kl}T_\Delta) \tag{2.56}$$

where C_{ijkl} and α_{ij} are Cartesian tensors. By symmetry considerations the maximum number of independent elastic constants that characterize C_{ijkl} is 21. The symmetric tensor α_{ij} is characterized by six independent coefficients of thermal expansion. This is the general form of anisotropy in linear elasticity.

Remark 2.3 When residual stresses are present in the reference configuration then equation (2.56) has to be modified. This point is discussed in Section 2.7.

3. **Equilibrium** Considering the dynamic equilibrium of a volume element, similar to that shown in Fig. 2.7, except that the edges are of length Δx, Δy, Δz, six equations of equilibrium are written: the resultants of the forces and moments must vanish. Assuming that the material is not loaded by distributed moments (body moments), consideration of moment equilibrium leads to the conclusion that $\tau_{xy} = \tau_{yx}$, $\tau_{yz} = \tau_{zy}$, $\tau_{zx} = \tau_{xz}$, i.e., the stress tensor is symmetric. Assuming further that the components of the stress tensor are continuous and differentiable, application of

---

8 Robert Hooke 1635–1703.
9 Gabriel Lamé 1795–1870.

d'Alembert's principle[10] and consideration of force equilibrium leads to three partial differential equations:

$$\frac{\partial \sigma_x}{\partial x} + \frac{\partial \tau_{xy}}{\partial y} + \frac{\partial \tau_{xz}}{\partial z} + F_x - \varrho \frac{\partial^2 u_x}{\partial t^2} = 0 \tag{2.57}$$

$$\frac{\partial \tau_{xy}}{\partial x} + \frac{\partial \sigma_y}{\partial y} + \frac{\partial \tau_{yz}}{\partial z} + F_y - \varrho \frac{\partial^2 u_y}{\partial t^2} = 0 \tag{2.58}$$

$$\frac{\partial \tau_{xz}}{\partial x} + \frac{\partial \tau_{yz}}{\partial y} + \frac{\partial \sigma_z}{\partial z} + F_z - \varrho \frac{\partial^2 u_z}{\partial t^2} = 0 \tag{2.59}$$

where F_x, F_y, F_z are the components of the body force vector (in N/m^3 units), ϱ is the specific density (in kg/m$^3 \equiv$ Ns^2/m^4 units). These equations are called the equations of motion. In index notation:

$$\sigma_{ij,j} + F_i = \varrho \frac{\partial^2 u_i}{\partial t^2}. \tag{2.60}$$

For elastostatic problems the time derivative is zero and the boundary conditions are independent of time. This yields the equations of static equilibrium:

$$\sigma_{ij,j} + F_i = 0. \tag{2.61}$$

2.4.1 The Navier equations

The equations of motion, called the Navier equations[11], are obtained by substituting eq. (2.54) into eq. (2.60). In elastodynamics the effects of temperature are usually negligible, hence we will assume $\mathcal{T}_\Delta = 0$:

$$Gu_{i,jj} + (\lambda + G)u_{j,ji} + F_i = \varrho \frac{\partial^2 u_i}{\partial t^2}. \tag{2.62}$$

In elastostatic problems we have:

$$Gu_{i,jj} + (\lambda + G)u_{j,ji} + F_i = (3\lambda + 2G)\alpha(\mathcal{T}_\Delta)_{,i}. \tag{2.63}$$

Exercise 2.23 Derive eq. (2.63) by substituting eq. (2.54) into eq. (2.61). Explain the rules under which the indices are changed to obtain eq. (2.63). Hint: $(\epsilon_{kk}\delta_{ij})_{,j} = (u_{k,k}\delta_{ij})_{,j} = u_{k,ki} = u_{j,ji}$.

Exercise 2.24 Derive the equilibrium equations from first principles.

Exercise 2.25 In deriving eq. (2.62) and (2.63) it was assumed that λ and G are constants. Formulate the analogous equations assuming that λ and G are smooth functions of $x_i \in \Omega$.

Exercise 2.26 Assume that Ω is the union of two or more subdomains and the material properties are constants on each subdomain but vary from subdomain to subdomain. Formulate the elastostatic problem for this case.

---

10 Jean Le Rond d'Alembert 1717–1783.
11 Claude Louis Marie Henri Navier 1785–1836.

2.4.2 Boundary and initial conditions

As in the case of heat conduction, we will consider three kinds of boundary conditions: prescribed displacements, prescribed tractions and spring boundary conditions. Tractions are forces per unit area acting on the boundary. Prescribed displacements and tractions are often specified in a normal-tangent reference frame.

1. **Prescribed displacement.** One or more components of the displacement vector is prescribed on all or part of the boundary. This is called a kinematic boundary condition.
2. **Prescribed traction.** One or more components of the traction vector is prescribed on all or part of the boundary. The definition of traction vector is given in Appendix K.1.
3. **Linear spring.** A linear relationship is prescribed between the traction and displacement vector components. The general form of this relationship is:

$$T_i = c_{ij}(d_j - u_j) \tag{2.64}$$

where T_i is the traction vector, c_{ij} is a positive-definite matrix that represents the distributed spring coefficients; d_j is a prescribed function that represents displacement imposed on the spring and u_j is the (unknown) displacement vector function on the boundary. The spring coefficients c_{ij} (in N/m³ units) may be functions of the position x_k but are independent of the displacement u_i. This is called a "Winkler spring[12]".

A schematic representation of this boundary condition on an infinitesimal boundary surface element is shown in Fig. 2.8 under the assumption that c_{ij} is a diagonal matrix and therefore three spring coefficients $c_1 \overset{\text{def}}{=} c_{11}, c_2 \overset{\text{def}}{=} c_{22}, c_3 \overset{\text{def}}{=} c_{33}$ characterize the elastic properties of the boundary condition.

Fig. 2.8 should be interpreted to mean that the imposed displacement d_i will cause a differential force ΔF_i to act on the centroid of the surface element. Suspending the summation rule, the magnitude of ΔF_i is

$$\Delta F_i = c_i \Delta A(d_i - u_i), \qquad i = 1, 2, 3$$

where u_i is the displacement of the surface element. The corresponding traction vector is:

$$T_i = \lim_{\Delta A \to 0} \frac{\Delta F_i}{\Delta A} = c_i(d_i - u_i), \qquad i = 1, 2, 3.$$

Figure 2.8 Spring boundary condition. Schematic representation.

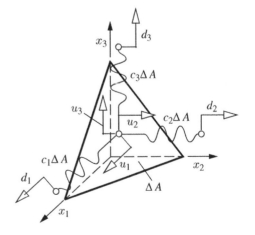

---

12 Emil Winkler 1835–1888.

The enumerated boundary conditions may occur in any combination. For example, the displacement vector component u_1, the traction vector component T_2 and a linear combination of T_3 and u_3 may be prescribed on a boundary segment.

In engineering practice boundary conditions are most conveniently prescribed in the normal-tangent reference frame. The normal is uniquely defined on smooth surfaces but the tangential coordinate directions are not. It is necessary to specify the tangential coordinate directions with respect to the reference frame used. The required coordinate transformations are discussed in Appendix K.

The boundary conditions are generally time-dependent. For time-dependent problems the initial conditions, that is, the initial displacement and velocity fields, denoted respectively by $U(x, y, z)$ and $V(x, y, z)$, have to be prescribed:

$$\mathbf{u}(x, y, z, 0) = \mathbf{U}(x, y, z) \quad \text{and} \quad \left(\frac{\partial \mathbf{u}}{\partial t} \right)_{(x, y, z, 0)} = \mathbf{V}(x, y, z).$$

Exercise 2.27 Assume that the following boundary conditions are given in the normal-tangent reference frame x_i', x_1' being coincident with the normal: $T_1' = c_1'(d_1' - u_1')$; $T_2' = T_3' = 0$. Using the transformation $x_i' = g_{ij} x_j$, determine the boundary conditions in the x_i coordinate system. (See the Appendix, Section K.2.)

2.4.3 Symmetry, antisymmetry and periodicity

Symmetry and antisymmetry of vectors in two dimensions with respect to the y axis is illustrated in Fig. 2.9

The definition of symmetry and antisymmetry of vectors in three dimensions is analogous: the corresponding vector components parallel to a plane of symmetry (resp. antisymmetry) have the same absolute value and the same (resp. opposite) sign. The corresponding vector components normal to a plane of symmetry (resp. antisymmetry) have the same absolute value and opposite (resp. same) sign.

In a plane of symmetry the normal displacement and the shearing traction components are zero. In a plane of antisymmetry the normal traction is zero and the in-plane components of the displacement vector are zero.

When the solution is periodic on Ω then a periodic sector of Ω has at least one periodic boundary segment pair denoted by $\partial \Omega_p^+$ and $\partial \Omega_p^-$. On corresponding points of a periodic boundary segment pair, $P^+ \in \partial \Omega_p^+$ and $P^- \in \partial \Omega_p^-$ the normal component of the displacement vector and the periodic in-plane components of the displacement vector have the same value. The normal component of the traction vector and the periodic in-plane components of the traction vector have the same absolute value but opposite sign.

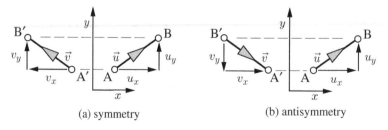

(a) symmetry (b) antisymmetry

Figure 2.9 Symmetry and antisymmetry of vectors in two dimensions.

Exercise 2.28 A homogeneous isotropic elastic body with Poisson's ratio zero occupies the domain $\Omega = \{x, y, z \mid |x| < a,\ |y| < b,\ |z| < c\}$. Define tractions on the boundaries of Ω such that the tractions satisfy equilibrium and the plane $z = 0$ is a plane of (a) symmetry, (b) antisymmetry.

Exercise 2.29 Consider the domain shown in Fig. 2.2. Assume that $T_n = 0$ and $T_t = \tau_o$ where τ_o is constant, is prescribed on $\partial\Omega^{(o)}$, $T_n = T_t = 0$ on the circular boundaries $\partial\Omega^{(k)}$ and $u_n = u_t = 0$ on $\partial\Omega^{(i)}$. Specify periodic boundary conditions on $\partial\Omega_p^+$ and $\partial\Omega_p^-$.

2.4.4 Dimensional reduction in linear elasticity

Owing to the complexity of three-dimensional problems in elasticity, dimensional reduction is widely used. Various kinds of dimensional reduction are possible in elasticity, such as planar, axisymmetric, shell, plate, beam and bar models. Each of these model types is sufficiently important to have generated a substantial technical literature. In the following models for planar and axially symmetric problems are discussed. Models for beams, plates and shells will be discussed separately.

Planar elastostatic models: Notation

We consider a prismatic body of length ℓ. The material points occupy the domain Ω_ℓ, defined as follows:

$$\Omega_\ell = \{(x, y, z) \mid (x, y) \in \omega,\ -\ell/2 < z < \ell/2,\ \ell > 0\} \tag{2.65}$$

where $\omega \in \mathbb{R}^2$ is a bounded domain. The lateral boundary of the body is denoted by

$$\Gamma_\ell = \{(x, y, z) \mid (x, y) \in \partial\omega,\ -\ell/2 < z < \ell/2,\ \ell > 0\} \tag{2.66}$$

and the faces are denoted by

$$\gamma_\pm = \{(x, y, z) \mid (x, y) \in \omega,\ z = \pm\ell/2\}. \tag{2.67}$$

The notation is shown in Fig. 2.10(a). The diameter of ω will be denoted by d_ω.

The material properties, the volume forces and temperature change acting on Ω_ℓ and tractions acting on Γ_ℓ will be assumed to be independent of z. Therefore the x, y plane is a plane of symmetry. The tangential direction will be understood to be with reference to contours parallel to $\partial\omega$.

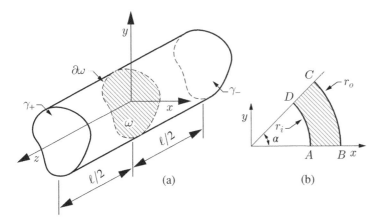

Figure 2.10 Notation.

Plane strain

When the boundary conditions on γ_+ and γ_- are $u_z = \tau_{zx} = \tau_{zy} = 0$ and on Γ_ℓ normal and tangential displacements and/or tractions are prescribed that are independent of z, then a plane strain model may be used. The stress-strain relationships for plane strain models are obtained from Hooke's law, see equations (2.47) to (2.52), by letting $\epsilon_z = \gamma_{yz} = \gamma_{zx} = 0$ to obtain

$$
\begin{Bmatrix} \sigma_x \\ \sigma_y \\ \tau_{xy} \end{Bmatrix} = \begin{bmatrix} \lambda + 2G & \lambda & 0 \\ \lambda & \lambda + 2G & 0 \\ 0 & 0 & G \end{bmatrix} \begin{Bmatrix} \epsilon_x \\ \epsilon_y \\ \gamma_{xy} \end{Bmatrix} - (3\lambda + 2G)\alpha T_\Delta \begin{Bmatrix} 1 \\ 1 \\ 0 \end{Bmatrix}. \tag{2.68}
$$

Observe that the boundary conditions prescribed on γ_+ and γ_- are periodic. Therefore the plane strain solution is extended by periodic repetition to $-\infty < \ell < \infty$ independent of what value is assigned to ℓ. This implies that $u_z = 0$ and hence $\epsilon_z = 0$. The solution is z-independent, therefore $\partial u_x/\partial z = \partial u_y/\partial z = 0$, hence the shearing strains γ_{xz} and γ_{yz} are zero. The 2D and 3D solutions are the same in the $x\,y$ plane. The stress σ_z can be computed from the plane strain solution using $\sigma_z = -\nu(\sigma_x + \sigma_y) - E\alpha T_\Delta$.

Note that the zero normal displacement condition on γ_+ and γ_- is an idealization that cannot be realized in practice.

Plane stress

When the boundary conditions on γ_+ and γ_- are zero normal and shearing tractions and on Γ_ℓ normal and tangential displacements and/or tractions are prescribed that are independent of z, then often a plane stress model is used. Letting $\sigma_z = \tau_{yz} = \tau_{zx} = 0$, equations (2.47) to (2.52) are simplified to

$$
\begin{Bmatrix} \sigma_x \\ \sigma_y \\ \tau_{xy} \end{Bmatrix} = \frac{E}{1 - \nu^2} \begin{bmatrix} 1 & \nu & 0 \\ \nu & 1 & 0 \\ 0 & 0 & \frac{1-\nu}{2} \end{bmatrix} \begin{Bmatrix} \epsilon_x \\ \epsilon_y \\ \gamma_{xy} \end{Bmatrix} - \frac{E\alpha T_\Delta}{1 - \nu} \begin{Bmatrix} 1 \\ 1 \\ 0 \end{Bmatrix}. \tag{2.69}
$$

The conditions $\sigma_z = \tau_{zx} = \tau_{zy} = 0$ on Ω_ℓ cannot be exactly satisfied by the three-dimensional solution in the general case, except in the limit when ℓ approaches zero. Therefore the solution of a plane stress model is an approximation of the corresponding 3D model, and the error of approximation depends on Poisson's ratio and ℓ.

Remark 2.4 When the material is homogeneous and isotropic and only tractions are specified on Γ_ℓ (which must satisfy the equations if equilibrium) and the body force is zero[13], then the problem is classified as a first fundamental boundary value problem of elasticity [60]. The stress field is completely determined by the boundary conditions and the compatibility condition. It is independent of the material properties E and ν and therefore independent of whether the model is plane stress or plane stain. The displacement field can be determined from the stress field up to rigid body displacements.

Exercise 2.30 Consider the prismatic body that has the annular cross-section $ABCD$ shown in Fig. 2.10(b). Assume that the faces γ_+ and γ_- are stress free. The boundary conditions for the other faces are listed in Table 2.1. Determine or estimate difference between the plane stress and fully three-dimensional solutions measured (a) in energy norm and (b) the maximum norm of the von Mises stress using the following data: $r_o = 200$ mm, $r_i = 175$ mm, $\ell = 40$ mm, $\alpha = 45°$, $E = 70.0$ GPa, $\nu = 0.3$, $q_n = -20$ MPa, $q_t = 100$ MPa.

---

13 It is possible to consider a more general case of body forces, but that is not addressed here.

Table 2.1 Exercise 2.30: Boundary conditions

Segment	Case 1	Case 2	Case 3
AB	symmetry	antisymmetry	$u_x = u_y = 0$
BC	$T_n = q_n, T_t = 0$	$T_n = 0, T_t = q_t$	$T_n = q_n, T_t = 0$
CD	symmetry	antisymmetry	symmetry
DA	$T_n = T_t = 0$	$T_n = 0, T_t = (r_o/r_i)^2 q_t$	$T_n = T_t = 0$

Partial solution: Case 1 is the classical problem of a cylindrical elastic pipe subjected to external pressure q. The exact stress distribution is known. The domain $ABCD$ represents a 45° segment of the pipe. The symmetry boundary conditions extend the domain by periodic repetition to the entire cross-section. Clearly, the external pressure satisfies the equations of equilibrium. The difference between the plane stress and three-dimensional solution is zero in energy norm, independent of ℓ, and also zero in the maximum norm of the von Mises stress. The maximum von Mises stress is 170.7 MPa.

The Navier equations

The equations of equilibrium are:

$$\frac{\partial \sigma_x}{\partial x} + \frac{\partial \tau_{xy}}{\partial y} + F_x = 0 \tag{2.70}$$

$$\frac{\partial \tau_{yx}}{\partial x} + \frac{\partial \sigma_y}{\partial y} + F_y = 0 \tag{2.71}$$

where F_x, F_y are the components of body force vector $\mathbf{F}(x, y)$ (in N/m^3 units).

The Navier equations are obtained by substituting the stress-strain and strain-displacement relationships into the equilibrium equations. For plane strain;

$$(\lambda + G)\frac{\partial}{\partial x}\left(\frac{\partial u_x}{\partial x} + \frac{\partial u_y}{\partial y}\right) + G\left(\frac{\partial^2 u_x}{\partial x^2} + \frac{\partial^2 u_x}{\partial y^2}\right) = \frac{E\alpha}{1 - 2v}\frac{\partial T_\Delta}{\partial x} - F_x \tag{2.72}$$

$$(\lambda + G)\frac{\partial}{\partial y}\left(\frac{\partial u_x}{\partial x} + \frac{\partial u_y}{\partial y}\right) + G\left(\frac{\partial^2 u_y}{\partial x^2} + \frac{\partial^2 u_y}{\partial y^2}\right) = \frac{E\alpha}{1 - 2v}\frac{\partial T_\Delta}{\partial y} - F_y. \tag{2.73}$$

The boundary conditions are most conveniently written in the normal-tangent ($n\,t$) reference frame illustrated in Fig. 2.4. The relationship between the $x\,y$ and $n\,t$ components of displacements and tractions is established by the rules of vector transformation, described in the Appendix, Section 2.4.2. The linear boundary conditions listed in Section 2.4.2 are applicable in two dimensions as well.

Exercise 2.31 Derive eq. (2.69) and eq. (2.68) from Hooke's law.

Exercise 2.32 Show that for plane stress the Navier equations are:

$$\frac{E}{2(1 - v)}\frac{\partial}{\partial x}\left(\frac{\partial u_x}{\partial x} + \frac{\partial u_y}{\partial y}\right) + G\left(\frac{\partial^2 u_y}{\partial x^2} + \frac{\partial^2 u_y}{\partial y^2}\right) = \frac{E\alpha}{1 - 2v}\frac{\partial T_\Delta}{\partial y} - F_y. \tag{2.74}$$

$$\frac{E}{2(1 - v)}\frac{\partial}{\partial y}\left(\frac{\partial u_x}{\partial x} + \frac{\partial u_y}{\partial y}\right) + G\left(\frac{\partial^2 u_y}{\partial x^2} + \frac{\partial^2 u_y}{\partial y^2}\right) = \frac{E\alpha}{1 - 2v}\frac{\partial T_\Delta}{\partial y} - F_y. \tag{2.75}$$

Exercise 2.33 Denote the components of the unit normal vector to the boundary by n_x and n_y. Show that

$$T_x = T_n n_x - T_t n_y$$

$$T_y = T_n n_y + T_t n_x$$

where T_n (resp. T_t) is the normal (resp. tangential) component of the traction vector.

Axisymmetric elastostatic models

The radial, circumferential and axial coordinates are denoted by r, θ and z respectively and the displacement, stress, strain and traction components are labeled with corresponding subscripts. The problem is formulated in terms of the displacement vector components $u_r(r, z)$ and $u_z(r, z)$.

1. **The linear strain-displacement relationships** in cylindrical coordinates are [105]:

$$\epsilon_r \overset{\text{def}}{=} \frac{\partial u_r}{\partial r} \tag{2.76}$$

$$\epsilon_\theta \overset{\text{def}}{=} \frac{u_r}{r} \tag{2.77}$$

$$\epsilon_z \overset{\text{def}}{=} \frac{\partial u_z}{\partial z} \tag{2.78}$$

$$\epsilon_{rz} = \frac{\gamma_{rz}}{2} \overset{\text{def}}{=} \frac{1}{2} \left(\frac{\partial u_r}{\partial z} + \frac{\partial u_z}{\partial r} \right). \tag{2.79}$$

2. **Stress-strain relationships.** For isotropic materials the stress-strain relationship is obtained from eq. (2.54):

$$\begin{Bmatrix} \sigma_r \\ \sigma_\theta \\ \sigma_z \\ \tau_{rz} \end{Bmatrix} = \begin{bmatrix} \lambda + 2G & \lambda & \lambda & 0 \\ \lambda & \lambda + 2G & \lambda & 0 \\ \lambda & \lambda & \lambda + 2G & 0 \\ 0 & 0 & 0 & G \end{bmatrix} \begin{Bmatrix} \epsilon_r \\ \epsilon_\theta \\ \epsilon_z \\ \gamma_{rz} \end{Bmatrix} - \frac{E\alpha \mathcal{T}_\Delta}{1 - 2\nu} \begin{Bmatrix} 1 \\ 1 \\ 1 \\ 0 \end{Bmatrix} \tag{2.80}$$

3. **Equilibrium.** The elastostatic equations of equilibrium are [105]:

$$\frac{1}{r} \frac{\partial(r\sigma_r)}{\partial r} + \frac{\partial \tau_{rz}}{\partial z} - \frac{\sigma_\theta}{r} + F_r = 0 \tag{2.81}$$

$$\frac{1}{r} \frac{\partial(r\tau_{rz})}{\partial r} + \frac{\partial \sigma_z}{\partial z} + F_z = 0. \tag{2.82}$$

Exercise 2.34 Write down the Navier equations for the axisymmetric model.

2.4.5 Incompressible elastic materials

When $\nu \to 1/2$ then $\lambda \to \infty$ therefore the relationship represented by eq. (2.54) breaks down. Referring to eq. (2.53), the sum of normal strain components is related to the sum of normal stress components by:

$$\epsilon_{kk} = \frac{1 - 2\nu}{E} \sigma_{kk} + 3\alpha \mathcal{T}_\Delta. \tag{2.83}$$

The sum of normal strain components is called the volumetric strain and will be denoted by $\epsilon_{\text{vol}} \overset{\text{def}}{=} \epsilon_{kk}$. Defining:

$$\sigma_0 \overset{\text{def}}{=} \frac{1}{3} \sigma_{kk} \tag{2.84}$$

we have

$$\epsilon_{\text{vol}} \equiv \epsilon_{kk} = \frac{3(1 - 2v)}{E}\sigma_0 + 3\alpha\mathcal{T}_\Delta \tag{2.85}$$

where the first term on the right is the mechanical strain, the second term is the thermal strain. For incompressible materials, that is when $v = 1/2$, $\epsilon_{\text{vol}} = 3\alpha\mathcal{T}_\Delta$ is independent of σ_0. Therefore σ_0 cannot be computed from the strains in the usual way. Substituting eq. (2.85) into eq. (2.54) and letting $v = 1/2$ we have:

$$\sigma_{ij} = \sigma_0\delta_{ij} + \frac{2E}{3}(\epsilon_{ij} - \alpha\mathcal{T}_\Delta\delta_{ij}). \tag{2.86}$$

Substituting into eq. (2.61), and assuming that E and α are constant;

$$(\sigma_0)_{,i} + \frac{2E}{3}(\epsilon_{ij,j} - \alpha(\mathcal{T}_\Delta)_{,i}) + F_i = 0.$$

Writing

$$\epsilon_{ij,j} = \frac{1}{2}(u_{i,j} + u_{j,i})_{,j} = \frac{1}{2}(u_{i,jj} + u_{j,ij})$$

and interchanging the order of differentiation in the second term, we have:

$$u_{j,ij} = u_{j,ji} = (u_{j,j})_{,i} = (\epsilon_{jj})_{,i} = 3\alpha(\mathcal{T}_\Delta)_{,i}.$$

Therefore, for incompressible materials,

$$\epsilon_{ij,j} = \frac{1}{2}u_{i,jj} + \frac{3}{2}\alpha(\mathcal{T}_\Delta)_{,i}.$$

The problem is to determine u_i such that

$$(\sigma_0)_{,i} + \frac{E}{3}\left(u_{i,jj} + \alpha(\mathcal{T}_\Delta)_{,i}\right) + F_i = 0 \tag{2.87}$$

subject to the condition of incompressibility, that is, the condition that volumetric strain can be caused by temperature change but not by mechanical stress,

$$u_{i,i} = 3\alpha\mathcal{T}_\Delta \tag{2.88}$$

and the appropriate boundary conditions. If displacement boundary conditions are prescribed over the entire boundary ($\partial\Omega_u = \partial\Omega$) then the problem has a solution only if the prescribed displacements are consistent with the incompressibility condition:

$$\int_{\partial\Omega} u_i n_i \, dS = 3\int_\Omega \alpha\mathcal{T}_\Delta \, dV.$$

This follows directly from integrating eq. (2.88) and applying the divergence theorem.

Exercise 2.35 An incompressible bar of constant cross-section and length ℓ is subjected uniform temperature change (i.e., $\mathcal{T}_\Delta$ is constant). The centroidal axis of the bar is coincident with the x_1 axis. The boundary conditions are: $u_1(0) = u_1(\ell) = 0$. The body force vector is zero. Explain how (2.86) and (2.88) are applied in this case to find that $\sigma_{11} = -E\alpha\mathcal{T}_\Delta$.

2.5 Stokes flow

The flow of viscous fluids at very low Reynolds numbers[14] ($Re < 1$) is modeled by the Stokes[15] equations. There is a close analogy between the Stokes equations and the equations of incompressible elasticity discussed in Section 2.4.5. In fluid mechanics the average compressive normal stress is the pressure p. The vector u_i represents the components of the velocity vector and the shear modulus of the incompressible elastic solid $E/3$ is replaced by the coefficient of dynamic viscosity μ (measured in Ns/m^2 units):

$$\mu u_{i,jj} = p_{,i} - F_i \tag{2.89}$$
$$u_{i,i} = 0. \tag{2.90}$$

Exercise 2.36 Write the Stokes equations in unabridged notation.

Exercise 2.37 Assume that velocities are prescribed over the entire boundary for a Stokes problem (i.e., $\partial\Omega_u = \partial\Omega$). What condition must be satisfied by the prescribed velocities?

Remark 2.5 In his chapter we treated the physical properties, such as coefficients of heat conduction, the surface coefficient, the modulus of elasticity and Poisson's ratio as given constants. Readers should be mindful of the fact that physical properties are empirical data inferred from experimental observations. Owing to variations in experimental conditions and other factors, these data are not known precisely and are always subject to restrictions. For example, stress is proportional to strain up to the proportional limit only, the flux is proportional to the temperature gradient only within a narrow range of temperatures. In fact, the coefficient of thermal conductivity (k) is typically temperature-dependent. For example, in the case of AISI 304 stainless steel k changes from about 15 to about 20 W/m°C in the temperature range of 0 to 400°C [18]. For a narrow range of temperature (say 100 to 200°C) the size of the uncertainty in k is about the same as the change in the mean value. Therefore, using a constant value for k, this range may be "good enough". Ignoring temperature-dependence for a much wider range of temperatures may lead to large errors, however.

Taking into account temperature-dependence of the coefficient of thermal conductivity leads to the formulation of non-linear problems, solutions for which are found by iteratively solving sequences of linear problems. This will be discussed in Section 9.1.2.

Analysts usually rely on data published in various handbooks. The published data can vary widely, however. For example, it was reported in [19] that published data on the coefficient of thermal conductivity of pure iron varies between 71.8 and 80.4 W/(mK).

2.6 Generalized formulation of problems of linear elasticity

Consider the equations of equilibrium (2.60):

$$\sigma_{ij,j} + F_i = 0. \tag{2.91}$$

Multiply (2.91) by a test function v_i, and integrate on Ω:

$$\int_\Omega \sigma_{ij,j} v_i \, dV + \int_\Omega F_i v_i \, dV = 0. \tag{2.92}$$

---

14 Osborne Reynolds 1842–1912.
15 George Gabriel Stokes 1819–1903.

Observe that if the equilibrium equation (2.91) is satisfied then eq. (2.92) holds for arbitrary v_i, subject to the condition that the indicated operations are defined. We write:

$$\int_\Omega \sigma_{ij,j} v_i \, dV = \int_\Omega (\sigma_{ij} v_i)_{,j} \, dV - \int_\Omega \sigma_{ij} v_{i,j} \, dV$$

and use the divergence theorem to obtain

$$\int_\Omega \sigma_{ij,j} v_i \, dV = \int_{\partial\Omega} \sigma_{ij} n_j v_i \, dS - \int_\Omega \sigma_{ij} v_{i,j} \, dV.$$

Noting that $\sigma_{ij} n_j = T_i$ (see eq. (K.3) in the Appendix), eq. (2.92) can be written as:

$$\int_\Omega \sigma_{ij} v_{i,j} \, dV = \int_\Omega F_i v_i \, dV + \int_{\partial\Omega} T_i v_i \, dS. \tag{2.93}$$

Observe that $\sigma_{ij} v_{i,j}$ equals the sum $\sigma_{11} v_{1,1} + \sigma_{22} v_{2,2} + \sigma_{33} v_{3,3}$ plus the sum of pairs like $\sigma_{12} v_{1,2} + \sigma_{21} v_{2,1}$. Since $\sigma_{ij} = \sigma_{ji}$, this can be written as:

$$\sigma_{12} v_{1,2} + \sigma_{21} v_{2,1} = \sigma_{12} \frac{1}{2}(v_{1,2} + v_{2,1}) + \sigma_{21} \frac{1}{2}(v_{2,1} + v_{1,2}) = \sigma_{12} \epsilon_{12}^{(v)} + \sigma_{21} \epsilon_{21}^{(v)}$$

where the superscript (v) indicates that these are the infinitesimal strain terms corresponding to the test function v_i, that is,

$$\epsilon_{ij}^{(v)} \stackrel{\text{def}}{=} \frac{1}{2}(v_{i,j} + v_{j,i}).$$

Therefore:

$$\sigma_{ij} v_{i,j} = \sigma_{ij} \epsilon_{ij}^{(v)}$$

and eq. (2.93) can be written as:

$$\int_\Omega \sigma_{ij} \epsilon_{ij}^{(v)} \, dV = \int_\Omega F_i v_i \, dV + \int_{\partial\Omega} T_i v_i \, dS. \tag{2.94}$$

On substituting eq. (2.56) into eq. (2.94) we have:

$$\int_\Omega C_{ijkl} \epsilon_{ij}^{(v)} \epsilon_{kl} \, dV = \int_\Omega F_i v_i \, dV + \int_{\partial\Omega} T_i v_i \, dS + \int_\Omega C_{ijkl} \epsilon_{ij}^{(v)} \alpha_{kl} T_\Delta \, dV. \tag{2.95}$$

Let $\partial\Omega_u$ denote the boundary region where $u_i = \hat{u}_i$ is prescribed; let $\partial\Omega_T$ denote the boundary region where $T_i = \hat{T}_i$ is prescribed and let $\partial\Omega_s$ denote the boundary region where $T_i = k_{ij}(d_j - u_j)$ (i.e., spring boundary condition) is prescribed. Let us define

$$B(\mathbf{u}, \mathbf{v}) \stackrel{\text{def}}{=} \int_\Omega C_{ijkl} \epsilon_{ij}^{(v)} \epsilon_{kl} \, dV + \int_{\partial\Omega_s} k_{ij} u_j v_i \, dS \tag{2.96}$$

and

$$F(\mathbf{v}) \stackrel{\text{def}}{=} \int_\Omega F_i v_i \, dV + \int_{\partial\Omega_T} \hat{T}_i v_i \, dS + \int_{\partial\Omega_s} k_{ij} d_j v_i \, dS + \int_\Omega C_{ijkl} \epsilon_{ij}^{(v)} \alpha_{kl} T_\Delta \, dV \tag{2.97}$$

where $\mathbf{u} \equiv u_i$ and $\mathbf{v} \equiv v_i$. In the interest of simplicity we will assume that the material constants E, v and α are piecewise smooth functions.

The space $E(\Omega)$, called the energy space, is defined by

$$E(\Omega) \stackrel{\text{def}}{=} \{\mathbf{u} \mid B(\mathbf{u}, \mathbf{u}) < \infty\} \tag{2.98}$$

and the norm

$$\|\mathbf{u}\|_E \overset{\text{def}}{=} \sqrt{\frac{1}{2}B(\mathbf{u}, \mathbf{u})} \tag{2.99}$$

is associated with $E(\Omega)$. The space of admissible functions is defined by:

$$\tilde{E}(\Omega) \overset{\text{def}}{=} \{u_i \mid u_i \in E(\Omega),\ u_i = \hat{u}_i \text{ on } \partial\Omega_u\}.$$

Note that this definition imposes a restriction on the prescribed displacement conditions: there has to be an $u_i \in E(\Omega)$ so that $u_i = \hat{u}_i$ on $\partial\Omega_u$.

The space of test functions is defined by:

$$E^0(\Omega) \overset{\text{def}}{=} \{u_i \mid u_i \in E(\Omega),\ u_i = 0 \text{ on } \partial\Omega_u\}.$$

The generalized formulation is stated as follows: "Find $\mathbf{u} \in \tilde{E}(\Omega)$ such that $B(\mathbf{u}, \mathbf{v}) = F(\mathbf{v})$ for all $\mathbf{v} \in E^0(\Omega)$".

2.6.1 The principle of minimum potential energy

By definition, the potential energy is the functional:

$$\pi(\mathbf{u}) \overset{\text{def}}{=} \frac{1}{2}B(\mathbf{u}, \mathbf{u}) - F(\mathbf{u}) \tag{2.100}$$

The principle of minimum potential energy states that the exact solution of the generalized formulation based on the principle of virtual work is the minimizer of the potential energy functional on the space of admissible functions:

$$\pi(\mathbf{u}_{EX}) = \min_{\mathbf{u} \in \tilde{E}(\Omega)} \pi(\mathbf{u}). \tag{2.101}$$

The proof given in Section 1.2.2 is directly applicable to the problem of elasticity.

The definition of the potential energy may be modified by the addition of an arbitrary constant. Specifically, referring to eq. (2.96) and (2.97), we define the potential energy for the problem of elasticity as follows:

$$\Pi(\mathbf{u}) \overset{\text{def}}{=} \frac{1}{2}\int_\Omega C_{ijkl}(\epsilon_{ij} - \alpha_{ij}T_\Delta)(\epsilon_{kl} - \alpha_{kl}T_\Delta)\,dV + \frac{1}{2}\int_{\partial\Omega_s} k_{ij}(u_i - d_i)(u_j - d_j)\,dS$$
$$- \int_\Omega F_i u_i\,dV - \int_{\partial\Omega_T} T_i u_i\,dS. \tag{2.102}$$

The advantage of this definition, over the definition given by eq. (2.100), is that in the special cases when a free body is subjected to a temperature change ($\epsilon_{ij} = \alpha_{ij}T_\Delta$), or a body with a spring boundary condition is given a rigid body displacement ($u_i = d_i$), then $\Pi(\mathbf{u}) = 0$, whereas $\pi(\mathbf{u}) \neq 0$.

In the finite element method $\tilde{E}(\Omega)$ is replaced by a finite-dimensional subspace $\tilde{S}$:

$$\Pi(\mathbf{u}_{FE}) = \min_{\mathbf{u} \in \tilde{S}(\Omega)} \Pi(\mathbf{u}). \tag{2.103}$$

When the sequence of finite element spaces is hierarchic (i.e., $\tilde{S}_1 \subset \tilde{S}_2 \subset \cdots$) then the potential energy converges monotonically.

Exercise 2.38 Compare the definition of $\pi(\mathbf{u})$ given by eq. (2.100), with the definition of $\Pi(\mathbf{u})$ given by eq. (2.102) and show that the two definitions differ by a constant, defined as follows:

$$\Pi(\mathbf{u}) - \pi(\mathbf{u}) = \frac{1}{2}\int_\Omega C_{ijkl}\alpha_{ij}\alpha_{kl}T_\Delta^2 \, dV + \frac{1}{2}\int_{\partial\Omega_s} k_{ij}d_id_j \, dS.$$

Isotropic elasticity

When the material is isotropic then we can substitute eq. (2.54) into eq. (2.93) to obtain:

$$\int_\Omega \left(\lambda\epsilon_{kk}\epsilon_{ii}^{(v)} + 2G\epsilon_{ij}\epsilon_{ij}^{(v)}\right) \, dV = \int_\Omega F_iv_i \, dV + \int_{\partial\Omega} T_iv_i \, dS +$$
$$\int_\Omega \frac{E}{1-2v}\alpha T_\Delta\epsilon_{ii}^{(v)} \, dV. \tag{2.104}$$

Define the differential operator matrix $[D]$ and the material stiffness matrix $[E]$ as follows:

$$[D] \stackrel{\text{def}}{=} \begin{bmatrix} \dfrac{\partial}{\partial x} & 0 & 0 \\[2mm] 0 & \dfrac{\partial}{\partial y} & 0 \\[2mm] 0 & 0 & \dfrac{\partial}{\partial z} \\[2mm] \dfrac{\partial}{\partial y} & \dfrac{\partial}{\partial x} & 0 \\[2mm] 0 & \dfrac{\partial}{\partial z} & \dfrac{\partial}{\partial y} \\[2mm] \dfrac{\partial}{\partial z} & 0 & \dfrac{\partial}{\partial x} \end{bmatrix} \qquad [E] \stackrel{\text{def}}{=} \begin{bmatrix} \lambda+2G & \lambda & \lambda & 0 & 0 & 0 \\ \lambda & \lambda+2G & \lambda & 0 & 0 & 0 \\ \lambda & \lambda & \lambda+2G & 0 & 0 & 0 \\ 0 & 0 & 0 & G & 0 & 0 \\ 0 & 0 & 0 & 0 & G & 0 \\ 0 & 0 & 0 & 0 & 0 & G \end{bmatrix}. \tag{2.105}$$

Furthermore, denote $\mathbf{u} \equiv \{u\} \stackrel{\text{def}}{=} \{u_x \, u_y \, u_z\}^T$ and $\mathbf{v} \equiv \{v\} \stackrel{\text{def}}{=} \{v_x \, v_y \, v_z\}^T$. It is left to the reader to show that eq. (2.104) can be written in the following form:

$$\int_\Omega ([D]\{v\})^T[E][D]\{u\} \, dV = \int_\Omega \{v\}^T\{F\} \, dV + \int_{\partial\Omega} \{v\}^T\{T\} \, dS$$
$$+ \int_\Omega \left\{ \frac{\partial v_x}{\partial x} \, \frac{\partial v_y}{\partial y} \, \frac{\partial v_z}{\partial z} \right\} \begin{Bmatrix} 1 \\ 1 \\ 1 \end{Bmatrix} \frac{E\alpha T_\Delta}{1-2v} \, dV \tag{2.106}$$

where $\{F\} \stackrel{\text{def}}{=} \{F_x \, F_y \, F_z\}^T$ is the body force vector and $\{T\} \stackrel{\text{def}}{=} \{T_x \, T_y \, T_z\}^T$ is the traction vector.

Remark 2.6 In the general anisotropic case, represented by eq. (2.95), we have

$$\int_\Omega ([D]\{v\})^T[E][D]\{u\} \, dV = \int_\Omega \{v\}^T\{F\} \, dV + \int_{\partial\Omega} \{v\}^T\{T\} \, dS$$
$$+ \int_\Omega ([D]\{v\})^T[E]\{\alpha\}T_\Delta \, dV \tag{2.107}$$

where the material stiffness matrix $[E]$ is a symmetric positive definite matrix with 21 independent coefficients and $\{\alpha\} \stackrel{\text{def}}{=} \{\alpha_{11} \, \alpha_{22} \, \alpha_{33} \, 2\alpha_{12} \, 2\alpha_{23} \, 2\alpha_{31}\}^T$.

Exercise 2.39 Show that for isotropic elastic materials with $v \neq 0$ $\Pi(\mathbf{u})$ can be written in the following form:

$$\Pi(\mathbf{u}) = \frac{1}{2} \int_\Omega \left\{ \lambda \left(\epsilon_{kk} - \frac{1+v}{v} \alpha \mathcal{T}_\Delta \right)^2 + 2G\epsilon_{ij}\epsilon_{ij} - \frac{E}{v} \left(\alpha \mathcal{T}_\Delta \right)^2 \right\} dV +$$

$$\frac{1}{2} \int_{\partial\Omega_s} k_{ij} \left(u_i - d_i \right) \left(u_j - d_j \right) dS - \int_\Omega F_i u_i dV - \int_{\partial\Omega_T} T_i u_i dS \tag{2.108}$$

and verify that when an unconstrained elastic body is subjected to a temperature change $\mathcal{T}_\Delta$ (i.e., $\epsilon_{ij} = \alpha \mathcal{T}_\Delta \delta_{ij}$) then $\Pi(\mathbf{u}) = 0$.

2.6.2 The RMS measure of stress

The root-mean-square (RMS) measure of stress is closely related to the energy norm. Using the notation

$$\epsilon \equiv \{\epsilon\} \stackrel{\text{def}}{=} [D]\{u\}, \quad \sigma \equiv \{\sigma\} = [E]\{\epsilon\}, \tag{2.109}$$

the RMS measure of stress is defined by

$$S(\sigma) \stackrel{\text{def}}{=} \left(\frac{1}{V} \int_\Omega \{\sigma\}^T \{\sigma\} \, dV \right)^{1/2} \tag{2.110}$$

where V is the volume. Therefore we can write

$$S^2(\sigma) = \frac{1}{V} \int_\Omega ([E]\{\epsilon\})^T [E]\{\epsilon\} \, dV = \frac{1}{V} \int_\Omega \{\epsilon\}^T [E][E]\{\epsilon\} \, dV \tag{2.111}$$

where we used $[E]^T = [E]$. When the boundary conditions do not include spring constraints then the strain energy is

$$\|\mathbf{u}\|_{E(\Omega)}^2 = \frac{1}{2} \int_\Omega \{\epsilon\}^T [E]\{\epsilon\} \, dV. \tag{2.112}$$

Using equations (2.111) and (2.112), we can write

$$\frac{2\Lambda_{\min}}{V} \|\mathbf{u}\|_{E(\Omega)}^2 \leq S^2(\sigma) \leq \frac{2\Lambda_{\max}}{V} \|\mathbf{u}\|_{E(\Omega)}^2 \tag{2.113}$$

where $\Lambda_{\max}$ (resp. $\Lambda_{\min}$) is the maximum (resp. minimum) eigenvalue of the matrix $[E]$. From this it follows that if $\mathbf{u}_{FE}$ converges to $\mathbf{u}_{EX}$ in energy norm then $S(\sigma_{FE})$ converges to $S(\sigma_{EX})$ at the same rate. Equivalently, the error $S(\sigma_{FE} - \sigma_{EX})$ is bounded from above and below by the error in energy norm:

$$\sqrt{\frac{2\Lambda_{\min}}{V}} \|\mathbf{u}_{FE} - \mathbf{u}_{EX}\|_{E(\Omega)} \leq S(\sigma_{FE} - \sigma_{EX}) \leq \sqrt{\frac{2\Lambda_{\max}}{V}} \|\mathbf{u}_{FE} - \mathbf{u}_{EX}\|_{E(\Omega)}. \tag{2.114}$$

This result indicates that when $\Lambda_{\max}$ is large then $S(\sigma_{FE} - \sigma_{EX})$ can be large even when the error in energy norm is small. Consider the potential energy expression given by eq (2.108) and assume that $\mathcal{T}_\Delta = 0$. The first term of the integrand is $\lambda \epsilon_{kk}^2$ where ϵ_{kk} is the infinitesimal volumetric strain. As $\lambda \to \infty$, the minimization of the potential energy results in $\epsilon_{kk} \to 0$. When computing σ_{kk} from Hooke's law we have

$$\sigma_{kk} = (3\lambda + 2G)\epsilon_{kk} = \frac{E}{1-2v} \epsilon_{kk} \tag{2.115}$$

that is, ϵ_{kk} is multiplied by a large number, which magnifies the error in ϵ_{kk}. The shear stresses and the differences of normal stresses are independent of λ and therefore can be computed directly from the strains. An indirect method for the computation of $\sigma_x + \sigma_y$ for plane strain problems is discussed in [98].

Remark 2.7 Using $\{\sigma\} = [E]\{\epsilon\}$ where $[E]$ is given by eq. (2.105), it is not difficult to see that the differences of normal stresses and shear stresses can be computed from the numerical simulation directly. Unlike in the case of the sum of normal stresses, the coefficients do not go to infinity as v approaches $1/2$.

Exercise 2.40 Find the eigenvalues of matrix $[E]$ for plane strain and plane stress. Partial answer: For plane strain

$$\Lambda_{\max} = 2(\lambda + G) \equiv \frac{E}{(1 - 2v)(1 + v)}, \quad \Lambda_{\min} = G. \tag{2.116}$$

Observe that $\Lambda_{\max} \to \infty$ as $v \to 1/2$.

2.6.3 The principle of virtual work

In many engineering textbooks eq. (2.93) is given a physical interpretation and treated as a fundamental principle of continuum mechanics, called the principle of virtual work. In this view, the test function v_i is understood to be some arbitrary displacement field, imposed by an agent that is independent of the applied body force F_i and traction T_i. For this reason v_i is called "virtual displacement".

The terms on the right-hand side of eq. (2.93) represent work by the body force and the traction forces acting on the body, collectively called "external forces", caused by the virtual displacement. The left-hand side represents virtual work done by the internal stresses. To see this, refer to Fig. 2.11 and assume that vertex A of the infinitesimal hexahedral element, the coordinates of which are x_i, is subjected to a virtual displacement v_i. Then, since v_i is continuous and differentiable, the face located at $x_1 + dx_1$ will be displaced, relative to point A, by $v_{i,1}dx_1$ and the virtual work done by σ_{11} is:

$$dW_{\sigma_{11}} = \underbrace{(\sigma_{11}dx_2dx_3)}_{\text{force}} \underbrace{(v_{1,1}dx_1)}_{\text{displacement}} = \sigma_{11}v_{1,1}dV$$

similarly, the virtual work done by σ_{13} is:

$$dW_{\sigma_{13}} = \underbrace{(\sigma_{13}dx_2dx_3)}_{\text{force}} \underbrace{(v_{3,1}dx_1)}_{\text{displacement}} = \sigma_{13}v_{3,1}dV$$

etc.

The principle of virtual work states that *the virtual work of external forces is equal to the virtual work of internal stresses*. Note that since this result is based on the equilibrium equations (2.91), it is independent of the material properties and therefore holds for any continuum.

Equation (2.93) is the generic form of the principle of virtual work. Particular statements of the principle of virtual work depend on the material properties and the boundary conditions.

Exercise 2.41 Starting from eq. (2.16) derive the counterpart of eq. (2.94) for the stationary heat conduction problem.

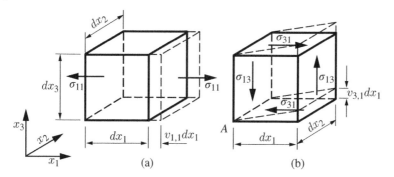

Figure 2.11 Virtual displacements corresponding to (a) σ_{11} and (b) σ_{13}.

2.6.4 Uniqueness

The generalized formulation based on the principle of virtual work is unique in the energy space $E(\Omega)$. The proof of uniqueness given by Theorem 1.1 is applicable to the elasticity problem in three dimensions.

Uniqueness in the energy space does not necessarily mean uniqueness of the displacement field **u**. When $\partial\Omega_u$ and $\partial\Omega_s$ are both empty then there are six linearly independent test functions in $E^0(\Omega) = E(\Omega)$ for which $\epsilon_{ij}^{(v)} = 0$ and hence $B(\mathbf{u}, \mathbf{v}) = 0$. Three of these functions correspond to rigid body displacements:

$$\epsilon_{11}^{(v)} = 0: \quad v_i^{(1)} = c_1 \{1\ 0\ 0\}^T$$
$$\epsilon_{22}^{(v)} = 0: \quad v_i^{(2)} = c_2 \{0\ 1\ 0\}^T$$
$$\epsilon_{33}^{(v)} = 0: \quad v_i^{(3)} = c_3 \{0\ 0\ 1\}^T$$

and three correspond to infinitesimal rigid body rotations:

$$\epsilon_{12}^{(v)} = 0: \quad v_i^{(4)} = c_4 \{-x_2\ x_1\ 0\}^T \quad \text{rotation about} \quad x_3$$
$$\epsilon_{23}^{(v)} = 0: \quad v_i^{(5)} = c_5 \{0 - x_3\ x_2\}^T \quad \text{rotation about} \quad x_1$$
$$\epsilon_{31}^{(v)} = 0: \quad v_i^{(6)} = c_6 \{x_3\ 0 - x_1\}^T \quad \text{rotation about} \quad x_2$$

where $c_1, c_2, \ldots, c_6$ are arbitrary constants. Consequently the body force vector and the surface tractions must satisfy the following six conditions:

$$F(\mathbf{v}) = 0: \quad \int_\Omega F_i v_i^{(k)}\, dV + \int_{\partial\Omega} T_i v_i^{(k)}\, dS = 0 \qquad k = 1, 2, \ldots, 6. \tag{2.117}$$

The physical interpretation of these conditions is that the body must be in equilibrium, i.e., the sum of forces and the sum of moments must be zero.

The solution is unique up to rigid body displacements and rotations. In order to ensure uniqueness of the solution, "rigid body constraints" are imposed, that is, functions that represent rigid body displacements and rotations are eliminated from the space of test functions.

$$E^0(\Omega) = \{v_i \mid v_i \in E(\Omega),\ v_i^{(k)} = 0,\ k = 1, 2, \ldots, 6\}. \tag{2.118}$$

The values of the rigid body displacements are arbitrary, therefore the space of admissible functions is:

$$\tilde{E}(\Omega) = \{u_i \mid u_i \in E(\Omega),\ u_i^{(k)} = \hat{u}_i^{(k)},\ k = 1, 2, \ldots, 6\}. \tag{2.119}$$

where $\hat{u}_i^{(k)}$ are arbitrary rigid body displacements, usually chosen to be zero.

Figure 2.12 Rigid body constraints. Notation.

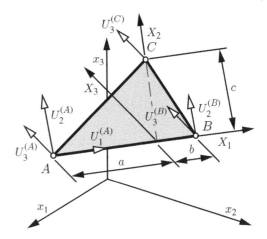

Rigid body constraints are enforced by setting six displacement components in at least three non-collinear points to arbitrary values. The usual procedure is as follows: Three non-collinear points, labeled A, B, C in Fig. 2.12, are selected arbitrarily. A Cartesian coordinate system is associated with these points such that points A and B lie on axis X_1, axis X_3 is perpendicular to the plane defined by the points ABC and axis X_2 is perpendicular to axes X_1, X_3. The displacement components $U_1^{(A)}$, $U_2^{(A)}$, $U_3^{(A)}$, $U_2^{(B)}$, $U_3^{(B)}$ and $U_3^{(C)}$, shown in Fig. 2.12, are assigned arbitrary values, typically zero. This will ensure that those modes of displacement that correspond to rigid body displacements and rotations are removed from the trial space.

For example, denoting the displacement vector components of the origin of the coordinate system X_1, X_2, X_3, shown in Fig. 2.12, by C_1, C_2, C_3 and the infinitesimal rotations about the X_1, X_2 and X_2 axes by C_4, C_5, C_6 respectively, we can set the displacement vector components in points A, B and C corresponding to rigid body displacements and rotations to zero:

$$\begin{Bmatrix} U_1^{(A)} \\ U_2^{(A)} \\ U_3^{(A)} \\ U_2^{(B)} \\ U_3^{(B)} \\ U_3^{(C)} \end{Bmatrix} = \begin{bmatrix} 1 & 0 & 0 & 0 & 0 & 0 \\ 0 & 1 & 0 & 0 & 0 & -a \\ 0 & 0 & 1 & 0 & a & 0 \\ 0 & 1 & 0 & 0 & 0 & b \\ 0 & 0 & 1 & 0 & -b & 0 \\ 0 & 0 & 1 & c & 0 & 0 \end{bmatrix} \begin{Bmatrix} C_1 \\ C_2 \\ C_3 \\ C_4 \\ C_5 \\ C_6 \end{Bmatrix} = 0 \tag{2.120}$$

where the determinant of the coefficient matrix is $c(a+b)^2$ which is nonzero for any choice of three non-collinear points. Here we set the displacements equal to zero, which is common practice; however, arbitrary values could have been assigned. The important point is that the coefficient matrix in eq. (2.120) must be nonsingular. An additional requirement is that the exact solution in the points where rigid body constraints are prescribed must be continuous.

Example 2.7 The thin elastic plate-like body with a circular hole, shown in Fig. 2.13, is loaded by constant normal tractions T_0. The equilibrium conditions (2.117) are obviously satisfied. In this case there are six rigid body modes and the space of admissible functions is given by eq. (2.119) and the space of test functions by eq. (2.118). Letting

$$u_1^{(A)} = u_2^{(A)} = u_3^{(A)} = u_2^{(B)} = u_3^{(B)} = u_3^{(C)} = 0$$

the rigid body constraints are enforced.

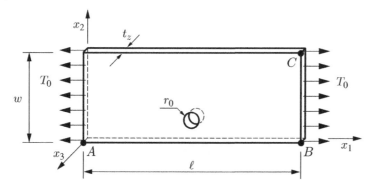

Figure 2.13 Example 2.7. Notation

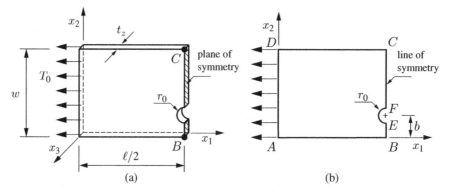

Figure 2.14 Notation.

In many cases $\partial\Omega_u$ and/or $\partial\Omega_s$ is not empty but the prescribed conditions do not provide sufficient number of constraints to prevent all rigid body displacements and/or rotations. In those cases it is necessary to provide a sufficient number of rigid body constraints to prevent rigid body displacement, as illustrated in the following example.

Example 2.8 If the center of the circular hole in the elastic plate-like body of Example 2.7 is located at $x_1 = \ell/2$ then the solution is symmetric with respect to the plane $x_1 = \ell/2$ and the problem may be formulated on the half domain shown in Fig. 2.14(a). On a plane of symmetry the normal displacement component u_n and the shearing stress components are zero.

Setting $u_n = u_1 = 0$ on the plane of symmetry prevents displacement in the x_1 direction and rotations about the x_2 and x_3 axes, but does not prevent displacements in the x_2 and x_3 coordinate directions, nor rotation about the x_1 axis. We may set

$$u_2^{(B)} = u_3^{(B)} = u_3^{(C)} = 0$$

to prevent rigid body motion allowed by the plane of symmetry.

A three-dimensional problem can be reduced to a planar problem only when there is a plane of symmetry. The domain of the planar problem lies in this plane of symmetry. Therefore there are at most three rigid body displacements: two in-plane displacement components and a rotation about an arbitrary point in the mid-plane.

For instance, the problem in Example 2.8 can be formulated as a planar problem. In that case the implied zero normal displacement of the mid-plane prevents displacement in the x_3 direction as well as rotation about the x_1 and x_2 axes. The line of symmetry shown in Fig. 2.14(b) prevents displacement in the x_1 direction and rotation about the x_3 axis. Therefore only one rigid body constraint has to be imposed, for example by letting $u_2^{(A)} = 0$. Note that the coordinate axes x_1, x_2 lie in the plane $x_3 = 0$.

Exercise 2.42   Refer to Fig. 2.13. Relocate point A to $\{0\ 0\ 0\}$ and point C to $\{\ell\ w\ 0\}$. Assume that the following constraints are specified:

$$u_1^{(A)} = u_2^{(A)} = u_3^{(A)} = u_2^{(B)} = u_3^{(B)} = u_3^{(C)} = 0.$$

Write down the system of equations analogous to eq. (2.120) and verify that the coefficient matrix is of full rank.

Remark 2.8   In the foregoing discussion it was assumed that the material is isotropic. If the material is not isotropic then symmetry, antisymmetry and periodicity cannot be used in general. An exception is when the material is orthotropic and the axes of material symmetry are aligned with the axes of geometric symmetry.

2.7   Residual stresses

Up to this point we tacitly assumed that an elastic body is stress free when the mechanical strain tensor is zero, see eq. (2.56).

Metal forming, such as forging, rolling and drawing, invariably induce residual stresses in metals[16]. These residual stresses, present in metal stock, are called bulk residual stresses. In addition, metal cutting operations, such as turning, drilling and milling, remove thin layers of metal by shear, leaving behind a layer of residual stresses that rapidly decay with respect to distance from the surface. The strongly affected boundary layer is typically between 0.2 and 1 mm wide [50]. Shot peening is used for inducing compressive residual stresses to increase durability and to remedy distortions caused by bulk and machining-induced residual stresses. Fastener holes in aircraft structures are often cold-worked to produce a layer of compressive residual stresses for improved durability.

In composite materials there is a large difference in the coefficients of thermal expansion of the fiber and matrix. Residual stresses develop when a part cools after curing.

We denote the reference configuration of an elastic body by Ω_0, its boundary points by $\partial\Omega_0$, and the residual stress field in the reference configuration by σ_{ij}^0. The residual stress field satisfies the equations of equilibrium and the stress-free boundary conditions:

$$\sigma_{ij,j}^{(0)} = 0 \ \text{ in } \ \Omega_0 \ \text{ and } \ \sigma_{ij}^{(0)} n_j = 0 \ \text{ on } \ \partial\Omega_0 \tag{2.121}$$

where n_j is the unit normal to the boundary. In the presence of residual stresses the generalized Hooke's law given by eq. (2.56) is modified to

$$\sigma_{ij} = \sigma_{ij}^{(0)} + C_{ijkl}(\epsilon_{kl} - \alpha_{kl}T_\Delta). \tag{2.122}$$

---

16 For example, certain types of aluminum plates are hot-rolled, quenched, over-aged and stretched by the imposition of 1.5–3.0% strain in the rolling direction.

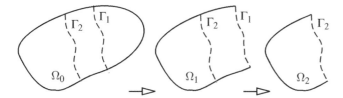

Figure 2.15 Notation.

The distribution of residual stresses is inferred from measurements of strains in non-destructive tests and measurements of displacements and strains in destructive and semi-destructive experiments. In destructive and semi-destructive experiments material is removed and the distribution and magnitude of residual stresses are inferred from displacements and strains caused by the redistribution of residual stresses. Therefore it is necessary to use numerical simulation for the interpretation of physical observations. See, for example, [63, 64].

Let us introduce a cut Γ_1 that will produce domain Ω_1 and a second cut Γ_2 that will produce domain Ω_2, see Fig. 2.15. Denote the displacement field following the first (resp. second) cut by $u_i^{(1)}$ (resp. $u_i^{(2)}$).

Assuming that the principle of superposition is applicable and the cuts do not cause residual stresses, we now show that $u_i^{(2)}$ depends on $\sigma_{ij}^{(0)}$ and Ω_2 but not on $u_i^{(1)}$ or Ω_1 Observe that, since the cuts create free surfaces, $u_i^{(1)}$ satisfies

$$B(u_i^{(1)}, v_i) = -\int_{\Gamma_1} \sigma_{ij}^{(0)} n_j v_i \, dS \quad \text{for all} \quad v_i \in E(\Omega_1). \tag{2.123}$$

We need to show that

$$B(u_i^{(2)}, v_i) = -\int_{\Gamma_2} \sigma_{ij}^{(0)} n_j v_i \, dS \quad \text{for all} \quad v_i \in E(\Omega_2) \tag{2.124}$$

independently of whether the body was the first cut at Γ_2 or first cut at Γ_1 and then at Γ_2.

Denote the stress field corresponding to $u_i^{(1)}$ on Ω_1 by s_{ij} and let w_i be the displacement field on Ω_2 corresponding to the traction s_{ij} acting on Γ_2. Therefore we have

$$B(w_i, v_i) = \int_{\Gamma_2} s_{ij} n_j v_i \, dS \quad \text{for all} \quad v_i \in E(\Omega_2). \tag{2.125}$$

The displacement field corresponding to the second cut is caused by the creation of a free surface on Γ_2:

$$B(u_i^{(2)} - w_i, v_i) = -\int_{\Gamma_2} \left(\sigma_{ij}^{(0)} + s_{ij} \right) n_j v_i \, dS \quad \text{for all} \quad v_i \in E(\Omega_2). \tag{2.126}$$

On adding equations (2.125) and (2.126) we find that $u_i^{(2)}$ satisfies eq. (2.124) which proves the statement that $u_i^{(2)}$ depends on $\sigma_{ij}^{(0)}$ and Ω_2 but not on $u_i^{(1)}$ or Ω_1.

Exercise 2.43 An annular aluminum plate with inner radius r_s and outer radius $r_s + b$, constant thickness d_a, was joined by shrink fitting to a stainless steel shaft. The configuration is shown in Fig. 2.16. Denote the mechanical properties of aluminum as follows: modulus of elasticity: E_a, modulus of rigidity: G_a, mass density: ϱ_a, coefficient of thermal expansion: α_a and the corresponding mechanical properties of stainless steel as $E_s, G_s, \varrho_s, \alpha_s$. Let $\ell_s = 80 \, \text{mm}, r_s = 17.5 \, \text{mm}, d_a = 15 \, \text{mm}, b = 150 \, \text{mm}, E_a = 72.0 \times 10^3 \, \text{MPa}, G_a = 28.0 \times 10^3 \, \text{MPa}, \varrho_a = 2800 \, \text{kg/m}^3, \alpha_a = 23.6 \times 10^{-6}/\text{K}, E_s = 190 \times 10^3 \, \text{MPa}, G_s = 75.0 \times 10^3 \, \text{MPa}, \varrho_s = 7920 \, \text{kg/m}^3, \alpha_s = 17.3 \times 10^{-6}/\text{K}.$

Figure 2.16 Notation for Exercise 2.7.

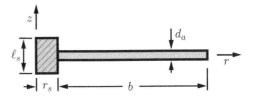

Consider the following conditions: (a) The shaft and the aluminum plate were heated to 220 °C, the shaft was inserted and then the assembly was cooled to 20 °C. Assume that at 220 °C the clearance between the shaft and the plate was zero and $\alpha_a > \alpha_s$. (b) The assembly is spinning about the z axis at an angular velocity ω.

Estimate the value of ω at which the membrane force in the aluminum plate, F_r, is approximately zero at $r = r_s$. By definition:

$$F_r = \int_{-d_a/2}^{d_a/2} \sigma_r \, dz.$$

Specify ω in units of cycles per second (Hertz). Note that, in order to have consistent units, kg/m^3 must be converted to Ns^2/mm^4.

2.8 Chapter summary

The formulation of mathematical models for linear problems in heat conduction and elasticity was described in strong and generalized forms.

Mathematical models for heat conduction are based on the conservation law and an empirical relationship between the derivatives of the temperature u and the flux vector q_i. We assumed that this relationship is linear. In reality, however, the coefficients of thermal conductivity depend on the value of u as well as the gradient of u. Only within a narrow range of u, and the gradient of u, should these coefficients be approximated by constants.

Mathematical models for elastic bodies are based on the conservation of momentum (in statics the equations of equilibrium) and an empirical linear relationship between the stress and strain tensors. This linear relationship holds for small displacements and strains only. It is important to bear in mind the limitations of mathematical models imposed by the assumptions incorporated in the models.

A mathematical model must never be confused with the physical reality that it was conceived to imitate. This important point will be addressed in greater detail in Chapter 5.

3

Implementation

This chapter is concerned with the algorithmic aspects of the finite element method. The finite element spaces, standard elements, the corresponding shape functions and mapping functions are described for two- and three-dimensional formulations. At the end of the solution process the coefficients of the shape functions, the mapping and the material properties are available at the element level. The quantities of interest are computed from this information either by direct or indirect methods.

3.1 Standard elements in two dimensions

Two-dimensional finite element meshes are comprised of triangular and quadrilateral elements. The standard quadrilateral element will be denoted by $\Omega_{st}^{(q)}$ and the standard triangular element by $\Omega_{st}^{(t)}$. The definition of standard elements is arbitrary. However, certain conveniences in mapping and assembly are achieved when the standard elements are defined as shown in Fig. 3.1. Note that the sides of the elements have the same length (2) as in one-dimension.

3.2 Standard polynomial spaces

The standard polynomial spaces in two and three dimensions are generalizations of the standard polynomial space $S^p(I_{st})$ defined in Section 1.3.1. Whereas the polynomial degree could be characterized by a single number (p) in one dimension, in higher dimensions more than one interpretation is possible.

3.2.1 Trunk spaces

Trunk spaces are polynomial spaces spanned by the set of monomials $\xi^i \eta^j$, $i, j = 0, 1, 2, \ldots, p$ subject to the restriction $i + j = 0, 1, 2, \ldots, p$. In the case of quadrilateral elements these are supplemented by one or two monomials of degree $p + 1$:

1. Triangles: The dimension of the space is $n(p) = (p + 1)(p + 2)/2$. For example, the space $S^6(\Omega_{st}^{(t)})$ is spanned by the 28 monomial terms indicated in Fig. 3.2. The space $S^p(\Omega_{st}^{(t)})$ comprises all polynomials of degree less than or equal to p.

Finite Element Analysis: Method, Verification and Validation, Second Edition. Barna Szabó and Ivo Babuška.
© 2021 John Wiley & Sons, Inc. Published 2021 by John Wiley & Sons, Inc.
Companion Website: www.wiley.com/go/szabo/finite_element_analysis

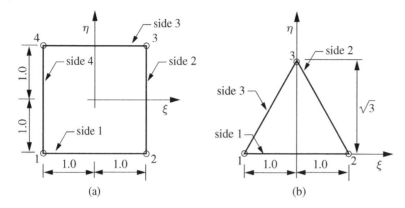

Figure 3.1 Standard quadrilateral and triangular elements $\Omega_{st}^{(q)}$ and $\Omega_{st}^{(t)}$.

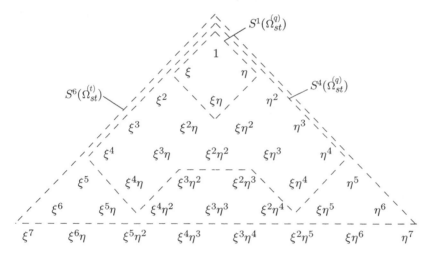

Figure 3.2 Trunk space. Illustration of spanning sets for $S^1(\Omega_{st}^{(q)})$, $S^4(\Omega_{st}^{(q)})$ and $S^6(\Omega_{st}^{(t)})$.

2. Quadrilaterals: Monomials of degree less than or equal to p are supplemented by $\xi\eta$ for $p = 1$ and by $\xi^p\eta$, $\xi\eta^p$ for $p \geq 2$. For example, the space $S^4(\Omega_{st}^{(q)})$ is spanned by the 17 monomial terms indicated in Fig. 3.2. The dimension of space $S^p(\Omega_{st}^{(q)})$ is

$$n(p) = \begin{cases} 4p & \text{for } p \leq 3 \\ 4p + (p-2)(p-3)/2 & \text{for } p \geq 4. \end{cases} \tag{3.1}$$

3.2.2 Product spaces

In two dimensions product spaces are spanned by the monomials 1, ξ, ξ^2, ..., ξ^p, 1, η, η^2, ..., η^q and their products. Thus the dimension of product spaces is $n(p, q) = (p + 1)(q + 1)$. Product spaces on triangles will be denoted by $S^{p,q}(\Omega_{st}^{(t)})$ and on quadrilaterals by $S^{p,q}(\Omega_{st}^{(q)})$. The spanning set of monomials for the space $S^{4,2}(\Omega_{st}^{(q)})$ is illustrated in Fig. 3.3.

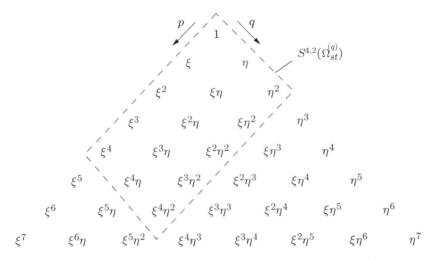

Figure 3.3 Product space. Illustration of spanning set for the space $S^{4,2}(\Omega_{st}^{(q)})$.

3.3 Shape functions

As in the one-dimensional case, we will discuss two types of shape functions: Shape functions based on Lagrange polynomials and hierarchic shape functions based on the integrals of Legendre polynomials. We will use the notation $N_i(\xi, \eta)$ $(i = 1, 2, \ldots, n)$ for both. The shape functions in two dimensions along the sides of each element are the same as the shape functions defined on the one-dimensional standard element I_{st}.

3.3.1 Lagrange shape functions

Elements with shape functions that span $S^p(\Omega_{st}^{(t)})$ and $S^p(\Omega_{st}^{(q)})$ with $p = 1$ and $p = 2$ are widely used in engineering practice. Each one of the Lagrange shape functions is unity in one of the node points and zero in the other node points. Therefore the approximating function u can be written as

$$u(\xi, \eta) = \sum_{i=1}^{n} u_i N_i(\xi, \eta)$$

where u_i is the value of u in the ith node point.

Quadrilateral elements

The shape functions of four-node quadrilateral elements span the space $S^1(\Omega_{st}^{(q)})$:

$$N_1(\xi, \eta) = \frac{1}{4}(1 - \xi)(1 - \eta) \tag{3.2}$$

$$N_2(\xi, \eta) = \frac{1}{4}(1 + \xi)(1 - \eta) \tag{3.3}$$

$$N_3(\xi, \eta) = \frac{1}{4}(1 + \xi)(1 + \eta) \tag{3.4}$$

$$N_4(\xi, \eta) = \frac{1}{4}(1 - \xi)(1 + \eta) \tag{3.5}$$

The shape functions of eight-node quadrilateral elements span the space $S^2(\Omega_{st}^{(q)})$. The shape functions corresponding to the vertex nodes are:

$$N_1(\xi, \eta) = \frac{1}{4}(1 - \xi)(1 - \eta)(-\xi - \eta - 1) \tag{3.6}$$

$$N_2(\xi, \eta) = \frac{1}{4}(1 + \xi)(1 - \eta)(\xi - \eta - 1) \tag{3.7}$$

$$N_3(\xi, \eta) = \frac{1}{4}(1 + \xi)(1 + \eta)(\xi + \eta - 1) \tag{3.8}$$

$$N_4(\xi, \eta) = \frac{1}{4}(1 - \xi)(1 + \eta)(-\xi + \eta - 1) \tag{3.9}$$

and the shape functions corresponding to the mid-side nodes are:

$$N_5(\xi, \eta) = \frac{1}{2}(1 - \xi^2)(1 - \eta) \tag{3.10}$$

$$N_6(\xi, \eta) = \frac{1}{2}(1 + \xi)(1 - \eta^2) \tag{3.11}$$

$$N_7(\xi, \eta) = \frac{1}{2}(1 - \xi^2)(1 + \eta) \tag{3.12}$$

$$N_8(\xi, \eta) = \frac{1}{2}(1 - \xi)(1 - \eta^2). \tag{3.13}$$

Observe that if we denote the coordinates of the vertices and the midpoints of the sides by (ξ_i, η_i), $i = 1, 2, \ldots, 8$, then $N_i(\xi_j, \eta_j) = \delta_{ij}$.

The shape functions of the nine-node quadrilateral element span $S^{2,2}(\Omega_{st}^{(q)})$ (i.e. the product space). In addition to the four nodes located in the vertices and the four nodes located in the mid-points of the sides, there is a node in the center of the element. Construction of these shape functions is left to the reader in the following exercise.

Exercise 3.1 Write down the shape functions for the nine-node quadrilateral element. Sketch the shape function associated with the node in the center of the element and one of the vertex shape functions and one of the side shape functions.

Triangular elements
The shape functions for triangular elements are usually written in terms of the triangular coordinates, defined as follows:

$$L_1 = \frac{1}{2}\left(1 - \xi - \frac{\eta}{\sqrt{3}}\right) \tag{3.14}$$

$$L_2 = \frac{1}{2}\left(1 + \xi - \frac{\eta}{\sqrt{3}}\right) \tag{3.15}$$

$$L_3 = \frac{\eta}{\sqrt{3}}. \tag{3.16}$$

Note that L_i is unity at node i and zero on the side opposite to node i. Also, $L_1 + L_2 + L_3 = 1$. The space $S^1(\Omega_{st}^{(t)})$ is spanned by the following shape functions:

$$N_i = L_i \quad i = 1, 2, 3. \tag{3.17}$$

These elements are called "three-node triangles". For the six-node triangles the shape functions are:

$$N_1 = L_1(2L_1 - 1) \tag{3.18}$$

$$N_2 = L_2(2L_2 - 1) \tag{3.19}$$

$$N_3 = L_3(2L_3 - 1) \tag{3.20}$$

$$N_4 = 4L_1 L_2 \tag{3.21}$$

$$N_5 = 4L_2 L_3 \tag{3.22}$$

$$N_6 = 4L_3 L_1 \tag{3.23}$$

which span $S^2(\Omega_{st}^{(t)})$.

3.3.2 Hierarchic shape functions

Hierarchic shape functions based on the integrals of Legendre polynomials are described for the nodes, sides and vertices of quadrilateral and triangular elements. The shape functions associated with the nodes and the sides are the same for the product and trunk spaces. Only the number of internal shape functions is different.

Quadrilateral elements
The nodal shape functions are the same as those for the four-node quadrilateral, given by equations (3.2) to (3.5).

The side shape functions are constructed by multiplying the shape functions N_3, N_4, ..., defined for the one-dimensional element (see Fig. 1.4), by linear blending functions. We define:

$$\phi_k(s) \overset{\text{def}}{=} \sqrt{\frac{2k-1}{2}} \int_{-1}^{s} P_{k-1}(t)\, dt, \quad k = 2, 3, \ldots \tag{3.24}$$

Note that the index k represents the polynomial degree. The shape functions of degree $p \geq 2$ are defined for the four sides as follows:

side 1: $\quad N_k^{(1)}(\xi, \eta) = \frac{1}{2}(1 - \eta)\phi_k(\xi) \tag{3.25}$

side 2: $\quad N_k^{(2)}(\xi, \eta) = \frac{1}{2}(1 + \xi)\phi_k(\eta) \tag{3.26}$

side 3: $\quad N_k^{(3)}(\xi, \eta) = \frac{1}{2}(1 + \eta)\phi_k(-\xi) \tag{3.27}$

side 4: $\quad N_k^{(4)}(\xi, \eta) = \frac{1}{2}(1 - \xi)\phi_k(-\eta) \tag{3.28}$

where $k = 2, 3, \ldots, p$. Thus there are $4(p-1)$ side shape functions. The argument of ϕ_k is negative for sides 3 and 4 because the positive orientation of the sides is counterclockwise. This will affect shape functions of odd degrees only.

The internal shape functions are zero on the sides. For the trunk space there are $(p-2)(p-3)/2$ internal shape functions ($p \geq 4$) constructed from the products of ϕ_k:

$$N_p^{(k,l)}(\xi, \eta) = \phi_k(\xi)\phi_l(\eta) \qquad k, l = 2, 3, \ldots, p, \quad k + l = 4, 5, \ldots, p. \tag{3.29}$$

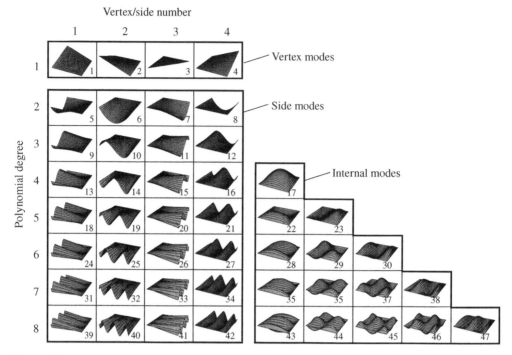

Vertex/side number

Polynomial degree

Figure 3.4 Hierarchic shape functions for quadrilateral elements. Trunk space, $p = 1$ to $p = 8$.

The shape functions are assigned unique sequential numbers, as shown in Fig. 3.4. For the product space there are $(p-1)(q-1)$ internal shape functions, defined for $p, q \geq 2$ by

$$N_{pq}^{(k,l)}(\xi, \eta) = \phi_k(\xi)\phi_l(\eta), \quad k = 2, 3, \ldots, p, \; l = 2, 3, \ldots, q, \; k + l \leq p + q. \tag{3.30}$$

Triangular elements

The nodal shape functions are the same as those for the three-node triangles, given by eq. (3.17). The side shape functions are constructed as follows: Define:

$$\tilde{\phi}_k(s) = 4\frac{\phi_k(s)}{1 - s^2} \quad k = 2, 3, \ldots \tag{3.31}$$

where $\phi_k(s)$ is the function defined by eq. (3.24). For example,

$$\tilde{\phi}_2(s) = -\sqrt{6}, \quad \tilde{\phi}_3(s) = -\sqrt{10}s, \quad \tilde{\phi}_4(s) = -\sqrt{\frac{7}{8}}(5s^2 - 1), \text{ etc.}$$

Using

$$s = \begin{cases} L_2 - L_1 & \text{for side 1} \\ L_3 - L_2 & \text{for side 2} \\ L_1 - L_3 & \text{for side 3} \end{cases} \tag{3.32}$$

the definition of the side shape functions is

$$\text{side 1:} \quad N_k^{(1)}(L_1, L_2, L_3) = L_1 L_2 \tilde{\phi}_k(L_2 - L_1) \tag{3.33}$$

$$\text{side 2:} \quad N_k^{(2)}(L_1, L_2, L_3) = L_2 L_3 \tilde{\phi}_k(L_3 - L_2) \tag{3.34}$$

$$\text{side 3:} \quad N_k^{(3)}(L_1, L_2, L_3) = L_3 L_1 \tilde{\phi}_k(L_1 - L_3) \tag{3.35}$$

where $k = 2, 3, \ldots, p$. Thus there are $3(p-1)$ side shape functions.

For the trunk space there are $(p-1)(p-2)/2$ internal shape functions $(p \geq 3)$ defined as follows:

$$N_p^{(k,l)} = L_1 L_2 L_3 P_k (L_2 - L_1) P_l (2L_3 - 1) \quad k,l = 0,1,2,\ldots,p-3 \tag{3.36}$$

where $k + l \leq p - 3$ and P_k is the kth Legendre polynomial.

Exercise 3.2 Sketch the shape function $N_3^{(2)}(L_1, L_2, L_3)$.

3.4 Mapping functions in two dimensions

The commonly used mapping procedures for the transformation of the standard element to the elements of the mesh are outlined in this section.

3.4.1 Isoparametric mapping

The term "isoparametric mapping" is meant to convey the idea that the same shape functions are used for providing topological descriptions for elements as for the element-level approximations. If the mapping is of lower (resp. higher) polynomial degree than the approximating functions then it is said to be subparametric (resp. superparametric).

Isoparametric mapping is based on the Lagrange shape functions described in Section 3.3.1. The most commonly used isoparametric mapping procedures are the linear and quadratic mappings.

Isoparametric mapping for quadrilateral elements
Linear mapping of quadrilateral elements from the standard quadrilateral element shown in Fig. 3.1(a) to the kth element is defined by

$$x = Q_x^{(k)}(\xi, \eta) = \sum_{i=1}^{4} N_i(\xi, \eta) X_i \tag{3.37}$$

$$y = Q_y^{(k)}(\xi, \eta) = \sum_{i=1}^{4} N_i(\xi, \eta) Y_i \tag{3.38}$$

where (X_i, Y_i) are the coordinates of vertex i of the kth element numbered in counterclockwise order and N_i are the shape functions defined by equations (3.2) through (3.5).

Quadratic mapping of quadrilateral elements from the standard quadrilateral element is defined by:

$$x = Q_x^{(k)}(\xi, \eta) = \sum_{i=1}^{8} N_i(\xi, \eta) X_i \tag{3.39}$$

$$y = Q_y^{(k)}(\xi, \eta) = \sum_{i=1}^{8} N_i(\xi, \eta) Y_i \tag{3.40}$$

where (X_i, Y_i) are the coordinates of the four vertices numbered in counterclockwise order and the mid-points of the four sides also numbered in counterclockwise order. The side between nodes 1 and 2 is the first side and N_i are the shape functions defined by equations (3.6) through (3.13). Quadratic isoparametric mapping of a quadrilateral element and typical numbering of the node points are illustrated in Fig. 3.5(a).

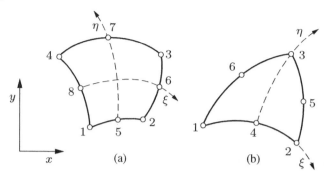

Figure 3.5 Isoparametric quadrilateral and triangular elements.

Isoparametric mapping for triangular elements

Linear mapping of triangles from the standard triangular element shown in Fig. 3.1(b) to the kth element is defined by

$$x = Q_x^{(k)}(L_1, L_2, L_3) = \sum_{i=1}^{3} L_i X_i \tag{3.41}$$

$$y = Q_y^{(k)}(L_1, L_2, L_3) = \sum_{i=1}^{3} L_i Y_i. \tag{3.42}$$

Quadratic isoparametric mapping of triangular elements from the standard quadrilateral element is given by

$$x = Q_x^{(k)}(L_1, L_2, L_3) = \sum_{i=1}^{6} N_i(L_1, L_2, L_3) X_i \tag{3.43}$$

$$y = Q_y^{(k)}(L_1, L_2, L_3) = \sum_{i=1}^{6} N_i(L_1, L_2, L_3) Y_i \tag{3.44}$$

where N_i are the shape functions defined by equations (3.18) through (3.23). The mapping of a triangular element and typical numbering of the node points are illustrated in Fig. 3.5(b).

Remark 3.1 The mapped shape functions are polynomials only in the special cases when the standard triangle is mapped into a straight-side triangle and when the standard quadrilateral element is mapped into a parallelogram. In general the mapped shape functions are not polynomials. The mapped shape functions are called "pull-back polynomials". The accuracy of finite element approximation is governed by the properties of the pull-back polynomials. Depending on exact solution, approximation by the pull-back polynomials can be better or worse than approximation by polynomials. In some cases the mapping is designed to improve the approximation.

Exercise 3.3 Show that quadratic parametric mapping applied to straight-side triangular and quadrilateral elements is identical to linear mapping.

Exercise 3.4 Consider the element shown in Fig. 3.6(b). Note that nodes 4 and 6 are located in the "quarter point position", i.e. they are located $1/4$ of the length of sides 1 and 3, respectively, from vertex 1. Introduce the polar coordinates r, θ and ρ, ϕ defined as shown in Fig. 3.6. The triangular

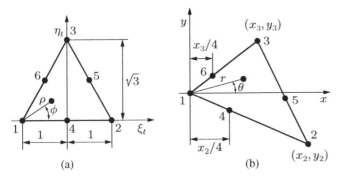

Figure 3.6 Notation for (a) the standard triangular element and (b) quarter-point mapping.

coordinates can be written in terms of ρ, ϕ as follows:

$$L_1 = \frac{1}{2}\left(2 - \rho\cos\phi - \frac{1}{\sqrt{3}}\rho\sin\phi\right)$$

$$L_2 = \frac{1}{2}\left(\rho\cos\phi \quad \frac{1}{\sqrt{3}}\,\rho\sin\phi\right)$$

$$L_3 = \frac{\rho\sin\phi}{\sqrt{3}}.$$

Show that ρ is proportional to $\sqrt{r}$ for any fixed ϕ when quadratic parametric mapping is used. The significance of this is that quarter point mapping modifies the element-level basis functions in such a way that in the neighborhood of vertex 1 the element level basis functions will contain the term $\sqrt{r}$. Quarter point mapping is frequently used in the solution of fracture mechanics problems.

Remark 3.2 The quarter point element is a special kind of singular element. Singular finite elements are formulated with the objective to reduce the errors of approximation caused by singular points. This is done by enlarging the finite element space through the addition of singular functions.

3.4.2 Mapping by the blending function method

To illustrate the method, let us consider a simple case where only one side (side 2) of a quadrilateral element is curved, as shown in Fig. 3.7. The curve $x = x_2(\eta)$, $y = y_2(\eta)$ is given in parametric form with $-1 \leq \eta \leq 1$. We can now write:

$$x = \frac{1}{4}(1-\xi)(1-\eta)X_1 + \frac{1}{4}(1+\xi)(1-\eta)X_2 + \frac{1}{4}(1+\xi)(1+\eta)X_3$$
$$+ \frac{1}{4}(1-\xi)(1+\eta)X_4 + \left(x_2(\eta) - \frac{1-\eta}{2}X_2 - \frac{1+\eta}{2}X_3\right)\frac{1+\xi}{2}. \tag{3.45}$$

Observe that the first four terms in this expression are the linear mapping terms given by eq. (3.37). The fifth term is the product of two functions: One function, the bracketed expression, represents the difference between $x_2(\eta)$ and the x-coordinates of the chord that connects points (X_2, Y_2) and (X_3, Y_3). The other function is the linear blending function $(1+\xi)/2$ which is unity along side 2 and zero along side 4. Therefore we can write:

$$x = \frac{1}{4}(1-\xi)(1-\eta)X_1 + \frac{1}{4}(1-\xi)(1+\eta)X_4 + x_2(\eta)\frac{1+\xi}{2}. \tag{3.46}$$

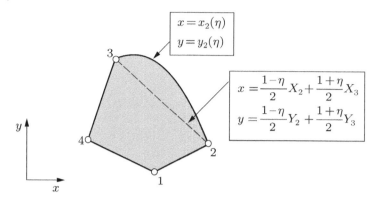

Figure 3.7 Quadrilateral element with one curved side.

Similarly:

$$y = \frac{1}{4}(1 - \xi)(1 - \eta)Y_1 + \frac{1}{4}(1 - \xi)(1 + \eta)Y_4 + y_2(\eta)\frac{1 + \xi}{2}. \tag{3.47}$$

In the general case all sides may be curved. We write the curved sides in parametric form:

$$x = x_i(s), \quad y = y_i(s), \quad -1 \le s \le +1 \quad \text{where } s = \begin{cases} \xi & \text{on sides 1 and 3} \\ \eta & \text{on sides 2 and 4} \end{cases}$$

and the subscripts represent the side numbers of the standard element. In this case the mapping functions are:

$$x = \frac{1}{2}(1 - \eta)x_1(\xi) + \frac{1}{2}(1 + \xi)x_2(\eta) + \frac{1}{2}(1 + \eta)x_3(\xi) + \frac{1}{2}(1 - \xi)x_4(\eta)$$
$$- \frac{1}{4}(1 - \xi)(1 - \eta)X_1 - \frac{1}{4}(1 + \xi)(1 - \eta)X_2 - \frac{1}{4}(1 + \xi)(1 + \eta)X_3$$
$$- \frac{1}{4}(1 - \xi)(1 + \eta)X_4, \tag{3.48}$$

$$y = \frac{1}{2}(1 - \eta)y_1(\xi) + \frac{1}{2}(1 + \xi)y_2(\eta) + \frac{1}{2}(1 + \eta)y_3(\xi) + \frac{1}{2}(1 - \xi)y_4(\eta)$$
$$- \frac{1}{4}(1 - \xi)(1 - \eta)Y_1 - \frac{1}{4}(1 + \xi)(1 - \eta)Y_2 - \frac{1}{4}(1 + \xi)(1 + \eta)Y_3$$
$$- \frac{1}{4}(1 - \xi)(1 + \eta)Y_4. \tag{3.49}$$

The inverse mapping, that is: $\xi = Q_\xi^{(k)}(x, y)$, $\eta = Q_\eta^{(k)}(x, y)$, cannot be given explicitly in general, but (ξ, η) can be computed very efficiently for any given (x, y) by means of the Newton-Raphson method or some other iterative procedure.

Exercise 3.5 Refer to Fig. 3.8(a). Show that the mapping of the quadrilateral element by the blending function method is:

$$x = r_i \cos(\theta_m + \eta\theta_d)\frac{1 - \xi}{2} + r_o \cos(\theta_m + \eta\theta_d)\frac{1 + \xi}{2}$$
$$y = r_i \sin(\theta_m + \eta\theta_d)\frac{1 - \xi}{2} + r_o \sin(\theta_m + \eta\theta_d)\frac{1 + \xi}{2}$$

where

$$\theta_m = \frac{\theta_1 + \theta_2}{2}, \quad \theta_d = \frac{\theta_2 - \theta_1}{2}.$$

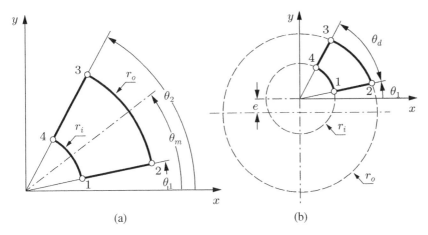

Figure 3.8 Quadrilateral elements bounded by circular segments.

Exercise 3.6 Refer to Fig. 3.8(b). A quadrilateral element is bounded by two circles. The centers of the circles are offset as shown. Write down the mapping by the blending function method in terms of the given parameters. Hint: Using the law of cosines, the radius of the arc between node 2 and node 3 is:

$$r_{2-3} = -e \sin \theta + \sqrt{r_o^2 - e^2 \cos^2 \theta}$$

where θ is the angle measured from the x-axis.

3.4.3 Mapping algorithms for high order elements

High order elements are usually mapped by the blending function method with the bounding curves approximated by polynomial functions, similar to isoparametric mapping. The reasons for this are that (a) the boundary curves are generally not available in analytical form and (b) standard treatment of all bounding curves is preferable from the point of view of implementation. For example, the boundaries of a domain are typically represented by a collection of splines in computer aided design (CAD) software products. In the interest of generality of implementation, the bounding curves are interpolated using the Lagrangian basis functions defined by eq. (1.50).

The quality of the approximation depends on the choice of the interpolation points. Specifically, the interpolation points must be defined so that, for the given polynomial degree of interpolation, the interpolation function is close to the best possible approximation of the boundary curve by polynomials in maximum norm. This is realized when the points are chosen so as to minimize the Lebesque constant. Details are available in Appendix F. In one dimension the abscissas of the Lobatto points[1] are close to the optimal interpolation points. For details we refer to [31].

Exercise 3.7 Refer to Appendix F. Approximate the semicircle of unit radius by polynomials of degree 5 using (a) the nodal set T_1^5, (b) the nodal set T_2^5 and (c) six uniformly distributed interpolation points. Compute the maximum relative error in the radius for each case.

Hint: The coordinates of the points on the perimeter of the semi-circle can be written as:

$$x_c = \cos(\pi(1 + \xi)/2), \quad y_c = \sin(\pi(1 + \xi)/2), \quad -1 \le \xi \le 1$$

---

1 See Appendix E.

Solution of this problem involves writing a short computer program. The maximum relative error in the radius in case (a) is 0.058%, in case (b) it is 0.061% and in case (c) it is 1.50%.

Rigid body rotations

In two-dimensional elasticity infinitesimal rigid body rotation is represented by the displacement vector $\mathbf{u} = C\{-y\,x\}^T$. Having introduced the mappings $x = Q_x^{(k)}(\xi, \eta)$, $y = Q_y^{(k)}(\xi, \eta)$, infinitesimal rigid body rotations are represented exactly by element k only when $Q_x^{(k)}(\xi, \eta) \in S^{p_k}(\Omega_{\mathrm{st}})$ and $Q_y^{(k)}(\xi, \eta) \in S^{p_k}(\Omega_{\mathrm{st}})$. Iso- and subparametric mappings satisfy this condition, hence rigid body rotation imposed on an element will not induce strains. Superparametric mappings and mapping by the blending function method when the sides are not polynomials, or are polynomials of degree higher than p_k, do not satisfy this condition. It has been argued that for this reason only iso- and subparametric mappings should be employed.

This argument is flawed, however. One should view this question in the following light: Errors are introduced when rigid body rotations are not represented exactly and also by the approximation of boundary curves. With the blending function method analytic curves, such as circles, are represented exactly, but the rigid body rotation terms are approximated. With iso- and subparametric mappings the boundaries are approximated but the rigid body rotation terms are represented exactly. In either case the errors of approximation will go to zero as the number of degrees of freedom is increased whether by mesh refinement or by increasing the polynomial degrees. This is illustrated by the following exercises.

Exercise 3.8   Refer to Fig. 3.8(a). Let $\theta_1 = 0$, $\theta_2 = 60°$, $r_i = 1.0$ and $r_o = 2.0$. Use $E = 200$ GPa, $\nu = 0.3$, plane strain. Impose nodal displacements consistent with rigid body rotation about the origin: $\mathbf{u} = C\{y - x\}$. For example let $u_x^{(1)} = 0$, $u_y^{(1)} = Cr_i$, $u_y^{(2)} = Cr_o$ where the superscripts indicate the node numbers and C is the angle of rotation (in radians) about the positive z axis. Let $C = 0.1$ and compute the maximum equivalent strain[2] for $p = 1, 2, \ldots, 6$. Very rapid convergence to zero will be seen[3].

Exercise 3.9   Repeat Exercise 3.8 using uniform mesh refinement and p fixed at $p = 1$ and $p = 2$. Plot the maximum equivalent strain vs. the number of degrees of freedom.

Remark 3.3   The constant function is in the finite element space $S^{p_k}(\Omega_{\mathrm{st}})$ independent of the mapping. Therefore rigid body displacements are represented exactly.

3.5   Finite element spaces in two dimensions

Finite element spaces are sets of continuous functions characterized by the finite element mesh Δ, a polynomial space defined on standard elements and the functions used for mapping the standard elements into the elements of the mesh. We have seen an example of this in Section 1.3.2 where finite element spaces in one dimension were described. There the standard element was the interval

---

2 The equivalent strain is proportional to the root-mean-square of the differences of principal strains and therefore it is an indicator of the magnitude of shearing strain.

3 Curves are approximated by polynomials of degree 5 in StressCheck. This is a default value that can be changed by setting a parameter. When the default value is used then the mapping is superparametric for $p \leq 4$, isoparametric at $p = 5$ and subparametric for $p \geq 6$.

$I = (-1, 1)$, the standard space, denoted by S^p, was a polynomial space of degree p, and the mapping function was the linear function given by eq. (1.63). Finite element spaces in two dimensions are defined analogously by

$$S \overset{\text{def}}{=} S(\Omega, \Delta, \mathbf{p}, \mathbf{Q}) =$$
$$\{\mathbf{u} \mid \mathbf{u} \in E(\Omega), \ \mathbf{u}(Q_x^{(k)}(\xi, \eta), \ Q_y^{(k)}(\xi, \eta)) \in S^{p_k}(\Omega_{st}), \ k = 1, 2, \ldots, M(\Delta)\} \tag{3.50}$$

where $\mathbf{p}$ and $\mathbf{Q}$ represent, respectively, the arrays of the assigned polynomial degrees and the mapping functions. The expression

$$\mathbf{u}(Q_x^{(k)}(\xi, \eta), \ Q_y^{(k)}(\xi, \eta)) \in S^{p_k}(\Omega_{st})$$

indicates that the basis functions defined on element Ω_k are mapped from the shape functions of a polynomial space defined on standard triangular and quadrilateral elements.

3.6 Essential boundary conditions

As we have seen in Chapter 1, the essential boundary conditions are enforced by restriction. If an essential boundary condition can be written as a linear combination of the basis functions then enforcement is straightforward: The coefficients of the basis functions are known, and enforcement is treated as in the one-dimensional case.

When the prescribed boundary conditions cannot be written as linear combinations of the basis functions then it is necessary to approximate the prescribed boundary conditions by the basis functions that are not zero on the element boundaries and then set the coefficients of the basis functions to the appropriate value.

Let us assume, for example, that side 1 of a quadrilateral element of degree p lies on a boundary on which a Dirichlet condition $u_{FE} = U(s)$ is to be enforced. Let the parameter values s_1 and s_2 correspond to nodes 1 and 2. We introduce the transformation

$$s = s(\xi) = \frac{1-\xi}{2} s_1 + \frac{1+\xi}{2} s_2$$

and define

$$u(\xi) \overset{\text{def}}{=} U(s(\xi)) - \frac{1-\xi}{2} U(s_1) - \frac{1+\xi}{2} U(s_2).$$

Using the ordinary least squares method, the function $u(\xi)$ is approximated by $u_{FE}(\xi)$ defined by

$$u_{FE}(\xi) \overset{\text{def}}{=} \sum_{i \in I_1(p)} a_i N_i(\xi, -1)$$

where $I_1(p)$ is the set of indices of the standard shape functions associated with side 1 that are zero in the nodal points. The coefficients of the nodal shape functions are $U(s_1)$ and $U(s_2)$.

3.7 Elements in three dimensions

Three-dimensional finite element meshes are comprised of hexahedral, tetrahedral, pentahedral elements, less frequently other types of elements, such as pyramid elements, are used. The standard hexahedral element, denoted by $\Omega_{st}^{(h)}$, is the set of points

$$\Omega_{st}^{(h)} \overset{\text{def}}{=} \{\xi, \eta, \zeta \mid -1 \le \xi, \eta, \zeta \le 1\}. \tag{3.51}$$

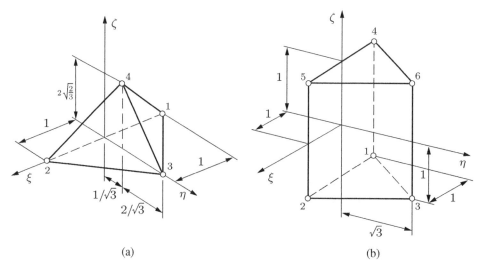

(a)   (b)

Figure 3.9 The standard tetrahedral and pentahedral elements $\Omega_{st}^{(th)}$ and $\Omega_{st}^{(p)}$.

The standard tetrahedral element, denoted by $\Omega_{st}^{(th)}$, and the standard pentahedral element, denoted by $\Omega_{st}^{(p)}$, are shown in Fig. 3.9. Note that the edges of the elements have the length 2.0, as in one and two dimensions.

The shape functions are analogous to those in one and two dimensions. For example, the eight-node hexahedral element has vertex shape functions such as:

$$N_1 = \frac{1}{8}(1 - \xi)(1 - \eta)(1 - \zeta).$$

The 20-node hexahedron is a generalization of the 8-node quadrilateral element to three dimensions. Similarly, the 4-node and 10-node tetrahedra are generalizations of the 3-node and 6-node triangles.

The hierarchic shape functions are also analogous to the shape functions defined in one and two dimensions. A detailed description of shape functions for hexahedral elements can be found in [35]. The shape functions associated with the edges and faces are the same along the edges and on the faces as in two dimensions. For example, the edge shape function of the pentahedral element, associated with the edge between nodes 1 and 2, corresponding to $p = 2$ is:

$$N_7 = L_1 L_2 \tilde{\phi}_2(L_2 - L_1)\frac{1 - \zeta}{2}$$

where $\tilde{\phi}_2$ is defined by eq. (3.31). The face shape function of the pentahedral element, associated with the face defined by nodes 1, 2 and 3, at $p = 3$ is:

$$N_{13} = L_1 L_2 L_3 \frac{1 - \zeta}{2}.$$

The internal shape functions of the pentahedral elements are the products of the internal shape functions defined for the triangular elements in eq. (3.36) and the function $\phi_k(\eta)$ defined by eq. (3.24).

Exercise 3.10 Write down the shape functions N_{12} for the standard tetrahedral, pentahedral and hexahedral elements.

3.7.1 Mapping functions in three dimensions

Mesh generators typically produce mapping by linear or quadratic interpolation. The coordinates of the interpolation points are determined from surface representations the generic form of which is $x_i = x_i(u, v)$ where u, v are the surface parameters.

A surface may be composed of one or more patches, each patch parameterized separately. In other words, the underlying geometry may be piecewise analytic. Various healing procedures are used for filling gaps that may exist between neighboring surfaces or patches. Element faces may intersect with more than one patch or surface.

The finite element representation approximates these surfaces by the element-level basis functions. Continuity of the basis functions on element boundaries is enforced. In the *h*-version of the finite element method the errors associated with the approximation of surfaces decreases as the number of elements is increased. This is not the case for the *p*-version where the number of elements is fixed. Therefore it is necessary to control the errors associated with mapping independently from the number of elements. Extension of the isoparametric mapping procedures to high order elements, combined with the blending function method, provides a satisfactory solution to this problem.

The mapping is based on Lagrange polynomials with the collocation points fixed on the standard element faces subject to the constraints that the mapping of the edges must be common to all element types and the interior points selected in such a way that the Lebesque constant is minimum [31]. The main points are summarized in Appendix F.

Example 3.1   A commercial automatic mesh generator[4] created 202 triangular elements on the spherical surface of unit radius shown in Fig. 3.10. The triangles were mapped by the optimal interpolation points for the standard triangle, $p = 5$. The error in maximum norm over all triangles is

$$\left|1 - \sqrt{x^2 + y^2 + z^2}\right|_{max} = 4.838 \times 10^{-7}.$$

If quadratic isoparametric mapping were used then this error would have been three orders of magnitude greater (2.875×10^{-4}).

Figure 3.10   Meshing of a spherical surface with 202 triangular elements.

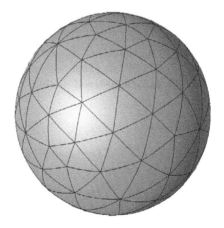

---

4 MeshSim®, a product of Simmetrix Inc.

Example 3.2 A spherical shell of radius r_s, wall thickness $t_s = r_s/100$, is subjected to an internal pressure q. By the membrane theory of shells the principal stress in the shell will be

$$\sigma_1 = \sigma_2 = \frac{1}{2}\frac{qr_s}{t_s}. \tag{3.52}$$

Letting $p = 2, r_s = 1, t_s = 0.01$, Poisson's ratio: $\nu = 0$ and using the same mesh as in Example 3.1, with (a) optimal interpolation points corresponding to $p = 5$ and (b) quadratic isoparametric mapping we get $\sigma_1 = 100.0$ (four significant digits accuracy) for (a) and $\sigma_1 = 102.4$ (two significant digits) for (b).

3.8 Integration and differentiation

In Section 1.3.3, the coefficients of the stiffness matrix, Gram matrix and the right hand side vector were computed on the standard element. In two and three dimensions the corresponding procedures are analogous; however, with the exception of some important special cases, the mappings are generally nonlinear.

The mapping functions

$$x = Q_x^{(k)}(\xi,\eta,\zeta), \quad y = Q_y^{(k)}(\xi,\eta,\zeta), \quad z = Q_z^{(k)}(\xi,\eta,\zeta) \tag{3.53}$$

map a standard element Ω_{st} onto the kth element Ω_k. In the following we will drop the superscript when it is clear that we refer to the mapping of the kth element. A mapping is said to be proper when the following three conditions are met: (a) The mapping functions Q_x, Q_y, Q_z are single valued functions of ξ, η, ζ and possess continuous first derivatives; and (b) the Jacobian determinant $|J|$ (defined in the next section) is positive in every point of Ω_{st}. The mapping functions used in FEA must meet these criteria.

3.8.1 Volume and area integrals

Volume integrals on the kth element are computed on the corresponding standard element. The volume integral of a scalar function $F(x,y,z)$ on Ω_k is:

$$\int_{\Omega_k} F(x,y,z)\,dxdydz = \int_{\Omega_{st}} \mathcal{F}(\xi,\eta,\zeta)|J|\,d\xi d\eta d\zeta \tag{3.54}$$

where $\mathcal{F}(\xi,\eta,\zeta) = F(Q_x(\xi,\eta,\zeta),\ Q_y(\xi,\eta,\zeta),\ Q_z(\xi,\eta,\zeta))$ and $|J|$ is the determinant of the Jacobian matrix[5], called the Jacobian determinant. The Jacobian determinant in eq. (3.54) arises from the definition of the differential volume: Let us denote the position vector of an arbitrary point P in the element $\overline{\Omega}_k$ by $\mathbf{r}$:

$$\mathbf{r} = x\mathbf{e}_x + y\mathbf{e}_y + z\mathbf{e}_z \tag{3.55}$$

where $\mathbf{e}_x, \mathbf{e}_y, \mathbf{e}_z$ are the orthogonal basis vectors of a (right-hand) Cartesian coordinate system. By definition, the differential volume is understood to be the scalar triple product:

$$dV = \left(\frac{\partial \mathbf{r}}{\partial x}dx \times \frac{\partial \mathbf{r}}{\partial y}dy\right)\cdot\frac{\partial \mathbf{r}}{\partial z}dz = (\mathbf{e}_x \times \mathbf{e}_y)\cdot\mathbf{e}_z\,dxdydz = dxdydz. \tag{3.56}$$

---

5 Carl Gustav Jacob Jacobi 1804–1851.

Given the change of variables of eq. (3.53), the analogous expression is:

$$dV = \left(\frac{\partial \mathbf{r}}{\partial \xi} d\xi \times \frac{\partial \mathbf{r}}{\partial \eta} d\eta \right) \cdot \frac{\partial \mathbf{r}}{\partial \zeta} d\zeta = \begin{vmatrix} \dfrac{\partial x}{\partial \xi} & \dfrac{\partial y}{\partial \xi} & \dfrac{\partial z}{\partial \xi} \\[2mm] \dfrac{\partial x}{\partial \eta} & \dfrac{\partial y}{\partial \eta} & \dfrac{\partial z}{\partial \eta} \\[2mm] \dfrac{\partial x}{\partial \zeta} & \dfrac{\partial y}{\partial \zeta} & \dfrac{\partial z}{\partial \zeta} \end{vmatrix} d\xi d\eta d\zeta \equiv |J| \, d\xi d\eta d\zeta. \tag{3.57}$$

From equations (3.56) and (3.57) we get: $dxdydz = |J| d\xi d\eta d\zeta$.

The vectors $\partial \mathbf{r}/\partial \xi$, $\partial \mathbf{r}/\partial \eta$, $\partial \mathbf{r}/\partial \zeta$ are a set of right-handed basis vectors, that is, their scalar triple product yields a positive number. If the Jacobian determinant is negative then the right-handed coordinate system is transformed into a left-handed one in which case the mapping is improper.

In two dimensions the mapping is

$$x = Q_x(\xi, \eta), \quad y = Q_y(\xi, \eta), \quad z = Q_z(\zeta) = \frac{t_z}{2}\zeta$$

where t_z is the thickness, see Fig. 2.4. When the thickness is constant then the integration in the transverse (ζ) direction can be performed explicitly and only an area integral needs to be evaluated.

$$dV = t_z \, dA = t_z \begin{vmatrix} \dfrac{\partial x}{\partial \xi} & \dfrac{\partial y}{\partial \xi} \\[2mm] \dfrac{\partial x}{\partial \eta} & \dfrac{\partial y}{\partial \eta} \end{vmatrix} d\xi d\eta \tag{3.58}$$

In the finite element method the integrations are performed by numerical quadrature, details of which are given in Appendix E. The minimum number of quadrature points depends on the polynomial degree of the shape functions: When the mapping is linear and the material properties are constant then the coefficients of the stiffness matrix should be exact (up to numerical round-off errors) otherwise the stiffness matrix may become singular. When the material properties vary, or the mapping is non-linear, then the number of integration points should be increased so that the errors in the stiffness coefficients are small. Experience has shown that increasing the number of integration points in each coordinate direction by two for curved elements is sufficient in most cases. Similar considerations apply to the load vector.

Exercise 3.11 Show that the Jacobian matrix of straight-side triangular elements, that is, triangular elements mapped by eq. (3.41) and eq. (3.42), is independent of L_1, L_2 and L_3 and show that the area of a triangle in terms of its vertex coordinates (X_i, Y_i), $i = 1, 2, 3$ is:

$$A = \frac{X_1(Y_2 - Y_3) + X_2(Y_3 - Y_1) + X_3(Y_1 - Y_2)}{2}.$$

Exercise 3.12 Show that for straight side quadrilaterals the Jacobian determinant is constant only if the quadrilateral element is a parallelogram.

3.8.2 Surface and contour integrals

Given the mapping functions, each face is parameterized by imposing the appropriate restriction on the mapping function. For example, if integration is to be performed on the face of a hexahedron that corresponds to $\zeta = 1$ then, referring to eq. (3.53), the parametric form of the surface becomes:

$$x = Q_x(\xi, \eta, 1), \quad y = Q_y(\xi, \eta, 1), \quad z = Q_z(\xi, \eta, 1) \tag{3.59}$$

and, using the definition of $\mathbf{r}$ given in (3.55), the surface integral of a scalar function $F(x, y, z)$ is:

$$\int\int_{(\partial\Omega_k)_{\zeta=1}} F(x, y, z) \, dS = \int_{-1}^{+1} \int_{-1}^{+1} \mathcal{F}(\xi, \eta, 1) \left| \frac{\partial \mathbf{r}}{\partial \xi} \times \frac{\partial \mathbf{r}}{\partial \eta} \right| d\xi d\eta$$

where $\mathcal{F}$ is obtained from F by replacing x, y, z with the mapping functions Q_x, Q_y, Q_z. The treatment of the other faces is analogous.

In two dimensions the contour integral of a scalar function $F(x, y)$ on the side of a quadrilateral element corresponding to $\eta = 1$ is:

$$\int_{(\partial\Omega_k)_{\eta=1}} F(x, y) \, ds = \int_{-1}^{+1} \mathcal{F}(\xi, 1) \left| \frac{d\mathbf{r}}{d\xi} \right| d\xi.$$

The other sides are treated analogously. The positive sense of the contour integral is counterclockwise.

3.8.3 Differentiation

The approximating functions, and hence the solution, are known in terms of the shape functions defined on standard elements. Therefore differentiation with respect to x, y and z has to be expressed in terms of differentiation with respect to ξ, η, ζ. Using the chain rule, we have:

$$\begin{Bmatrix} \dfrac{\partial}{\partial \xi} \\[2ex] \dfrac{\partial}{\partial \eta} \\[2ex] \dfrac{\partial}{\partial \zeta} \end{Bmatrix} = \begin{bmatrix} \dfrac{\partial x}{\partial \xi} & \dfrac{\partial y}{\partial \xi} & \dfrac{\partial z}{\partial \xi} \\[2ex] \dfrac{\partial x}{\partial \eta} & \dfrac{\partial y}{\partial \eta} & \dfrac{\partial z}{\partial \eta} \\[2ex] \dfrac{\partial x}{\partial \zeta} & \dfrac{\partial y}{\partial \zeta} & \dfrac{\partial z}{\partial \zeta} \end{bmatrix} \begin{Bmatrix} \dfrac{\partial}{\partial x} \\[2ex] \dfrac{\partial}{\partial y} \\[2ex] \dfrac{\partial}{\partial z} \end{Bmatrix}. \tag{3.60}$$

On multiplying by the inverse of the Jacobian matrix, we have the expression used for computing the derivatives of the shape functions defined on standard elements:

$$\begin{Bmatrix} \dfrac{\partial}{\partial x} \\[2ex] \dfrac{\partial}{\partial y} \\[2ex] \dfrac{\partial}{\partial z} \end{Bmatrix} = \begin{bmatrix} \dfrac{\partial x}{\partial \xi} & \dfrac{\partial y}{\partial \xi} & \dfrac{\partial z}{\partial \xi} \\[2ex] \dfrac{\partial x}{\partial \eta} & \dfrac{\partial y}{\partial \eta} & \dfrac{\partial z}{\partial \eta} \\[2ex] \dfrac{\partial x}{\partial \zeta} & \dfrac{\partial y}{\partial \zeta} & \dfrac{\partial z}{\partial \zeta} \end{bmatrix}^{-1} \begin{Bmatrix} \dfrac{\partial}{\partial \xi} \\[2ex] \dfrac{\partial}{\partial \eta} \\[2ex] \dfrac{\partial}{\partial \zeta} \end{Bmatrix}. \tag{3.61}$$

Computation of the first derivatives from a finite element solution in a given point is discussed in Section 3.11.

3.9 Stiffness matrices and load vectors

The algorithms for the computation of stiffness matrices and load vectors for three-dimensional elasticity are outlined in the following. Their counterparts for two-dimensional elasticity and heat conduction are analogous. The algorithms are based on eq. (2.107). However, the integrals are evaluated element by element:

$$\int_{\Omega_k} ([D]\{v\})^T [E][D]\{u\} \ dV = \int_{\Omega_k} \{v\}^T \{F\} \ dV + \int_{\partial\Omega_k \cap \partial\Omega_T} \{v\}^T \{T\} \ dS$$
$$+ \int_{\Omega_k} ([D]\{v\})^T [E]\{\alpha\} \mathcal{T}_\Delta \ dV \qquad (3.62)$$

where the differential operator matrix $[D]$ and the material stiffness matrix $[E]$ are as defined by eq. (2.105). The kth element is denoted by Ω_k. The second term on the right-hand side represents the virtual work of tractions acting on boundary segment $\partial\Omega_T$. This term is present only when one or more of the boundary surfaces of the element lie on $\partial\Omega_T$. For the sake of simplicity we assume that the number of degrees of freedom is the same for all three fields on Ω_k. We denote the number of degrees of freedom per field by n and define the $3 \times 3n$ matrix $[N]$ as follows:

$$[N] = \begin{bmatrix} N_1 & N_2 & \cdots & N_n & 0 & 0 & \cdots & 0 & 0 & 0 & \cdots & 0 \\ 0 & 0 & \cdots & 0 & N_1 & N_2 & \cdots & N_n & 0 & 0 & \cdots & 0 \\ 0 & 0 & \cdots & 0 & 0 & 0 & \cdots & 0 & N_1 & N_2 & \cdots & N_n \end{bmatrix}.$$

The jth column of $[N]$, denoted by $\{N_j\}$, is the jth shape function vector. We write the trial and test functions as linear combinations of the shape function vectors:

$$\{u\} = \sum_{j=1}^{3n} a_j \{N_j\} \quad \text{and} \quad \{v\} = \sum_{i=1}^{3n} b_i \{N_i\}. \qquad (3.63)$$

3.9.1 Stiffness matrices

The elements of the stiffness matrix k_{ij} can be written in the form:

$$k_{ij}^{(k)} = \int_{\Omega_k} ([D]\{N_i\})^T [E][D]\{N_j\} \ dV. \qquad (3.64)$$

We take advantage of the fact that two elements of $\{N_i\}$ are zero. The position of the zero elements depends on the value of the index i. For example, when $1 \le i \le n$ then:

$$[D]\{N_i\} = \begin{Bmatrix} \partial/\partial x \\ 0 \\ 0 \\ \partial/\partial y \\ 0 \\ \partial/\partial z \end{Bmatrix} N_i = \underbrace{\begin{bmatrix} 1 & 0 & 0 \\ 0 & 0 & 0 \\ 0 & 0 & 0 \\ 0 & 1 & 0 \\ 0 & 0 & 0 \\ 0 & 0 & 1 \end{bmatrix}}_{[M_1]} \begin{Bmatrix} \partial/\partial x \\ \partial/\partial y \\ \partial/\partial z \end{Bmatrix} N_i = [M_1][J_k]^{-1} \begin{Bmatrix} \partial/\partial \xi \\ \partial/\partial \eta \\ \partial/\partial \zeta \end{Bmatrix} N_i$$

where $[J_k]^{-1}$ is the inverse of the Jacobian matrix corresponding to element k, see eq. (3.61), $[M_1]$ is a logical matrix. When the index i changes, only $[M_1]$ has to be replaced. Specifically, when

$(n + 1) \leq i \leq 2n$ then $[M_1]$ is replaced by $[M_2]$; when $(2n + 1) \leq i \leq 3n$ then $[M_1]$ is replaced by $[M_3]$ which are defined as follows:

$$[M_2] = \begin{bmatrix} 0 & 0 & 0 \\ 0 & 1 & 0 \\ 0 & 0 & 0 \\ 1 & 0 & 0 \\ 0 & 0 & 1 \\ 0 & 0 & 0 \end{bmatrix} \qquad [M_3] = \begin{bmatrix} 0 & 0 & 0 \\ 0 & 0 & 0 \\ 0 & 0 & 1 \\ 0 & 0 & 0 \\ 0 & 1 & 0 \\ 1 & 0 & 0 \end{bmatrix}.$$

We define $\{D\} = \{\partial/\partial\xi\ \partial/\partial\eta\ \partial/\partial\zeta\}^T$ and write eq. (3.64) in a form suitable for evaluation by numerical integration which is described in Appendix E:

$$k_{ij}^{(k)} = \int_{\Omega_{st}} \left([M_\alpha][J_k]^{-1}\{D\}N_i\right)^T [E][M_\beta][J_k]^{-1}\{D\}N_j|J_k|\ d\xi d\eta d\zeta. \tag{3.65}$$

The domain of integration is a standard hexahedral, tetrahedral or pentahedral element. The indices α and β take on the values 1, 2, 3 depending on the range of the indices i and j. Therefore the element stiffness matrix $[K^{(k)}]$ consists of six blocks $[K_{\alpha\beta}^{(k)}]$:

$$[K^{(k)}] = \begin{bmatrix} [K_{11}^{(k)}] & [K_{12}^{(k)}] & [K_{13}^{(k)}] \\ & [K_{22}^{(k)}] & [K_{23}^{(k)}] \\ \text{sym.} & & [K_{33}^{(k)}] \end{bmatrix}. \tag{3.66}$$

Exercise 3.13 Assume that the mapping functions are given for the kth element. For example, in two dimensions:

$$x = \alpha Q_x^{(k)}(\xi, \eta), \quad y = \alpha Q_y^{(k)}(\xi, \eta)$$

where $\alpha > 0$ is some real number. Assume further that the element stiffness matrix has been computed for $\alpha = 1$. How will the elements of the stiffness matrix change as functions of α in one, two and three dimensions? Assume that in two dimensions the thickness is independent of α.

Exercise 3.14 Refer to equations (2.76) to (2.80). Develop an expression for the computation of the terms of the stiffness matrix, analogous to $k_{ij}^{(k)}$ given by eq. (3.65), for axisymmetric elastostatic models.

3.9.2 Load vectors

The computation of element level load vectors corresponding volume forces, surface tractions and thermal loading is based on the corresponding terms on the right-hand side of eq. (3.62).

Volume forces

Refer to the first term on the right-hand side of eq. (3.62). The computation of the load vector corresponding to volume force $\{F\}$ acting on element k is a straightforward application of eq. (3.54):

$$r_i^{(k)} = \int_{\Omega_{st}} \{N_i\}^T \{F\}\ |J_k|\ d\xi d\eta d\zeta \qquad i = 1, 2, \ldots, 3n. \tag{3.67}$$

Surface tractions

Refer to the second term on the right-hand side of eq. (3.62). Assume that traction vectors are acting on a hexahedral element on the face $\zeta = 1$. In this case the ith term of the load vector is:

$$r_i^{(k)} = \int_{-1}^{+1} \int_{-1}^{+1} \{N_i\}^T \{T\} \left| \frac{\partial \mathbf{r}}{\partial \xi} \times \frac{\partial \mathbf{r}}{\partial \eta} \right|_{\zeta=1} d\xi d\eta \qquad (3.68)$$

where the range of i is the set of indices of shape functions associated with the face $\zeta = 1$. The other faces are treated analogously.

Thermal loading

Refer to the third term on the right-hand side of eq. (3.62). The expression for $r_i^{(k)}$ corresponding to thermal loading is:

$$r_i^{(k)} = \int_{\Omega_{st}} \left([M_\beta][J_k]^{-1}[D]\{N_i\}\right)^T [E]\{\alpha\}\mathcal{T}_\Delta \, |J_k| \, d\xi d\eta d\zeta \qquad i = 1, 2, \ldots, 3n \qquad (3.69)$$

where $\beta = 1$ when $1 \leq i \leq n$, $\beta = 2$ when $(n+1) \leq i \leq 2n$ and $\beta = 3$ when $(2n+1) \leq i \leq 3n$. The matrices $[M_\beta]$ and the operator $\{D\}$ are defined in Section 3.9.1.

Summary of the main points

A finite element space is characterized by a finite element mesh and the polynomial degrees and mapping functions assigned to the elements of the mesh. The polynomial degrees identify a polynomial space defined on a standard element. The polynomial space is spanned by basis functions, called shape functions. Two kinds of shape functions, called Lagrange and hierarchic shape functions, were described for quadrilateral and triangular elements. The finite element space is spanned by the mapped shape functions subject to the continuity requirements described in Section 1.3.2.

Unless the mappings of all elements are polynomial functions of degree equal to or less than the polynomial degree of elements, rigid body rotation will not be represented exactly by the finite element solution. Nevertheless, rapid convergence to the correct solution will occur as the finite element space is progressively enlarged by h-, p- or hp-extension. The mapping functions used in FEA must be such that the Jacobian determinant is positive in every point within the element.

3.10 Post-solution operations

Following assembly (outlined in Sections 1.3.5) and solution operations, the finite element solution is stored in the form of data sets that contain the coefficients of the shape functions, the mapping functions and indices that identify the polynomial space associated with each element.

Some of the data of interest, such as temperature, displacement, flux, strain, stress, can be computed from the finite element solution either by direct or indirect methods while others, such as stress intensity factors, can be computed by indirect methods only. In this chapter the techniques used for the computation and verification of engineering data are described.

3.11 Computation of the solution and its first derivatives

If one is interested in the value of the solution in a point (x_0, y_0, z_0) then the domain has to be searched to identify the element in which that point lies. Suppose that the point lies in the *k*th element. The next step is to find the standard coordinates (ξ_0, η_0, ζ_0) from the mapping functions.

$$x_0 = Q_x^{(k)}(\xi_0, \eta_0, \zeta_0), \quad y_0 = Q_y^{(k)}(\xi_0, \eta_0, \zeta_0), \quad z_0 = Q_z^{(k)}(\xi_0, \eta_0, \zeta_0). \tag{3.70}$$

Unless the mapping of the *k*th element happens to be linear, the inverse of the mapping function is not known explicitly. Therefore this step involves a root finding procedure, such as the Newton-Raphson method.

The next step is to look up the parameters that identify the standard space associated with the element and the computed coefficients of the basis functions. With this information the solution and its derivatives can be computed. For example, let us assume that the solution is a scalar function and the standard space $S^{p,q}(\Omega_{st}^{(q)})$ is associated with the *k*th element, denoted by Ω_k. Then the finite element solution in the point $(x_0, \ y_0) \in \Omega_k$ is:

$$u_{FE}(x_0, y_0) = \sum_{i=1}^{n} a_i^{(k)} N_i(\xi_0, \eta_0) \tag{3.71}$$

where $n = (p+1)(q+1)$ is the number of shape functions that span $S^{p,q}(\Omega_{st}^{(q)})$, $N_i(\xi, \eta)$ are the shape functions and $a_i^{(k)}$ are the corresponding coefficients. When the solution is a vector function then each component of $\mathbf{u}_{FE}$ is in the form of eq. (3.71).

Computation of the first derivatives of u_{FE} in the point $(x_0, \ y_0)$ involves the computation of the inverse of the Jacobian matrix in the corresponding point (ξ_0, η_0) and multiplying the derivatives of the finite element solution with respect to the standard coordinates. Referring to eq. (3.61);

$$\left\{ \begin{array}{c} \dfrac{\partial u_{FE}}{\partial x} \\[3mm] \dfrac{\partial u_{FE}}{\partial y} \end{array} \right\}_{(x_0, y_0)} = \left[\begin{array}{cc} \dfrac{\partial x}{\partial \xi} & \dfrac{\partial y}{\partial \xi} \\[3mm] \dfrac{\partial x}{\partial \eta} & \dfrac{\partial y}{\partial \eta} \end{array} \right]_{(\xi_0, \eta_0)}^{-1} \sum_{i=1}^{n} a_i^{(k)} \left\{ \begin{array}{c} \dfrac{\partial N_i}{\partial \xi} \\[3mm] \dfrac{\partial N_i}{\partial \eta} \end{array} \right\}_{(\xi_0, \eta_0)} \tag{3.72}$$

where $x = Q_x^{(k)}(\xi, \eta), y = Q_y^{(k)}(\xi, \eta)$. The flux vector (resp. stress tensor) is computed by multiplying the temperature gradient (resp. the strain tensor) by the thermal conductivity matrix (resp. material stiffness matrix). The transformation of vectors and tensors is described in Appendix K.

The derivatives of the finite element solution are discontinuous along inter-element boundaries. Therefore if the point selected for the evaluation of fluxes, stresses, etc. is a node point, or a point on an inter-element boundary, then the computed value depends on the element selected for the computation. The degree of discontinuity in the normal and shearing stresses, or the normal flux component, at inter-element boundaries is an indicator of the quality of the approximation. In implementations of the *h*-version the derivatives are typically evaluated in the integration points and are interpolated over the elements. In graphical displays in the form of contour plots discontinuities of the derivatives at element boundaries are often masked by means of smoothing the contour lines through averaging.

In the *p*-version the standard element is subdivided so as to produce a uniform grid, called display grid, and the solution and its derivatives are evaluated in the grid points. Since the standard coordinates of the grid points are known, inverse mapping is not involved. The quality of contour plots depends on the quality of data being displayed and on the fineness of the display grid.

Search for a maximum or minimum value also involves search on a uniform grid defined on the standard elements. The fineness of the grid, and hence the number of points searched for the minimum or maximum, is controlled by a parameter. In conventional implementations of the *h*-version the search grid is typically defined by the integration points or the node points.

Exercise 3.15 Consider two plane elastic elements that have a common edge. Assume that different material properties were assigned to the elements. Show that the normal and shearing stresses corresponding to the exact solution have to be the same along the common edge and hence the normal and shearing strains will be discontinuous. Hint: Consider equilibrium in the coordinate system defined in the normal and tangential directions.

3.12 Nodal forces

Recall the definition of nodal forces $\{f^{(k)}\}$ in Section 1.4.1

$$\{f^{(k)}\} = [K^{(k)}]\{a^{(k)}\} - \{\overline{r}^{(k)}\} \qquad k = 1, 2, \ldots, M(\Delta) \tag{3.73}$$

where $[K^{(k)}]$ is the stiffness matrix, $\{a^{(k)}\}$ is the solution vector and $\{\overline{r}^{(k)}\}$ is the load vector corresponding to volume forces and thermal loads acting on element k.

3.12.1 Nodal forces in the *h*-version

When solving problems of elasticity using finite element analysis based on the *h*-version, nodal forces are treated in the same way as concentrated forces are treated in statics. Typical uses of nodal forces are: (a) isolating some region of interest from a larger structure and treating the isolated region as if it were a free body held in equilibrium by the nodal forces and (b) computing stress resultants. The underlying assumption is that nodal forces reliably represent the load path, that is, the distribution of internal forces in a statically indeterminate structure. This assumption is usually justified by the argument that nodal forces satisfy the equations of static equilibrium for any element or group of elements.

The following discussion will show that satisfaction of equilibrium is related to the rank deficiency of unconstrained stiffness matrices. Consequently equilibrium of nodal forces should not be interpreted as an indicator of the quality of the finite element solution and does not guarantee that the nodal forces are reliable approximations of the internal forces in a statically indeterminate structure. On the other hand, nodal forces are useful for the computation of stress resultants.

Let us assume, for example, that Ω_k is an 8-node quadrilateral element. The number of degrees of freedom per field, denoted by n, is 8. The notation is shown on Fig. 3.11.

In expanded notation the elements of the nodal force vector $\{f^{(k)}\}$ are:

$$f_x^{(k,i)} = f_i^{(k)}, \quad f_y^{(k,i)} = f_{n+i}^{(k)}, \quad i = 1, 2, \ldots, n. \tag{3.74}$$

Similarly, the elements of the vector $\{\overline{r}\}$ are written as follows:

$$\overline{r}_x^{(k,i)} = \overline{r}_i^{(k)}, \quad \overline{r}_y^{(k,i)} = \overline{r}_{n+i}^{(k)}, \quad i = 1, 2, \ldots, n. \tag{3.75}$$

On adapting the notation used in eq. (3.65) to problems of planar elasticity we write:

$$f_x^{(k,i)} = \int_{\Omega_{\mathrm{st}}^{(q)}} \left([M_1][J_k]^{-1}\{\mathcal{D}\}N_i\right)^T [E] \sum_{j=1}^{2n} \{\varepsilon_j^{(k)}\}a_j^{(k)} \, |J_k| \, d\xi d\eta - \overline{r}_x^{(k,i)} \tag{3.76}$$

and

$$f_y^{(k,i)} = \int_{\Omega_{\mathrm{st}}^{(q)}} \left([M_2][J_k]^{-1}\{\mathcal{D}\}N_i\right)^T [E] \sum_{j=1}^{2n} \{\varepsilon_j^{(k)}\}a_j^{(k)} \, |J_k| \, d\xi d\eta - \overline{r}_y^{(k,i)} \tag{3.77}$$

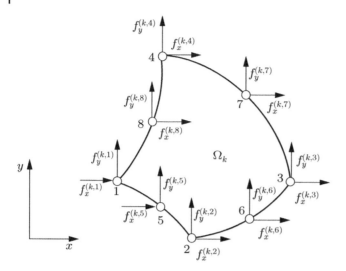

Figure 3.11 Nodal forces associated with the 8-node quadrilateral element. Notation.

where $[J_k]$ is the Jacobian matrix and

$$[M_1] = \begin{bmatrix} 1 & 0 \\ 0 & 0 \\ 0 & 1 \end{bmatrix} \quad [M_2] = \begin{bmatrix} 0 & 0 \\ 0 & 1 \\ 1 & 0 \end{bmatrix} \quad \{D\} = \left\{ \begin{array}{c} \dfrac{\partial}{\partial \xi} \\[2mm] \dfrac{\partial}{\partial \eta} \end{array} \right\}$$

and

$$\{\varepsilon_j^{(k)}\} = \begin{cases} [M_1][J_k]^{-1}\{D\}N_j & \text{for } j = 1, 2, \ldots, n \\ [M_2][J_k]^{-1}\{D\}N_{j-n} & \text{for } j = n+1, \ n+2, \ \ldots, \ 2n. \end{cases}$$

The elements of the load vector corresponding to body forces and thermal loads are:

$$\bar{r}_x^{(k,i)} = \int_{\Omega_{st}^{(q)}} N_i F_x |J_k| \ d\xi d\eta + \int_{\Omega_{st}^{(q)}} \left([M_1][J_k]^{-1}\{D\}N_i\right)^T [E]\{\alpha\} T_\Delta |J_k| \ d\xi d\eta \tag{3.78}$$

and

$$\bar{r}_y^{(k,i)} = \int_{\Omega_{st}^{(q)}} N_i F_y |J_k| \ d\xi d\eta + \int_{\Omega_{st}^{(q)}} \left([M_2][J_k]^{-1}\{D\}N_i\right)^T [E]\{\alpha\} T_\Delta |J_k| \ d\xi d\eta. \tag{3.79}$$

The equations of static equilibrium are:

$$\sum_{i=1}^{n} f_x^{(k,i)} + \int_{\Omega_k} F_x \ dxdy = 0 \tag{3.80}$$

$$\sum_{i=1}^{n} f_y^{(k,i)} + \int_{\Omega_k} F_y \ dxdy = 0 \tag{3.81}$$

and

$$\sum_{i=1}^{n} \left(X_i f_y^{(k,i)} - Y_i f_x^{(k,i)}\right) + \int_{\Omega_k} (xF_y - yF_x) \ dxdy = 0 \tag{3.82}$$

where X_i, Y_i are the coordinates of the ith node.

Satisfaction of eq. (3.80) and eq. (3.81) follows from the fact that

$$\sum_{i=1}^{n} N_i(\xi,\eta) = 1, \text{ therefore } \{D\}\sum_{i=1}^{n} N_i(\xi,\eta) = 0.$$

Satisfaction of eq. (3.82) follows from the mapping functions (3.39) and (3.40) and the fact that infinitesimal rigid body rotations do not cause strain:

$$\left([M_1][J_k]^{-1}\{D\}\sum_{i=1}^{n} Y_i N_i\right)^T \equiv \left\{\frac{\partial}{\partial x}\ 0\ \frac{\partial}{\partial y}\right\} y = \{0\ 0\ 1\} \tag{3.83}$$

$$\left([M_2][J_k]^{-1}\{D\}\sum_{i=1}^{n} X_i N_i\right)^T \equiv \left\{0\ \frac{\partial}{\partial y}\ \frac{\partial}{\partial x}\right\} x = \{0\ 0\ 1\}. \tag{3.84}$$

On substituting equations (3.83) and (3.84) into the expressions for $f_x^{(k,i)}, f_y^{(k,i)}, \bar{r}_x^{(k,i)}$ and $\bar{r}_y^{(k,i)}$ eq. (3.82) is obtained.

Note that the conditions for static equilibrium are satisfied independently of $\{a^{(k)}\}$. Therefore equilibrium of nodal forces is unrelated to the finite element solution.

Exercise 3.16 Consider the problem of heat conduction in two dimensions. Assume that the nodal fluxes were computed analogously to eq. (3.73). Define the term which is analogous to $\{\bar{r}\}$ and show that the sum of nodal fluxes plus the integral of the source term over an element is zero. Use an 8-node quadrilateral element and a 6-node triangular element to illustrate this point.

3.12.2 Nodal forces in the *p*-version

We have seen in the case of the 8-node quadrilateral element that equilibrium of nodal forces was related to the facts that the sum of the shape functions is unity and the functions x and y could be expressed as linear combinations of the shape functions. When the hierarchic shape functions based on the integrals of Legendre polynomials are used, such as those illustrated in Fig. 3.4, then the sum of the first four shape functions is unity independently of p. Therefore in the equilibrium equations (3.80) to (3.82) $n = 4$.

Example 3.3 A rectangular domain representing an elastic body of constant thickness is subjected to the boundary conditions $u_n = 0$, $u_t = \delta$ on boundary segments BC and DA shown in Fig. 3.12(a). The subscripts n and t refer to the normal and tangent directions respectively. Boundary segments AB and CD are traction-free.

We solve this as a plane stress problem using one element, represented by the shaded part of the domain, and product spaces. Antisymmetry condition is applied at $x = \ell/2$. The node numbering is shown in Fig. 3.12(b). Since only one element is used, the superscript that identifies the element number is dropped. Letting $\ell = 1000$ mm, $d = 50$ mm, $b = 20$ mm, $E = 200$ GPA, $v = 0.295$, $\delta = 5$ mm, the computed values of the nodal forces are shown in Table 3.1. In this example $\{\bar{r}\} = 0$. Therefore the nodal forces were computed from $\{f\} = [K]\{a\}$. The results shown in Table 3.1 indicate that equilibrium of nodal forces is satisfied at every p-level, independent of the accuracy of the finite element solution.

This problem could have been solved on the shaded domain shown in Fig. 3.13 in which case antisymmetry condition would be prescribed on the x axis. Using the notation shown in Fig. 3.13(b), the computed nodal forces are shown in Table 3.2 for $p = 8$. It is seen that the nodal forces once again satisfy the condition of static equilibrium.

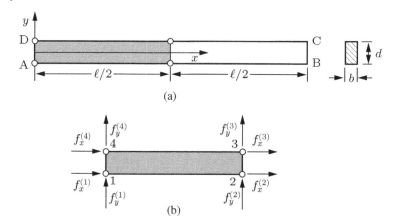

Figure 3.12 Example 3.3. Notation.

Table 3.1 Example 3.3. Nodal forces (kN). The notation is shown in Fig. 3.12(b). Product spaces.

p	$f_x^{(1)}$	$f_x^{(2)}$	$f_x^{(3)}$	$f_x^{(4)}$	$f_y^{(1)}$	$f_y^{(2)}$	$f_y^{(3)}$	$f_y^{(4)}$
1	−1971.3	0.000	0.000	1971.3	−98.564	98.564	98.564	−98.564
2	−64.678	0.000	0.000	64.678	−3.234	3.234	3.234	−3.234
3	−50.546	0.000	0.000	50.546	−2.527	2.527	2.527	−2.527
4	−50.190	0.000	0.000	50.190	−2.509	2.509	2.509	−2.509
5	−50.010	0.000	0.000	50.010	−2.500	2.500	2.500	−2.500
6	−49.907	0.000	0.000	49.907	−2.495	2.495	2.495	−2.495
7	−49.843	0.000	0.000	49.843	−2.492	2.492	2.492	−2.492
8	−49.802	0.000	0.000	49.802	−2.490	2.490	2.490	−2.490

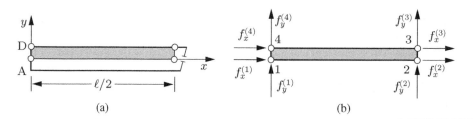

Figure 3.13 Example 3.3. The smallest solution domain.

Exercise 3.17 The nodal forces shown in Table 3.1 (resp. Table 3.2) were computed on the shaded domain shown in Fig. 3.12(a) (resp. Fig. 3.13(a)). Show that the stress resultants acting on boundary segments BC and DA are the same. Hint: Sketch and label the values of the nodal forces on the element shown in Fig. 3.13(b) and its antisymmetric pair.

Table 3.2 Example 3.3. Nodal forces (kN). Solution on the shaded domain shown in Fig. 3.13(a). The notation is shown in Fig. 3.13(b). Product space.

p	$f_x^{(1)}$	$f_x^{(2)}$	$f_x^{(3)}$	$f_x^{(4)}$	$f_y^{(1)}$	$f_y^{(2)}$	$f_y^{(3)}$	$f_y^{(4)}$
8	-12.454	-37.353	0.000	49.806	4.922	1.144	1.346	-7.412

3.12.3 Nodal forces and stress resultants

The nodal forces are related to the extraction functions for stress resultants. We illustrate this on the basis of Example 3.3. We have:

$$\int_\Omega ([D]\{v\})^T [E][D]\{u_{FE}\} b \ dxdy = \int_{\partial\Omega} (T_x^{(FE)} v_x + T_y^{(FE)} v_y) b \ ds \tag{3.85}$$

where Ω is the domain of the element shown in Fig. 3.12(b) and b is the thickness. If we are interested in the shear force acting on the side between nodes 2 and 3, denoted by $V_{2,3}$, then we select $v_x = 0$ on Ω and v_y a smooth function of Ω such that $v_y = 1$ on the side between nodes 2 and 3 and $v_y = 0$ on the side between nodes 4 and 1. Specifically if we select $v_y = N_2(\xi, \eta) + N_3(\xi, \eta)$ then we have:

$$V_{2,3} = \int_{\text{node }2}^{\text{node }3} T_y^{(FE)} b \ dy = \sum_{j=1}^{2n} (k_{n+2,j} + k_{n+3,j}) a_j = f_y^{(2)} + f_y^{(3)} \tag{3.86}$$

where n is the number of degrees of freedom per field. In Example 3.3 product spaces were used therefore $n = (p+1)^2$.

Exercise 3.18 Solve the problem of Example 3.3 and compute the stress resultants $V_{4,1}$ and $M_{4,1}$ on the side between nodes 4 and 1 by direct integration. Keep refining the mesh until satisfactory convergence is observed. Compare the results with those computed from the nodal forces in Table 3.1. By definition:

$$M_{4,1} = \int_{-d/2}^{+d/2} T_x y b \ dy.$$

This exercise shows that integration of stresses is much less efficient than extraction. The presence of singularities in points A and D influences the accuracy of the numerical integration.

3.13 Chapter summary

In order to meet the requirement of solution verification, it is necessary to show that the errors in the data of interest do not exceed stated tolerances. In practical problems the exact solution is typically unknown and it is not possible to determine the errors of approximation with certainty. It is possible however to show that necessary conditions are satisfied for the errors in the data of interest to be small.

Error estimation is based on the a priori knowledge that the data of interest corresponding to the exact solution are finite and independent of the discretization parameter. Therefore a necessary condition for the error to be small is that the computed data should exhibit convergence to a limit value as the number of degrees of freedom is increased. An efficient and robust way to achieve this is to use properly designed meshes and increase the polynomial degree. This will be discussed further in Chapter 4

4

Pre- and postprocessing procedures and verification

The questions of how to choose an effective finite element discretization scheme, given a set of input data, and how to extract the quantities of interest from the finite element solution and estimate their relative errors, are addressed in this chapter.

Preprocessing is concerned with the collection and verification of input data and the formulation of a discretization scheme with the goal to approximate the quantities of interest in an efficient manner. Based on the problem statement, which includes specification of the domain, the material properties, boundary conditions, the quantities of interest and the acceptable error tolerances, analysts are called upon to take into consideration the technical capabilities of the software tool(s) available to them and formulate a discretization scheme. This requires an understanding of the relationship between problem definition and the regularity of the underlying exact solution. For that reason, this chapter begins with a discussion on the regularity of functions. The main results of the relevant mathematical theorems are summarized.

Postprocessing is concerned with the extraction of the quantities of interest from the finite element solution and estimation of their relative errors. Should it be found that the error tolerances were exceeded, the discretization has to be revised and a new solution obtained. This chapter also covers postprocessing procedures for the extraction of flux and stress intensity factors.

4.1 Regularity in two and three dimensions

A rich and elaborate mathematical theory exists on the convergence properties of the finite element method. The details are beyond the scope of this book; however, an understanding of the meaning and implications of those theorems is essential in the practice of finite element analysis. The main point is that the solutions of elliptic boundary value problems on polygonal and polyhedral domains are typically piecewise analytic functions. Through a judicious combination of local mesh refinement and increase of polynomial degree, it is possible to achieve exponential rates of convergence and estimate the errors of approximation from the resulting sequence of finite element solutions.

Of particular interest are singular points and curves. Singularities are usually caused by simplifications introduced in the formulation of mathematical problems, which involves the definition of the solution domain, assignment of material properties and specification of the boundary conditions. In engineering applications it is generally useful to divide the solution domain into domains of primary and secondary interest. The QoIs are extracted from the domain(s) of primary interest.

Singularities in the domain of secondary interest are nuisances that tend to slow convergence and may perturb the computed quantities of interest. Singularities may also occur in the domain of primary interest. For example, in linear elastic fracture mechanics the quantity of interest is the

Finite Element Analysis: Method, Verification and Validation, Second Edition. Barna Szabó and Ivo Babuška.
© 2021 John Wiley & Sons, Inc. Published 2021 by John Wiley & Sons, Inc.
Companion Website: www.wiley.com/go/szabo/finite_element_analysis

stress intensity factor. An understanding of why singularities occur and how singularities affect the numerical solution is essential for the understanding and proper application of the finite element method.

The regularity of functions is measured by how many derivatives are square integrable. In one dimension the regularity of the exact solution depends on the smoothness of the coefficients κ and c and the forcing function f. In two and three dimensions regularity also depends on the vertex angles, material properties, boundary conditions and source functions. In three dimensions the regularity of the solution also depends on the angles at which boundary surfaces intersect.

4.2 The Laplace equation in two dimensions

We consider the solution of the Laplace equation in the neighborhood of a corner point of a two-dimensional domain, such as point B in Fig. 4.1(a). We assume that the boundary condition on Γ_{AB} and Γ_{BC} is either $u = 0$ or $\partial u / \partial n = 0$. The Laplace equation in polar coordinates $(r, \ \theta)$ is

$$\Delta u \equiv \frac{\partial^2 u}{\partial r^2} + \frac{1}{r} \frac{\partial u}{\partial r} + \frac{1}{r^2} \frac{\partial^2 u}{\partial \theta^2} = 0. \tag{4.1}$$

In particular, we are interested in solutions of the form: $u = r^\lambda F(\theta)$ with $\lambda \neq 0$. Such solutions are typically associated with geometric singularities, boundary conditions and intersections of material interfaces. Substituting into eq. (4.1), we have:

$$F'' + \lambda^2 F = 0$$

the general solution of which is

$$F = a \cos \lambda\theta + b \sin \lambda\theta \tag{4.2}$$

where a, b are arbitrary constants. Therefore the solution can be written as:

$$u = r^\lambda (a \cos \lambda\theta + b \sin \lambda\theta). \tag{4.3}$$

Consider, for example, the problem with the boundary condition $u = 0$ on boundary segments Γ_{AB} and Γ_{BC} (i.e. $\theta = \pm\alpha/2$):

$$a \cos \lambda\alpha/2 + b \sin \lambda\alpha/2 = 0$$
$$a \cos \lambda\alpha/2 - b \sin \lambda\alpha/2 = 0.$$

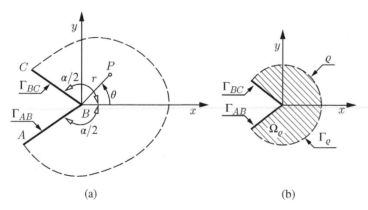

(a)   (b)

Figure 4.1 Reentrant corner. Notation

Adding and subtracting the two equations we find:

$$\cos \lambda\alpha/2 = 0 \quad \text{therefore} \quad \lambda\alpha/2 = \pm(2m-1)\pi/2, \quad m = 1, 2, \dots \tag{4.4}$$

$$\sin \lambda\alpha/2 = 0 \quad \text{therefore} \quad \lambda\alpha/2 = \pm n\pi, \quad n = 1, 2, \dots \tag{4.5}$$

Note that the values of λ that satisfy eq. (4.4) and eq. (4.5) can be either positive or negative. However, the function u, given by eq. (4.3), lies in the energy space only when $\lambda \geq 0$. Therefore, having excluded $\lambda = 0$ from consideration, we will be concerned with $\lambda > 0$ only. Denoting

$$\lambda_m^{(s)} \stackrel{\text{def}}{=} \frac{(2m-1)\pi}{\alpha}, \quad \lambda_n^{(a)} \stackrel{\text{def}}{=} \frac{2n\pi}{\alpha} \tag{4.6}$$

where $m, n = 1, 2, \dots$ the solution can be written in the following form:

$$u = \sum_{m=1}^{\infty} A_m r^{\lambda_m^{(s)}} \cos \lambda_m^{(s)}\theta + \sum_{n=1}^{\infty} B_n r^{\lambda_n^{(a)}} \sin \lambda_n^{(a)}\theta, \quad r \leq r_c \tag{4.7}$$

where r_c is the radius of convergence of the infinite series.

Observe that the first sum in eq. (4.7) is a symmetric function with respect to the x axis, the second is an antisymmetric function. If the solution is symmetric (with respect to the x axis) then $B_n = 0$, if it is antisymmetric then $A_m = 0$. Note that when $\alpha = \pi/k$ where $k = 1, 2, \dots$ then the powers of r are integers and u is an analytic function. For other values of α the solution u is not analytic in the corner point and when $\alpha > \pi$ then the first derivative of u with respect to r is infinity in the corner point, provided that $a_1 \neq 0$, and u lies in Sobolev space $H^{1+\lambda_1^{(s)}-\epsilon}(\Omega)$ where $\epsilon > 0$ is arbitrarily small.

Exercise 4.1 Consider the solution of $\Delta u = 0$ in the neighborhood of corner point B shown in Fig. 4.1(a), and let $u = 0$ on Γ_{BC} and the normal derivative $\partial u/\partial n = 0$ on Γ_{AB}. Show that u can be written as

$$u = \sum_{n=1}^{\infty} A_n r^{\lambda_n} (\cos \lambda_n\theta + (-1)^n \sin \lambda_n\theta) \quad \text{where} \quad \lambda_n = \frac{(2n-1)\pi}{2\alpha}.$$

Hint: The condition: $\partial u/\partial n = 0$ on Γ_{AB} is equivalent to $\partial u/\partial \theta = 0$.

Exercise 4.2 Construct the series expansion analogous to eq. (4.7) for the problem $\Delta u = 0$ in the neighborhood of the corner point B shown in Fig. 4.1(a), given that $\partial u/\partial n = 0$ on Γ_{AB} and Γ_{BC}.

Exercise 4.3 Show that $u(r, \theta)$, defined by eq. (4.3), is not in the energy space when $\lambda < 0$.

Exercise 4.4 Consider functions $u = r^\lambda F(\theta, \phi)$ where r, θ, ϕ are spherical coordinates centered on a corner point and F is an analytic function. Show that $\partial u/\partial r$ is square integrable in three dimensions when $\lambda > -1/2$.

4.2.1 2D model problem, $u_{EX} \in H^k(\Omega)$, $k - 1 > p$

We consider the L-shaped domain modified by removing a circular sector of radius r_0 centered on the origin. We denote this circular sector by Ω_0. The solution domain is defined by

$$\Omega \stackrel{\text{def}}{=} \{(X, Y) \mid (X, Y) \in (-1, \ 1)^2 \setminus [0, \ 1)^2 \setminus \Omega_0\}$$

where the backslash ($\setminus$) is the set operator for subtraction. The solution domain and notation are shown in Fig. 4.2(a).

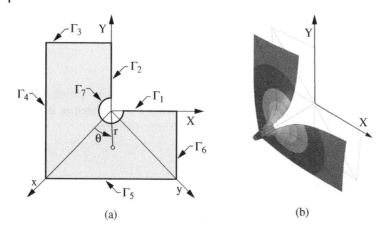

Figure 4.2 The L-shaped domain with a circular cut-out. (a) Notation. (b) The function $Q \overset{\text{def}}{=} (q_x + q_y)_{FE}$ corresponding to $r_0 = 0.05$ and $p = 8$.

The boundary conditions are defined so as to correspond to the exact solution:

$$u_{EX} = r^{2/3} \cos 2\theta/3. \tag{4.8}$$

Specifically, the boundary conditions are as follows: $u = 0$ on Γ_1 and Γ_2. On $\Gamma_q \overset{\text{def}}{=} \Gamma_3 \cup \Gamma_4 \cup \Gamma_5 \cup \Gamma_6 \cup \Gamma_7$ flux q_n is specified:

$$q_n = -\nabla u_{EX} \cdot \mathbf{n}_j \tag{4.9}$$

where $\mathbf{n}_j$ is the unit normal to the boundary segment Γ_j ($j = 3, 4, 5, 6$). On the circular boundary segment Γ_7 eq. (4.9) takes the following form:

$$q_n = \frac{\partial u_{EX}}{\partial r}.$$

In this problem the singular point is outside of the solution domain hence the exact solution is an analytic function on the entire domain. The smoothness of the solution increases with increasing r_0. The exact value of the potential energy is computed from

$$\pi_{\text{exact}} = -\frac{1}{2} \int_{\Gamma_q} q_n u_{EX} \, ds. \tag{4.10}$$

For $r_0 = 0.05$ we get $\pi_{\text{exact}} = -0.90364617$. We examine two discretization schemes in the following:

1. A sequence of uniform meshes characterized by $h = 1, \ 1/2, 1/3, \ \ldots$ with $p = 2$ (product space) assigned to all elements. The a priori estimate for h-convergence on a uniform mesh is given by eq. (1.91). In this case $k - 1 > p$ and hence

$$\| u_{EX} - u_{FE} \|_{E(\Omega)} \leq C(k, p) h^p \approx \frac{C(k, p)}{N^{p/2}} \tag{4.11}$$

   where $C(k, p)$ is a positive constant, independent of h, and we made use of the fact that in two dimensions $N \propto h^{-2}$. Therefore the asymptotic rate of convergence is $\beta = p/2 = 1$. As seen in Fig. 4.3, this value is approached from above.

2. Uniform mesh of 6 elements with p ranging from 2 to 8 (product space). Since u_{EX} is an analytic function, the rate of convergence will be exponential. Plotting the relative error in energy norm vs. N on log-log scale, the absolute value of the slope increases with N. This is visible in Fig. 4.3.

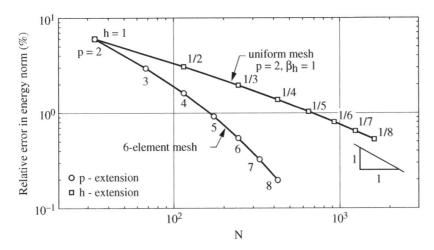

Figure 4.3 The L-shaped domain with a circular cut-out ($r_0 = 0.05$). Comparison of two discretization schemes.

The function

$$Q \overset{\text{def}}{=} (q_x + q_y)_{FE} = - \left(\frac{\partial u_{FE}}{\partial x} + \frac{\partial u_{FE}}{\partial y} \right)$$

corresponding to $r_0 = 0.05$, computed on a 6-element mesh using $p = 8$ (product space), is shown in Fig. 4.2(b). The exact solution is an analytic function, hence it can be expanded into a Taylor series everywhere within the domain and on the boundaries of the domain. The error term depends on the $(p + 1)$th derivative of u_{EX}.

4.2.2 2D model problem, $u_{EX} \in H^k(\Omega)$, $k - 1 \leq p$

We define our second model problem so that the first term of eq. (4.7), defined on the L-shaped domain shown in Fig. 4.4(a), is the exact solution u_{EX}. This is analogous to the one-dimensional model problem given by equation (1.5) with the solution $u_{EX} = x^\alpha(1 - x)$ on the interval $I = (0, 1)$.

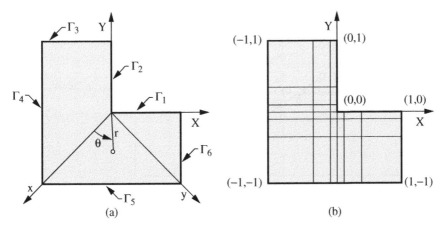

Figure 4.4 The L-shaped domain. (a) Notation, (b) radically graded 27-element mesh corresponding to the parameters $M = 3$ and $\gamma = 2$.

We assign the following boundary conditions: $u = 0$ on Γ_1 and Γ_2. On $\Gamma_q \stackrel{\text{def}}{=} \Gamma_3 \cup \Gamma_4 \cup \Gamma_5 \cup \Gamma_6$ flux q_n is specified consistent with the solution given by eq. (4.8), see eq. (4.9). The exact value of the potential energy is computed from

$$\pi_{\text{exact}} = -\frac{1}{2} \int_{\Gamma_q} q_n u_{EX} \, ds = -0.9181133309. \tag{4.12}$$

Except for the vertex located in the origin of the coordinate system, all derivatives of the solution exist. The solution lies in Sobolev space $H^{5/3-\epsilon}(\Omega)$, see Section A.2.4. Since the exact value of the potential energy is known, the relative error in energy norm can be computed for each discretization from eq. (1.100) and the realized rates of convergence can be computed from eq. (1.102). We examine three discretization schemes in the following:

1. A sequence of uniform meshes with $p = 2$ (product space). The size of the elements is given by

$$h_i = \frac{1}{3i}, \quad i = 1, 2, \ldots \tag{4.13}$$

The a priori estimate for h-convergence on a uniform mesh is given by eq. (1.91). In this case $k - 1 < p$ and hence

$$\| u_{EX} - u_{FE} \|_{E(\Omega)} \le C(k, p) h^{k-1} \approx \frac{C(k, p)}{N^{1/3}} \tag{4.14}$$

where $C(k, p)$ is a positive constant, independent of h, and we made use of the fact that in two dimensions $N \propto h^{-2}$. Therefore the asymptotic rate of convergence is $\beta = 1/3$.

2. Uniform mesh, $h = 1/3$ (27 elements), with p ranging from 2 to 8 (product space). Since the singular point is a vertex point, the asymptotic rate of convergence of the p-version in energy norm is twice that of the h-version, therefore $\beta = 2/3$. As seen in Fig. 4.5, this is indeed realized.

3. A fixed mesh of 27 elements with radical grading and p ranging from 1 to 8 (product space). The grading of the radical mesh is controlled by a parameter $\gamma > 0$. The node points on the positive

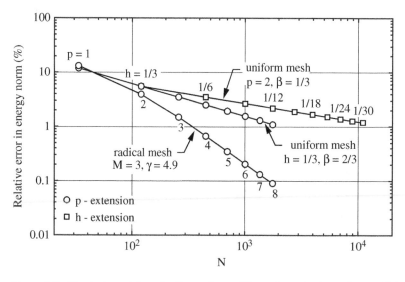

Figure 4.5 The L-shaped domain. Comparison of three discretization schemes. Relative error in energy norm (%).

x axis are

$$x_i = \left(\frac{i-1}{M}\right)^{\gamma}, \quad i = 1, 2, \ldots, M+1 \tag{4.15}$$

where M is the number of elements that have boundaries on the positive x axis. This partitioning is extended over the whole domain, as shown in Fig. 4.4(b), where the mesh corresponding to the parameters $M = 3$, $\gamma = 2$ is shown. We selected $\gamma = 4.9$ because this value minimizes the potential energy for the highest polynomial degree allowed by the software, in this case $p = 8$. The results for the radical mesh are shown in Table 4.1.

The entry in the last column of Table 4.1 is the effectivity index of the error estimator. The effectivity index, denoted by θ, is defined as the ratio of the estimated relative error to the exact relative error. The effectivity index can be computed only for those problems for which the exact solution is known. Observe that the relative errors are slightly overestimated. This occurs because the estimates are based on the assumption that the rate of convergence is algebraic, see eq. (1.92); however, in the pre-asymptotic range it is stronger than algebraic.

The results for the three discretization schemes considered here are shown in Fig. 4.5. It is seen that p-extension on a radically graded mesh is much more efficient than h-extension using a sequence of uniform meshes or p-extension using a fixed uniform mesh. The reason for this is that for the range of polynomial degrees $1 \leq p \leq 8$ (supported by the software) radical meshing overrefines the neighborhood of the singular corner and therefore the error is essentially coming from the smooth part of the solution where the rate of convergence of p-extensions is exponential. This is representative of the pre-asymptotic behavior of the p-version when strongly graded meshes are used. As the p-level is increased, the convergence curve will slow to the asymptotic rate, in this case $\beta = 2/3$. However, at that point the relative error in energy norm will be small.

This example illustrates an important point: in the finite element method the objective in selecting a discretization scheme is to achieve the desired accuracy in an efficient fashion. This is best accomplished when the mesh is laid out in such a way that, for the range of p-values supported by the software, p-extension is in the pre-asymptotic range.

Table 4.1 L-shaped domain, radical mesh ($\gamma = 4.9$), 27 elements, product space. Estimated and exact relative errors in energy norm.

p	N	π_p	β		$(e_r)_E$ (%)		θ
			Est.'d	Exact	Est.'d	Exact	
1	33	−0.90505595	−	−	11.93	11.93	1.00
2	120	−0.91529501	0.593	0.594	5.54	5.54	1.00
3	261	−0.91698177	0.586	0.587	3.52	3.51	1.00
4	456	−0.91753087	0.593	0.595	2.53	2.52	1.00
5	705	−0.91776880	0.598	0.603	1.95	1.94	1.01
6	1,008	−0.91789036	0.602	0.609	1.57	1.56	1.01
7	1,365	−0.91795963	0.603	0.614	1.31	1.29	1.01
8	1,776	−0.91800230	0.603	0.618	1.11	1.10	1.01
extrapolated:		−0.91811642	−	0.667	−	0	−

Dirichlet boundary condition

Let us impose Dirichlet boundary conditions on all boundary segments. Since u_{EX} is a sinusoidal function and our basis functions are the mapped shape functions, it is necessary to first approximate u_{EX} by the basis functions on the element boundaries and then enforce the Dirichlet condition on the coefficients of the basis functions. This procedure was outlined in Section 3.6. We are interested in the size of the error coming from approximation of the essential boundary conditions.

Remark 4.1 The a posteriori estimator of error described in Section 1.5.3 is based on the assumption that the h- or p-extension is in the asymptotic range, that is, h is sufficiently small or p is sufficiently large to justify replacement of the $\leq$ sign with the $\approx$ sign in eq. (1.91). As we have seen in this section, the pre-asymptotic rate of convergence is stronger than algebraic.

Exercise 4.5 Explain why the potential energy π_p is negative when Neumann conditions are specified on Γ_q but positive when Dirichlet conditions are specified on Γ.

4.2.3 Computation of the flux vector in a given point

The computation flux vectors from the finite element solution in a given point is straightforward. First, the standard coordinates (ξ_0, η_0) of the given point are determined, then the flux vector is evaluated in point (ξ_0, η_0) from

$$\left\{ \begin{matrix} q_x^{(FE)} \\ q_y^{(FE)} \end{matrix} \right\} = - \begin{bmatrix} k_x & k_{xy} \\ k_{xy} & k_y \end{bmatrix} [J]^{-1} \left\{ \begin{matrix} \partial u_{FE}/\partial \xi \\ \partial u_{FE}/\partial \eta \end{matrix} \right\} \tag{4.16}$$

where $[J]$ is the Jacobian matrix defined in Section 3.8.1.

Let us consider, for example the L-shaped domain with a circular cut-out shown in Fig. 4.6(a) with Dirichlet boundary conditions applied on each boundary segment. The boundary conditions correspond to the exact solution $u = r^{2/3} \cos(2\theta/3)$ (in the $x\,y$ coordinate system).

Table 4.2 L-shaped domain, radical mesh ($\gamma = 4.9$), 27 elements, product space. Estimated and exact relative errors in energy norm.

p	N	π_p	β		$(e_r)_E$ (%)		θ
			Est.'d	Exact	Est.'d	Exact	
1	16	0.95282287	0.000		19.444	19.444	1.00
2	85	0.91965811	0.932	0.932	4.102	4.102	1.00
3	208	0.91831905	1.127	1.126	1.497	1.497	1.00
4	385	0.91815493	1.300	1.298	0.672	0.673	1.00
5	616	0.91812462	1.394	1.388	0.349	0.351	1.00
6	901	0.91811717	1.442	1.419	0.202	0.204	0.99
7	1240	0.91811490	1.464	1.403	0.126	0.131	0.97
8	1633	0.91811408	1.467	1.330	0.084	0.091	0.93
extrapolated:		0.91811343				0.033	–

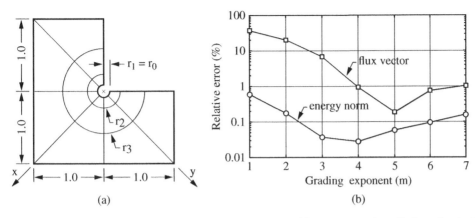

Figure 4.6 The L-shaped domain with a circular cut-out. (a) 18-element mesh, radical grading in the radial direction. (b) Relative errors.

The components of exact flux vector in the x, y coordinate system are

$$q_x = -\frac{2}{3}r^{-1/3}\cos(\theta/3), \quad q_y = -\frac{2}{3}r^{-1/3}\sin(\theta/3).$$

We define the relative error in the flux vector as

$$e_{\mathrm{fv}} \overset{\text{def}}{=} 100 \,\frac{\sqrt{\left(q_x - q_x^{(FE)}\right)^2 + \left(q_y - q_y^{(FE)}\right)^2}}{2r_1^{-1/3}/3} \tag{4.17}$$

where the denominator is the maximum of the absolute value of the flux vector over the domain. Specifically, we will be interested in the relative error of the flux vector in the point $x = r_0, y = 0$. Since this point lies on a line of symmetry, the flux vector is in the direction of the line of symmetry which is normal to the circular boundary.

We will use the 18-element mesh shown in Fig. 4.6(a) with the radii of the circles defined as follows:

$$r_i = r_1 + (1 - r_1)\left(\frac{i-1}{3}\right)^m, \quad i = 1, 2, 3$$

i.e. radical grading, controlled by the grading exponent m, is used in the radial direction. Letting $r_1 = r_0 = 0.001$ and $m = 1, 2, \ldots, p = 8$ (product space) we compute the relative errors in the flux vector in the point $(r_0, 0)$ and the energy norm[1].

The results of computation are shown in Fig. 4.6(b). It is seen that the smallest error in energy norm occurs at approximately $m = 4$ and the smallest error in the flux vector (e_{fv}) at approximately $m = 5$. At $m < 4$ the error in energy norm is small, whereas the error e_{fv} is large. The reason for this is that the error in energy norm depends on the square integral of the error in the first derivatives evaluated over the entire domain, whereas e_{fv} is evaluated in a particular point. In this example the error in the derivatives is large in a small neighborhood of the circular cut-out when $m < 4$; however, this error influences the energy norm of the error by a small amount only.

For $m < 4$ the error in the flux vector is coming mainly from under-refinement of the neighborhood of the small cut-out, the region of primary interest, whereas for $m > 5$ the error is mainly from the region of secondary interest.

---

1 The exact value of the potential energy is known: $\pi_{EX} = 0.91803479$ for $r_0 = 0.001$.

Remark 4.2 It is not difficult to find the value of the grading exponent that yields the minimum error in energy norm. In general, this is not the grading exponent that is optimal for the quantity of interest. It is reasonable to expect, however, that the error of approximation in the QoI will not be far from that minimum. In this example this is the case indeed, as seen in Fig. 4.6(b).

4.2.4 Computation of the flux intensity factors

An algorithm for the computation of the coefficients of the asymptotic expansions from finite element solutions is described in the following. The algorithm is based on (a) the existence of a path-independent integral and (b) the orthogonality of the eigenfunctions. The coefficients are called flux intensity factors.

Consider a two-dimensional domain Ω with boundary Γ. For any two functions in $E(\Omega)$ we have:

$$\int_{\Omega} \Delta u \, v \, dxdy = \oint_{\Gamma} (\nabla u \cdot \mathbf{n})v \, ds - \oint_{\Gamma} (\nabla v \cdot \mathbf{n})u \, ds + \int_{\Omega} \Delta v \, u \, dxdy$$

where we have applied the divergence theorem twice. When both u and v satisfy the Laplace equation, that is $\Delta u = 0$ and $\Delta v = 0$, then this equation becomes:

$$\oint_{\Gamma} (\nabla u \cdot \mathbf{n})v \, ds = \oint_{\Gamma} (\nabla v \cdot \mathbf{n})u \, ds. \tag{4.18}$$

This equation is applicable to Ω and any subdomain of Ω.

Path-independent integral

Now consider a subdomain $\Omega^{\star}$ in the neighborhood of a corner point, represented by the shaded region in Fig. 4.7. Assume that either $u = 0$ or $\nabla u \cdot \mathbf{n} = 0$ and either $v = 0$ or $\nabla v \cdot \mathbf{n} = 0$ on $\Gamma_2^{\star}$ and $\Gamma_4^{\star}$. Then eq. (4.18) becomes:

$$\int_{\Gamma_1^{\star}} (\nabla u \cdot \mathbf{n})v \, ds + \int_{\Gamma_3^{\star}} (\nabla u \cdot \mathbf{n})v \, ds = \int_{\Gamma_1^{\star}} (\nabla v \cdot \mathbf{n})u \, ds + \int_{\Gamma_3^{\star}} (\nabla v \cdot \mathbf{n})u \, ds$$

which is equivalent to

$$\int_{\Gamma_1^{\star}} (\nabla u \cdot \mathbf{n})v \, ds - \int_{\Gamma_1^{\star}} (\nabla v \cdot \mathbf{n})u \, ds = -\int_{\Gamma_3^{\star}} (\nabla u \cdot \mathbf{n})v \, ds + \int_{\Gamma_3^{\star}} (\nabla v \cdot \mathbf{n})u \, ds.$$

Observe that integration along $\Gamma_1^{\star}$ is clockwise about the corner point whereas integration along $\Gamma_3^{\star}$ is counterclockwise. Reversing the direction of integration along $\Gamma_1^{\star}$ so that both integrals are counterclockwise about the corner point, we find that the two integrals are equal, and

Figure 4.7 Definition of $\Omega^{\star}$.

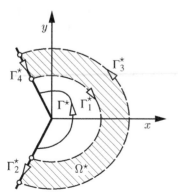

since $\Omega^\star$ is arbitrary, we may select an arbitrary counterclockwise path $\Gamma^\star$ and the integral expression

$$I_{\Gamma^\star} \overset{\text{def}}{=} -\int_{\Gamma^\star} (\nabla u \cdot \mathbf{n})v \, ds + \int_{\Gamma^\star} (\nabla v \cdot \mathbf{n})u \, ds \tag{4.19}$$

will be path-independent.

Orthogonality

Let $u = r^{\lambda_i}\phi_i(\theta)$; $v = r^{\lambda_j}\phi_j(\theta)$, and $\Gamma^\star = \Gamma_\varrho$ where Γ_ϱ is a circular path of radius ϱ, centered on the corner point. Then on Γ_ϱ we have:

$$\nabla u \cdot \mathbf{n} = \left(\frac{\partial u}{\partial r}\right)_{r=\varrho} = \lambda_i \varrho^{\lambda_i - 1}\phi_i(\theta)$$

$$\nabla v \cdot \mathbf{n} = \left(\frac{\partial v}{\partial r}\right)_{r=\varrho} = \lambda_j \varrho^{\lambda_j - 1}\phi_j(\theta).$$

Using $ds = \varrho \, d\theta$, eq. (4.19) can be written as

$$I_{\Gamma_\varrho} = (\lambda_j - \lambda_i)\varrho^{\lambda_i + \lambda_j} \int_{-\alpha/2}^{\alpha/2} \phi_i(\theta)\phi_j(\theta) \, d\theta.$$

Since I_{Γ_ϱ} is path-independent, the integral expression must be zero when $\lambda_j \neq \pm\lambda_i$. Note that since $\Omega^\star$ does not include the corner point, solutions corresponding to the negative eigenvalues are in the energy space. In the following we will denote

$$C_{ij} \overset{\text{def}}{=} \int_{-\alpha/2}^{\alpha/2} \phi_i(\theta)\phi_j(\theta) \, d\theta \quad \text{and} \quad C_{ij}^- \overset{\text{def}}{=} \int_{-\alpha/2}^{\alpha/2} \phi_i(\theta)\phi_j^-(\theta) \, d\theta \tag{4.20}$$

where $\phi_j^-(\theta)$ is the eigenfunction corresponding to $-\lambda_j$. The eigenfunctions are orthogonal in the sense that $C_{ij} = 0$ when $\phi_j \neq \phi_i$ and $\phi_j \neq \phi_i^-$.

Remark 4.3   In the case of the Laplace operator all eigenvalues are real and simple.

Exercise 4.6   Show that for the asymptotic expression given by eq. (4.7)

$$C_{ij} = \begin{cases} \alpha/2 & \text{if } i = j \\ 0 & \text{if } i \neq j. \end{cases}$$

Extraction of A_k

Using the orthogonality property of the eigenfunctions, it is possible to extract approximate values of the coefficients A_k, $k = 1, 2, \ldots$ from a finite element solution. Let us consider the asymptotic expansion

$$u_{EX} = \sum_{i=1}^{\infty} A_i r^{\lambda_i}\phi_i(\theta). \tag{4.21}$$

Suppose that we are interested in computing A_k. We then define the extraction function w_k as follows:

$$w_k \overset{\text{def}}{=} r^{-\lambda_k}\phi_k^-(\theta)$$

and evaluate the path-independent integral on a circular path Γ_ϱ centered on the corner point:

$$I_{\Gamma_\varrho}(u_{EX}, w_k) = -\int_{\Gamma_\varrho} (\nabla u_{EX} \cdot \mathbf{n})w_k \, ds + \int_{\Gamma_\varrho} (\nabla w_k \cdot \mathbf{n})u_{EX} \, ds. \tag{4.22}$$

It is now left to the reader to show that, utilizing the orthogonality property of the eigenfunctions, we get:

$$A_k = -\frac{1}{2C_{kk}^- \lambda_k} I_{\Gamma_\varrho}(u_{EX}, w_k). \tag{4.23}$$

In the finite element method u_{EX} is replaced by u_{FE} to obtain an approximate value for A_k. This method, called the contour integral method, is very efficient, as illustrated by the following example.

Example 4.1 Let us consider the L-shaped domain problem shown in Fig. 4.4 with the boundary conditions $u = 0$ on Γ_1 and Γ_2. On all other boundary segments q_n corresponding to the exact solution

$$u_{EX} = r^{2/3} \cos(2\theta/3) \tag{4.24}$$

is prescribed. Note that this is the leading term of the symmetric part of the expansion given by eq. (4.7). This example illustrates that if we substitute u_{FE} for u_{EX} in eq. (4.22) then we will get a good approximation to the exact value of the coefficient A_1 which, in this example, was chosen to be unity.

The extraction function is

$$w_1(\varrho, \theta) = \varrho^{-2/3} \cos(2\theta/3) \tag{4.25}$$

and

$$(\nabla w_1 \cdot \mathbf{n})_{r=\varrho} = -\frac{2}{3}\varrho^{-5/3} \cos(2\theta/3) \tag{4.26}$$

where ϱ is the radius of the circle used for extraction.

$$C_{11}^- = \int_{-3\pi/4}^{3\pi/4} \cos^2(2\theta/3) \, d\theta = 3\pi/4.$$

The approximate value of A_1 is computed from eq. (4.22) which in this case takes the form

$$A_1^{(FE)} = -\frac{1}{2C_{11}^- \lambda_1} \left(\varrho^{1/3} \int_{-3\pi/4}^{3\pi/4} (\nabla u_{FE} \cdot \mathbf{n}) \cos(2\theta/3) \, d\theta + \right.$$
$$\left. \frac{2}{3} \varrho^{-2/3} \int_{-3\pi/4}^{3\pi/4} \cos(2\theta/3) u_{FE} \, d\theta \right). \tag{4.27}$$

We compute u_{FE} and $\nabla u_{FE} \cdot \mathbf{n}$ along an arbitrary circular path of radius ϱ in integration points and compute $A_1^{(FE)}$ numerically. The integrands are smooth functions therefore the error associated with numerical integration can be controlled very efficiently. For this example we used a uniform mesh of 27 elements, $h = 1/3$ and an extraction circle of radius $\varrho = 0.2$. The results of computation for $p = 1$ to $p = 5$ (product space) are shown in Table 4.3

It is seen that the relative error in $A_1^{(FE)}$ decreases much faster than the relative error in energy norm. For that reason, the contour integral method is said to be superconvergent.

Table 4.3 Example 4.1: The results of extraction by the contour integral method using $\varrho = 0.2$.

polynomial degree (p)	1	2	3	4	5
$A_1^{(FE)}$	1.0242	1.0075	0.9920	1.0027	0.9987
Rel. error of $A_1^{(FE)}$ (%)	2.42	0.75	0.80	0.27	0.13
Rel. error in energy norm (%)	11.93	5.54	3.52	2.53	1.95

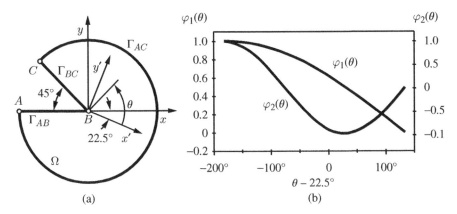

Figure 4.8 Example 4.2: The first two normalized eigenfunctions.

Example 4.2 We refer to the results of Exercise 4.1 and construct a model problem on the domain shown in Fig. 4.8(a) so that the exact solution is a linear combination of the first two terms of the asymptotic expansion:

$$u_{EX} = a_1 r^{\lambda_1} \underbrace{(\cos \lambda_1 \theta - \sin \lambda_1 \theta)}_{\phi_1(\theta)} + a_2 r^{\lambda_2} \underbrace{(\cos \lambda_2 \theta + \sin \lambda_2 \theta)}_{\phi_2(\theta)}$$

where $\lambda_1 = \pi/(2\alpha)$, $\lambda_2 = 3\pi/(2\alpha)$. Let $\alpha = 7\pi/4 = 315°$. The boundary conditions are: $\partial u/\partial n = 0$ on Γ_{AB}; $u = 0$ on Γ_{BC}, and on Γ_{AC}, i.e., the flux corresponding to u_{EX} is specified:

$$q_n = -a_1 \lambda_1 r^{\lambda_1 - 1}(\cos \lambda_1 \theta - \sin \lambda_1 \theta) - a_2 \lambda_2 r^{\lambda_2 - 1}(\cos \lambda_2 \theta + \sin \lambda_2 \theta).$$

We normalize the eigenfunctions as follows. Let θ_i be the angle where the absolute value of $\phi_i(\theta)$ is maximum:

$$\theta_i = \arg \max_{\theta \in I_\alpha} |\phi_i(\theta)| \quad \text{where } I_\alpha \overset{\text{def}}{=} \{\theta \mid -\alpha/2 \le \theta \le +\alpha/2\}.$$

The normalized eigenfunctions are defined such that the maximum value of $\varphi_i(\theta)$ on I_α is unity:

$$\varphi_i(\theta) \overset{\text{def}}{=} \phi_i(\theta)/\phi_i(\theta_i). \tag{4.28}$$

The functions $\varphi_1(\theta)$ and $\varphi_2(\theta)$ are shown in Fig. 4.8(b). If we let $a_1 = 1.0$, $a_2 = 0$ then, using a 16-element mesh with one layer of geometrically graded elements around the corner point, trunk space, the estimated relative error in energy norm is 17.21% at $p = 8$. Using the method of extraction described in this section, the computed value of the coefficient of the first normalized eigenfunction is $A_1 = 1.348$. Its exact value is 1.414, therefore the relative error is 4.66%.

Exercise 4.7 For the problem described in Example 4.2 let $a_1 = 0$ and $a_2 = 1.0$. Determine the approximate value of the coefficient of the second normalized eigenfunction A_2 and estimate the relative error. The exact value is -1.414.

4.2.5 Material interfaces

A schematic view of the material interface problem in two dimensions is shown in Fig. 4.9. The shadings represent different materials of constant thermal conductivity.

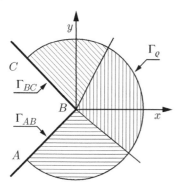

Figure 4.9 Multi-material interface, notation.

The intersection of material interfaces with boundaries are singular points. The solution in the neighborhood of these singular points is characterized by functions that are of the form $u = r^\lambda F(\theta)$, as before. However, here $F(\theta)$ is a piecewise analytic function.

Since the solution has to satisfy the Laplace equation on each sector, the solution on the kth sector can be written as

$$u^{(k)} = r^\lambda (a_k \cos(\lambda\theta) + b_k \sin(\lambda\theta)), \quad k = 1, 2, \dots, n \tag{4.29}$$

where n is the number of sectors. We assume that homogeneous Dirichlet or Neumann boundary conditions are prescribed on Γ_{AB} and Γ_{BC}. Imposing continuity on $u^{(k)}$, and on the flux normal to the material interface i.e. $u^{(k-1)}(r, \theta_k) = u^{(k)}(r, \theta_k)$ and $q_n^{k-1}(r, \theta_k) = -q_n^k(r, \theta_k)$ and enforcing the homogeneous boundary conditions, results in a system of $2n$ homogeneous equations in a_k and b_k. The procedure is illustrated by the following example.

Example 4.3 An aluminum plate with coefficient of thermal conductivity $k_{al} = 202$ W/(mK) is bonded to a chrome-nickel steel plate with $k_{cn} = 16.3$ W/(mK). The edges are offset, forming the corner shown in Fig. 4.10(a). We assume that the boundaries that lie on the x and y axes are perfectly insulated.

Four equations are necessary for the enforcement of the homogeneous boundary conditions and the continuity of the temperature and flux at the material interface:

$$\begin{bmatrix} -\sin(\lambda\pi/2) & 0 & \cos(\lambda\pi/2) & 0 \\ 0 & -\sin(2\lambda\pi) & 0 & \cos(2\lambda\pi) \\ \cos(\lambda\pi) & \cos(\lambda\pi) & \sin(\lambda\pi) & \sin(\lambda\pi) \\ -k_{al}\sin(\lambda\pi) & -k_{cn}\sin(\lambda\pi) & k_{al}\cos(\lambda\pi) & k_{cn}\cos(\lambda\pi) \end{bmatrix} \begin{Bmatrix} a_1 \\ a_2 \\ b_1 \\ b_2 \end{Bmatrix} = 0. \tag{4.30}$$

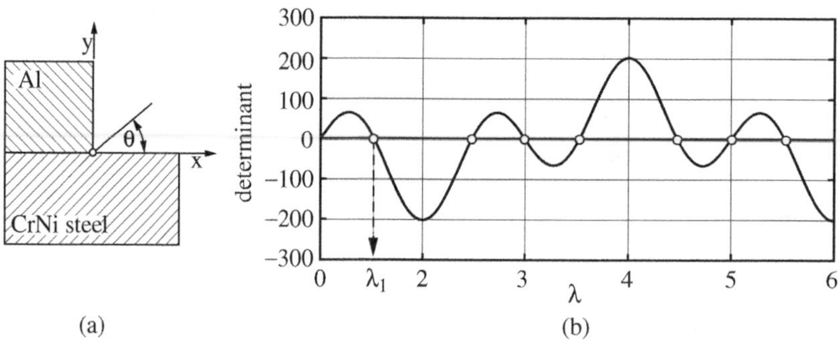

(a)

(b)

Figure 4.10 Example 4.3. (a) Notation. (b) Value of the determinant of the matrix in eq. (4.30).

These equations have a non-trivial solution only if the determinant of the coefficient matrix is zero. There are infinitely many eigenvalues. The determinant is plotted as a function of λ in the interval $0 < \lambda < 5$ in Fig. 4.10(b) where the first seven roots are indicated by open circles. The first eigenvalue is $\lambda_1 = 0.5238$.

Exercise 4.8 Plot the eigenfunction corresponding to λ_1 in Example 4.3. Scale the eigenfunction so that its maximum value is unity.

Exercise 4.9 Determine the second eigenvalue and the corresponding eigenfunction for the problem of Example 4.3. Partial answer: $\lambda_2 = 1.4762$.

Exercise 4.10 Find the first and second eigenvalues for the problem in Example 4.3 modified so that on the boundaries that lie on the x- and y-axes have zero temperature prescribed. Partial answer: $\lambda_1 = 0.8762$.

The Steklov method

It is possible to determine the eigenpairs numerically using a method known as the Steklov method[2]. This method is based on the idea that on a circular boundary of radius ϱ, shown in Fig. 4.1(b), the normal derivative of the function $u = r^\lambda F(\theta)$ is:

$$\left(\frac{\partial u}{\partial n}\right)_{\Gamma_\varrho} = \left(\frac{\partial u}{\partial r}\right)_{r=\varrho} = \lambda \varrho^{\lambda-1} F(\theta) = \frac{\lambda}{\varrho}\, u. \tag{4.31}$$

Consider problems where two or more materials are bonded and the material interfaces[3] intersect the boundary in the origin, see Fig. 4.1(b). Homogeneous natural or essential boundary conditions are prescribed on Γ_{AB} and Γ_{BC}.

The generalized form of eq. (2.34), subject to the assumptions that $\overline{Q} = 0$, steady state conditions exist, and either $u = 0$ or $\partial u/\partial n = 0$ on Γ_{AB} and Γ_{BC}, is:

$$\int_{\Omega_\varrho} \mathrm{grad}\, v\, [\kappa]\, \mathrm{grad}\, u\ dxdy = \frac{\lambda}{\varrho} \int_{\Gamma_\varrho} uv\ ds \quad \text{for all } v \in E^0(\Omega_\varrho) \tag{4.32}$$

where $[\kappa(x,y)]$ is a positive-definite matrix of material properties which are discontinuous at the material interfaces[4] and Ω_ϱ is defined in Fig. 4.1(b). Equation (4.32) is an eigenvalue problem. The eigenvalues are real numbers. On solving eq. (4.32) by the finite element method, the eigenfunctions are approximations of $F(\theta)$ by the (mapped) piecewise polynomial functions. The size of the numerical characteristic value problem can be reduced to the number of degrees of freedom associated with the contour Γ_ϱ. For additional discussion on the Steklov method and illustrative examples we refer to [112].

4.3 The Laplace equation in three dimensions

In this section we consider an extension of the L-shaped domain problem to three dimensions. The domain, shown in Fig. 4.11, is known as the Fichera cube[5]. The Fichera cube is extensively used in

---

2 Vladimir Andreevich Steklov 1864–1926.
3 The treatment of curved interfaces is not considered here.
4 The elements of the matrix $[\kappa]$ are usually piecewise constant functions.
5 Gaetano Fichera 1922–1996.

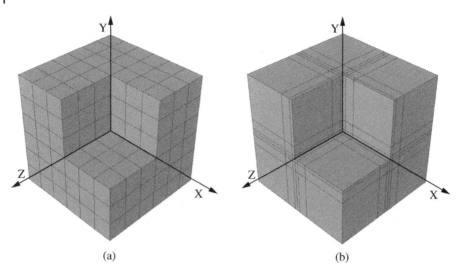

(a)   (b)

Figure 4.11 The Fichera domain, 189-element mesh. (a) Uniform mesh: $M = 3, \gamma = 1, h = 1/3$. (b) Radical grading with parameters $M = 3, \gamma = 2.5$.

benchmarking various discretization and error control procedures. The domain is defined by

$$\Omega \overset{\text{def}}{=} \{(X,Y,Z) \mid (X,Y,Z) \in (-1, \ 1)^3 \setminus [0, \ 1)^3\}. \tag{4.33}$$

The boundary surfaces lie in the planes $X = 0, Y = 0, Z = 0$ and $X = \pm 1, Y = \pm 1, Z = \pm 1$. On the boundaries $X = 0$ and $Z = 0$ homogeneous Dirichlet boundary condition ($u = 0$) is prescribed, on boundary $Y = 0$ homogeneous Neumann condition ($q_n = 0$), on the L-shaped boundaries $X = 1$, $Y = 1, Z = 1$, the Neumann condition $q_n = 0.1$ and on the boundaries $X = -1, Y = -1, Z = -1$ the condition $q_n = -0.5$ are prescribed.

In two dimensions it was possible to characterize the regularity of u_{EX} by a single parameter λ. This is not possible in three dimensions, nevertheless, one can expect that the singularities on the edges coincident with the coordinate axes are similar to the vertex singularity of the L-shaped domain, the coefficients of the expansion given by eq. (4.7) being functions of the edge coordinate, and convergence will be characterized by those singularities. See Remark 4.4. Theoretical analysis of the singularities on domains with piecewise smooth boundaries is addressed in [62].

The exact solution of this problem is not known. The reference value of the potential energy was estimated by extrapolation, from the results of p-extension on a 189-element radially graded mesh using the procedure described in Section 1.5.3. The data are listed in Table 4.4.

The grading parameter for the radical mesh was $\gamma = 6.2$. This parameter was chosen because it minimizes the potential energy at $p = 8$.

For h-extension, a sequence of uniform mesh was used and the rate of convergence was estimated with reference to the extrapolated value of the potential energy in Table 4.4. The results are listed in Table 4.5.

For p-extension, a 189-element uniform mesh was used and the rate of convergence was estimated with reference to the extrapolated value of the potential energy in Table 4.4. The results are listed in Table 4.6.

The estimated relative error in energy norm vs. degrees of freedom curves are plotted on log-log scale for h-extension and p-extension on uniform meshes and p-extension on a radical mesh in Fig. 4.12.

Table 4.4 The Fichera domain. p-convergence, radical mesh $\gamma = 6.2$, $M(\Delta) = 189$, product space.

p	N	π_p	β_p	$(e_r)_E$ (%)
1	288	-3.96897295	–	32.78
2	1,890	-4.34971145	0.423	14.78
3	5,940	-4.41536773	0.492	8.42
4	13,572	-4.43381736	0.533	5.42
5	25,920	-4.44057198	0.562	3.77
6	44,118	-4.44342415	0.565	2.79
7	69,300	-4.44482967	0.577	2.15
8	102,600	-4.44557782	0.577	1.71
extrapolated:		-4.44688294	–	–

Table 4.5 Fichera domain. h-convergence, uniform mesh refinement, $p = 2$ (product space).

h	$M(\Delta)$	N	π_h	k_h	β_h	$(e_r)_E$ (%)
1/3	189	1,890	-4.21669217	0	0	22.75
1/6	1,512	13,572	-4.30253820	1.337	0.118	18.02
1/9	5,103	44,118	-4.33699192	1.336	0.116	15.72
1/12	12,096	102,600	-4.35632002	1.336	0.115	14.27
1/15	23,625	198,090	-4.36893882	1.336	0.114	13.24
1/18	40,824	339,660	-4.37793385	1.336	0.114	12.45
1/21	64,827	536,382	-4.38472583	1.336	0.113	11.82
1/24	96,768	797,328	-4.39006756	1.336	0.113	11.30
1/27	137,781	1,131,570	-4.39439827	1.337	0.113	10.86
1/30	189,000	1,548,180	-4.39799291	1.337	0.113	10.49
1/33	251,559	2,056,230	-4.40103312	1.337	0.113	10.15
1/36	326,592	2,664,792	-4.40364413	1.337	0.113	9.86
1/39	415,233	3,382,938	-4.40591530	1.337	0.113	9.60
reference:			-4.44688294	–	–	–

For the h-version, quasiuniform meshes, we can write eq. (1.91) as

$$\sqrt{\pi_h - \pi} \approx C_h(k, p, u_{EX})h^{k-1} \tag{4.34}$$

where π_h is the potential energy corresponding to the mesh characterized by h, π is the (unknown) exact value of the potential energy, C_h is a constant, independent of h, and k is the index of the Sobolev space in which u_{EX} lies. Here we used the reference value of $\pi_{\text{ref}} = -4.44688294$ (see Table 4.4) to approximate the exact value of π.

Table 4.6 The Fichera domain. p-convergence, uniform mesh, $M(\Delta) = 189$, product space.

p	N	π_p	β_p	$(e_r)_E$ (%)
1	288	−3.96101944	–	33.05
2	1890	−4.21669217	0.199	22.75
3	5940	−4.30016800	0.197	18.16
4	13572	−4.34150404	0.200	15.39
5	25920	−4.36584861	0.203	13.50
6	44118	−4.38173819	0.205	12.10
7	69300	−4.39284714	0.207	11.02
8	102600	−4.40100653	0.209	10.16
reference:		−4.44688294	–	–

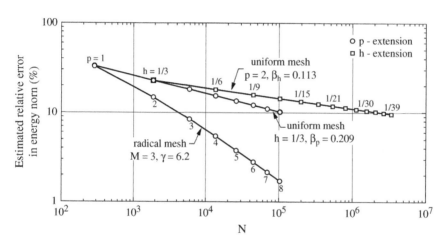

Figure 4.12 The Laplace problem on the Fichera domain. Comparison of three discretization schemes.

We estimate the index k from the sequence of finite element solutions of the Fichera problem shown in Table 4.5 where the column under the heading k_h shows the estimated values of k obtained by the following formula based on eq. (4.34):

$$k_h^{(i)} = 1 + \frac{1}{2} \frac{\log(\pi(h_i) - \pi_{\mathrm{ref}}) - \log(\pi(h_{i-1}) - \pi_{\mathrm{ref}})}{\log(h_i) - \log(h_{i-1})} \tag{4.35}$$

where the index i refers to the ith entry in the table. The results indicate that

$$\lim_{h \to 0} k_h \approx 1.337. \tag{4.36}$$

The rates of convergence computed using the method described in Section 1.5.3 are listed in column β_h. The estimate of β_h is based on eq. (1.92) which can be written as

$$\sqrt{\pi_h - \pi} \approx \frac{C}{N^{\beta_h}}. \tag{4.37}$$

In three dimensions we have the proportionality $N \propto h^{-3}$ therefore we expect

$$\beta_h \approx \frac{k_h - 1}{3} \approx 0.113. \tag{4.38}$$

As seen in Table 4.5, this is indeed the case.

Referring to Table 4.6, we see that the estimated rate of p-convergence is $\beta_p \approx 0.209$ and increasing. From this data we may conjecture either that in three dimensions, when all singular arcs are coincident with element edges, a condition that is usually satisfied in applications of the finite element method, $\beta_h < \beta_p < 2\beta_h$, or that $\beta_p \to 2\beta_h$, as in two dimensions; however, this will occur at very high values of N. Additional empirical data developed for a problem of elasticity on the Fichera domain suggests that the second conjecture is more likely to be true. See Appendix C.

Remark 4.4 In Section 4.2 we considered asymptotic expansions of the form given by eq. (4.3). The coefficients a and b were chosen so as to satisfy the boundary conditions on the intersecting edges. In three dimensions we assume the same functional form along the intersection of two plane surfaces; however, the coefficients will be functions of the edgewise coordinates. For example, referring to Fig. 4.11, along the X axis we have

$$u = r^\lambda (a(X) \cos(\lambda\theta) + b(X) \sin(\lambda\theta)) \tag{4.39}$$

which has to satisfy the strong form of the Laplace equation, written in cylindrical form;

$$\Delta u \equiv \frac{\partial^2 u}{\partial r^2} + \frac{1}{r}\frac{\partial u}{\partial r} + \frac{1}{r^2}\frac{\partial^2 u}{\partial \theta^2} + \frac{\partial^2 u}{\partial X^2} = 0. \tag{4.40}$$

This occurs only if $a(X)$ and $b(X)$ are linear functions. If $a(X)$ and $b(X)$ are polynomials of degree greater than 1 then the functional form of eq. (4.39) has to be augmented by functions, called shadow functions, defined so that eq. (4.39) is satisfied. Importantly, the shadow functions are smoother than the corresponding eigenfunctions for the L-shaped domain. Therefore the shadow functions do not affect the rate of convergence. For details we refer to [112].

4.4 Planar elasticity

The analysis of corner singularities in two-dimensional elasticity is analogous to the way the corner singularities of the Laplace problem were treated in Section 4.2. However, it is complicated by the fact that not all eigenvalues are real or simple. Details are given in Appendix G.

4.4.1 Problems of elasticity on an L-shaped domain

The analysis of the corner singularity of an L-shaped domain problem of two-dimensional elasticity, assuming that the intersecting boundary segments are stress-free, is outlined in Section G.2.3 in the appendix. The lowest positive eigenvalue is $\lambda_1 = 0.54448374$. Therefore the exact solution lies in the Sobolev space $H^{1+\lambda_1-\epsilon}(\Omega)$, hence we expect that the rate of h-convergence on uniform mesh will be $\beta_h = \lambda_1/2$ and the rate of p-convergence will be $\beta_p = \lambda_1$.

The stress field corresponding to the first term of the asymptotic expansion is given by eq. (G.28) and the displacement field is given by eq. (G.29). From this information the exact value of the potential energy is determined from the integral

$$\pi(\mathbf{u}_{EX}) = -\frac{1}{2}\oint \left[u_x(\sigma_x n_x + \tau_{xy} n_y) + \tau_{xy} n_x + \sigma_y n_y)\right] t_z \, ds$$

$$= -4.15454423 \frac{a_1^2 \ell^{2\lambda_1} t_z}{E} \tag{4.41}$$

where n_x, n_y are the unit normals to the boundary in the x, y coordinate system shown in Fig. 4.4, ℓ is a length scale, t_z is the thickness (constant) and E is the modulus of elasticity. Plane strain conditions were assumed with $\nu = 0.3$.

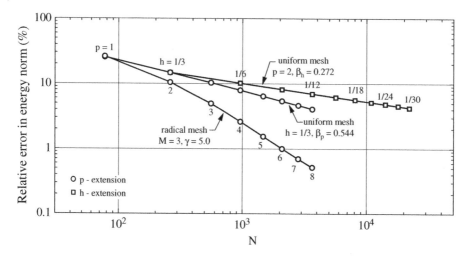

Figure 4.13 A problem of elasticity on an L-shaped domain. Comparison of three discretization schemes. Plane strain, $v = 0.3$, stress-free boundary conditions on the re-entrant edges.

The results computation for (a) *h*-extension using a sequence of uniform meshes with *h* ranging from 1/3 to 1/30, $p = 2$, product space, (b) *p*-extension using a 27-element uniform mesh with *p* ranging from 1 to 8, product space, and (c) *p*-extension on a 27-element radical mesh with $\gamma = 5.0$, product space, are shown in Fig. 4.13. The realized rates of convergence, β_h and β_p are very close to the theoretical estimates.

The estimated and exact convergence rates and relative errors in energy norm are listed in Table 4.7 for *p*-extension on the 27-element radical mesh. The exact values were computed using the exact value of the potential energy given by eq. (4.41). The estimated values are based on the extrapolated value of the potential energy. In the last row the error in the extrapolated value is shown.

Table 4.7 L-shaped domain, free-free boundary conditions, plane strain, $v = 0.3$, radical mesh ($\gamma = 5.0$), 27 elements, product space. Estimated and exact relative errors in energy norm.

p	N	$\dfrac{\pi_p E}{a_1^2 \ell^{2\lambda_1} t_z}$	β Est.'d	β Exact	$(e_r)_E$ (%) Est.'d	$(e_r)_E$ (%) Exact	θ
1	77	−3.87071791	−	−	26.137	26.138	1.00
2	263	−4.10911566	0.746	0.746	10.455	10.457	1.00
3	557	−4.14439620	0.999	0.999	4.939	4.942	1.00
4	959	−4.15172126	1.181	1.177	2.600	2.607	1.00
5	1469	−4.15355459	1.241	1.229	1.531	1.543	0.99
6	2087	−4.15412877	1.267	1.236	0.981	1.000	0.98
7	2813	−4.15434288	1.283	1.213	0.669	0.696	0.96
8	3647	−4.15443336	1.283	1.149	0.480	0.517	0.93
extrapolated:		−4.15454423				0.192	−

Exercise 4.11 Construct a 27-element mesh for the L-shape domain, similar to the mesh shown in Fig. 4.4. However, use geometric grading: locate nodal points along the intersecting boundary segments using eq. (1.57) and extend the mesh to the entire domain. Let $q = 0.07$. Compare the relative errors with those in Table 4.7. You will find that the two grading patterns produce similar results.

4.4.2 Crack tip singularities in 2D

Crack tip singularities have a great practical importance. This is because in damage-tolerant design it is assumed that small cracks are present at critical locations of a structure at the beginning of its service life. When the structure is subjected to cyclic loads then the cracks grow. The objective of design is to ensure that the cracks will not grow by more than a fraction of their critical lengths between inspection intervals.

The rate of crack growth is correlated with the coefficients of the leading terms of the asymptotic expansion. This is justified by the idea that the highly nonlinear process of crack growth is driven by the surrounding elastic stress field, and furthermore, the volume in which the nonlinear processes take place is sufficiently small so that only the leading terms of the asymptotic expansion have to be considered. The coefficients of the leading terms are proportional to the stress intensity factors which are the quantities of interest in linear elastic fracture mechanics (LEFM). This section is concerned with the extraction of stress intensity factors by the contour integral method, which is analogous to the method described in Section 4.2.4.

For cracks ($\alpha = 2\pi$) the equations (G.18) and (G.19) in the appendix reduce to one equation:

$$\sin 2\lambda\pi = 0 \quad \text{therefore} \quad \lambda_n = \pm\frac{n}{2}, \ n = 1, 2, 3, \ldots$$

hence all roots are real and simple. The goal is to compute the coefficients of the first terms of the symmetric (mode I) and antisymmetric (mode II) expansions.

In the engineering literature it is customary write the Cartesian components of the Mode I stress tensor corresponding in the following form:

$$\sigma_x = \frac{K_I}{\sqrt{2\pi r}} \cos\frac{\theta}{2}\left(1 - \sin\frac{\theta}{2}\sin\frac{3\theta}{2}\right) + T + O(r^{3/2}) \tag{4.42}$$

$$\sigma_y = \frac{K_I}{\sqrt{2\pi r}} \cos\frac{\theta}{2}\left(1 + \sin\frac{\theta}{2}\sin\frac{3\theta}{2}\right) + O(r^{3/2}) \tag{4.43}$$

$$\tau_{xy} = \frac{K_I}{\sqrt{2\pi r}} \sin\frac{\theta}{2}\cos\frac{\theta}{2}\cos\frac{3\theta}{2} + O(r^{3/2}) \tag{4.44}$$

where $-\pi \leq \theta \leq \pi$, T is a constant, called the T-stress, see Section H.1. The constant K_I is called the mode I stress intensity factor.

The antisymmetric (Mode II) stress tensor components are usually written in the following form:

$$\sigma_x = -\frac{K_{II}}{\sqrt{2\pi r}} \sin\frac{\theta}{2}\left(2 + \cos\frac{\theta}{2}\cos\frac{3\theta}{2}\right) + O(r^{3/2}) \tag{4.45}$$

$$\sigma_y = \frac{K_{II}}{\sqrt{2\pi r}} \sin\frac{\theta}{2}\cos\frac{\theta}{2}\cos\frac{3\theta}{2} + O(r^{3/2}) \tag{4.46}$$

$$\tau_{xy} = \frac{K_{II}}{\sqrt{2\pi r}} \cos\frac{\theta}{2}\left(1 - \sin\frac{\theta}{2}\sin\frac{3\theta}{2}\right) + O(r^{3/2}) \tag{4.47}$$

where K_{II} is called the mode II stress intensity factor.

Computation of stress intensity factors

The contour integral method (CIM) for the Navier equations, analogously to the CIM for the Laplace equation outlined in Section 4.2.4, is based on the existence of a path-independent integral and the orthogonality of eigenfunctions. Details on the formulation the 2D elasticity problem are available in [96].

We will consider the Navier equation under the assumption that the volume forces are zero and the temperature is constant. The path-independent contour integral is:

$$I_\Gamma \stackrel{\text{def}}{=} \int_\Gamma (T_x^{(\mathbf{u})} w_x + T_y^{(\mathbf{u})} w_y)\, ds - \int_\Gamma (T_x^{(\mathbf{w})} u_x + T_y^{(\mathbf{w})} u_y)\, ds \tag{4.48}$$

where the superscript $\mathbf{u}$ (resp. $\mathbf{w}$) represents the exact solution (resp. a test function that satisfies the Navier equation and the homogeneous boundary conditions on the edges that intersect in the singular point), Γ is an arbitrary contour that begins on one edge and runs in the counterclockwise direction to the other, as shown in Fig. 4.7. We will assume that Γ is a circular contour, the radius of which is arbitrary, and we will be interested in the computation of K_I, K_{II} and T. Details are presented in Appendix H.

Under special conditions the stress intensity factors can be determined from the energy release rate. The procedure for computing stress intensity factors from the energy release rate is described in Appendix H.

Example 4.4 The plane strain fracture toughness, denoted by K_{Ic}, is defined as the value of the stress intensity factor at which the propagation of cracks becomes rapid and unbounded. It is considered to be a material property that quantifies the resistance of materials to crack propagation.

Procedures for the measurement of K_{Ic} are governed by various standards[6]. Compact tension (CT) specimens and single edge notched bend (SENB) specimens are commonly used. A typical CT specimen with an initial crack of length a is shown in Fig. 4.14(a). The initial crack is produced through stable fatigue crack growth.

In this example we consider the following question: Since plane strain cannot be realized in an experiment, what is the difference between K_I of a plane strain model and K_I along the crack front of a CT specimen.

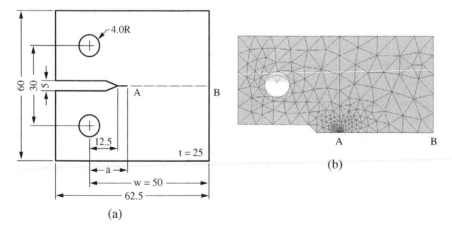

(a)

(b)

Figure 4.14 (a) A typical compact tension test specimen. (b) A typical finite element mesh.

---

6 See ASTM E399: Standard test method for linear-elastic plane-strain fracture toughness K_{Ic} of metallic materials. See also ASTM E1820, ISO 12737, ISO 12135.

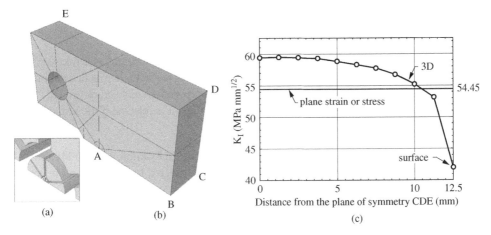

Figure 4.15 One quarter of a compact tension test specimen. (a) Mesh detail, exploded view. (b) 27 element mesh, geometric grading, $q = 0.15$, (c) computed values of K_I for 1 kN applied force and $a = 25$ mm.

The material is an aluminum alloy with the elastic properties $E = 71.7$ GPa, $v = 0.333$. Letting $a = 25$ mm and using the contour integral method, for an applied force of 1 kN we find that $K_I = 54.45$ MPa mm$^{1/2}$. This result was confirmed using a 27-element geometrically graded finite element mesh.

The 27-element mesh was extruded to obtain a three-dimensional representation of the CT specimen, see Fig. 4.15 (b). Since the CT specimen has two planes of symmetry, it is sufficient to use the quarter model shown in Fig. 4.15(b). Symmetry is prescribed on the rectangular area, the three vertices A, B, C of which are visible, and on the plane CDE. The cylindrical surface was loaded by distributed sinusoidal normal tractions, the resultant of which is 1 kN. The other surfaces are traction-free.

The stress intensity factor K_I was computed at 11 equally spaced points along the crack front by the contour integral method. At each point a sequence of 6 extraction circles were used with the radii ranging from 0.6 mm (just outside of the innermost elements shown in Fig. 4.15(a)) to 0.8 mm. The resulting values were extrapolated to zero radius by linear regression. In three dimensions the finite element solution is affected by the shadow functions (see Remark 4.4). Extrapolation to zero radius serves to remove the perturbations caused by shadow functions.

The extrapolated values are plotted in Fig. 4.15(c). The maximum value (59.4 MPa mm$^{1/2}$) occurs in the plane of symmetry. It is 8.4 % greater than the value obtained by the plane strain model. The singularity at the point where the crack front intersects the stress-free surface is different from the crack tip singularity in the case of plane strain. This is because the zero traction conditions have to be satisfied not only on the faces of the crack but also on the boundary surface.

Remark 4.5 The ASTM Standard E1820-01[7] provides the following formula for the computation of K_I:

$$K_I \approx \frac{P}{t\sqrt{w}} \frac{2 + a/w}{(1 - a/w)^{3/2}} (0.886 + 4.64(a/w)$$
$$- 13.32(a/w)^2 + 14.72(a/w)^3 - 5.6(a/w)^4)$$

---

7 Standard Test Method for Measurement of Fracture Toughness.

where P is the applied force. Using the data in Example 4.4 we get $K_I \approx 54.64$ MPa mm$^{1/2}$ which is a close approximation to the plane strain result. □

Remark 4.6 We assumed that the crack front is a straight line. This assumption was necessary in order to make comparison between the plane strain and 3D models possible. In practice the initial crack is induced by subjecting the specimen to cyclic loads that produce stable crack growth. This process results in a curved crack front. In standard tests a test result is rejected if the initial crack length varies by more than 5% along the crack front. □

Exercise 4.12 Assume that the fracture toughness of the material in Example 4.4 is 20 MPa m$^{1/2}$. Estimate the applied load at which rapid crack propagation is expected to occur when the crack length is 30 mm.

4.4.3 Forcing functions acting on boundaries

The regularity of the solution is influenced by the forcing function. We consider (a) a concentrated force and (b) a step function acting on a semi-infinite planar body in the following. The notation is shown in Fig. 4.16

Concentrated force
In this case the stress function is

$$U = -\frac{F_0 r}{\pi}\theta \sin\theta \tag{4.49}$$

see, for example, [105]. The corresponding stress components are:

$$\sigma_r = \frac{1}{r}\frac{\partial U}{\partial r} + \frac{1}{r^2}\frac{\partial^2 U}{\partial\theta^2} = -\frac{2F_0}{\pi}\frac{\cos\theta}{r} \tag{4.50}$$

$$\sigma_\theta = \frac{\partial^2 U}{\partial r^2} = 0 \tag{4.51}$$

$$\tau_{r\theta} = -\frac{\partial}{\partial r}\left(\frac{1}{r}\frac{\partial U}{\partial\theta}\right) = 0 \tag{4.52}$$

and the displacement components, up to rigid body displacement, with the symmetry constraint implied on $\theta = 0$, are:

$$u_r = -\frac{2F_0}{\pi E}\cos\theta \ln r - \frac{(1-\nu)F_0}{\pi E}\theta \sin\theta + C\cos\theta \tag{4.53}$$

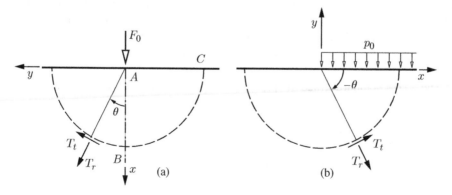

Figure 4.16 (a) Loading by a concentrated force. (b) Loading by a step function.

$$u_\theta = \frac{2vF_0}{\pi E} \sin\theta + \frac{2F_0}{\pi E} \ln r \sin\theta + \frac{(1-v)F_0}{\pi E}(\sin\theta - \theta\cos\theta) - C\sin\theta \qquad (4.54)$$

where C is an arbitrary constant. If we select an arbitrary reference point on the positive x axis, say $x = d$, and set $u_r(d, 0) = 0$ then from eq. (4.53) we get $C = 2F_0 \ln d/(\pi E)$. In this case $u_x(r, \theta)$ is measured from this point. Note that in the origin $u_r(0, \theta) = \infty$ and $\partial u_r/\partial r$ is not square integrable on any subdomain that includes the origin. Hence u_r is not in the energy space. Concentrated forces and point constraints are not admissible in two- and three-dimensional elasticity. This point will be discussed and illustrated by an example in Section 5.2.8.

Step function

The step function is shown in Fig. 4.16(b). The point where the normal traction changes from zero to $-p_0$ is a singular point. The stress function is:

$$U = -\frac{p_0 r^2}{2\pi}\left(\pi + \theta - \frac{1}{2}\sin 2\theta\right) - \pi \le \theta \le 0 \qquad (4.55)$$

and the stress components are:

$$\sigma_r = \frac{1}{r}\frac{\partial U}{\partial r} + \frac{1}{r^2}\frac{\partial^2 U}{\partial\theta^2} = -\frac{p_0}{\pi}\left(\pi + \theta + \frac{1}{2}\sin 2\theta\right) \qquad (4.56)$$

$$\sigma_\theta = \frac{\partial^2 U}{\partial r^2} = -\frac{p_0}{\pi}\left(\pi + \theta - \frac{1}{2}\sin 2\theta\right) \qquad (4.57)$$

$$\tau_{r\theta} = -\frac{\partial}{\partial r}\left(\frac{1}{r}\frac{\partial U}{\partial\theta}\right) = \frac{p_0}{2\pi}(1 - \cos 2\theta). \qquad (4.58)$$

In this case the stress components are finite but not single-valued in the singular point. Observe that along the x axis ($\theta = 0$) $\tau_{r\theta} = 0$ in the origin but along the y axis ($\theta = -\pi/2$) $\tau_{r\theta} = p_0/\pi$. In the origin it is multi-valued. Similarly, the sum of the normal stresses $\sigma_r + \sigma_\theta = -2p_0(1 + \theta/\pi)$ ranges from $-2p_0$ to 0 in the origin, depending on the direction from which the origin is approached.

Exercise 4.13 Refer to Fig. 4.16(b). Let $p_0 = 1$ MPa and the radius of the circle 100 mm. Specify tractions corresponding to the stresses given by equations (4.56) to (4.58) on the circular boundary. Plot the sum of normal stresses and report the error in energy norm.

4.5 Robustness

A numerical scheme for the approximation of a parameter-dependent problem is said to be robust if it is uniformly convergent for all admissible values of the parameter. In this section we consider robustness with Poisson's ratio as the parameter.

As Poisson's ratio v approaches $1/2$, the rate of h-convergence of low order elements, measured in energy norm, slows. Furthermore, the error in the first stress invariant, computed from the finite element solution using Hooke's law, increases. This is called Poisson ratio locking or volumetric locking.

To explain why locking occurs, we show that for an arbitrary mesh of regular quadrilateral elements, $p = 1$, the number of degrees of freedom is independent of the number of elements when the material is incompressible and hence convergence cannot occur at all.

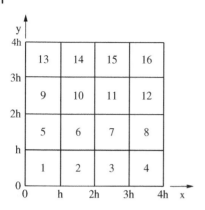

Figure 4.17 Poisson ratio locking. Notation.

Consider the uniform mesh of quadrilateral elements shown in Fig. 4.17. The displacement components for the ith element are:

$$u_x^{(i)} = a_1^{(i)} + a_2^{(i)}x + a_3^{(i)}y + a_4^{(i)}xy \tag{4.59}$$

$$u_y^{(i)} = a_5^{(i)} + a_6^{(i)}x + a_7^{(i)}y + a_8^{(i)}xy. \tag{4.60}$$

For incompressible elastic materials the volumetric mechanical strain $\epsilon_{kk} - 3\alpha\mathcal{T}_\Delta$ is zero, independent of the hydrostatic stress σ_0 defined by eq. (2.84). Therefore in the case of plane strain the displacement components satisfy the constraint

$$\epsilon_x^{(i)} + \epsilon_y^{(i)} \equiv \frac{\partial u_x^{(i)}}{\partial x} + \frac{\partial u_y^{(i)}}{\partial y} = 0. \tag{4.61}$$

Consequently $a_4^{(i)} = a_8^{(i)}$ and $a_7^{(i)} = -a_2^{(i)}$ and hence

$$u_x^{(i)} = a_1^{(i)} + a_2^{(i)}x + a_3^{(i)}y \tag{4.62}$$

$$u_y^{(i)} = a_5^{(i)} + a_6^{(i)}x - a_2^{(i)}y. \tag{4.63}$$

Consider now continuity between elements 1 and 2:

$$a_1^{(1)} + a_2^{(1)}h + a_3^{(1)}y = a_1^{(2)} + a_2^{(2)}h + a_3^{(2)}y \tag{4.64}$$

$$a_5^{(1)} + a_6^{(1)}h - a_2^{(1)}y = a_5^{(2)} + a_6^{(2)}h - a_2^{(2)}y \tag{4.65}$$

therefore

$$a_3^{(1)} = a_3^{(2)} \quad \text{and} \quad a_2^{(1)} = a_2^{(2)}$$

which is equivalent to

$$\frac{\partial u_x^{(1)}}{\partial x} = \frac{\partial u_x^{(2)}}{\partial x} \quad \text{and} \quad \frac{\partial u_x^{(1)}}{\partial y} = \frac{\partial u_x^{(2)}}{\partial y}.$$

Thus both derivatives of u_x are continuous on the interface of elements 1 and 2. The same argument can be used to show that both derivatives are continuous on the interfaces of elements 2 and 3, 3 and 4, etc. as well as on the interfaces of elements 5 and 6 and so on. Since u_x is continuous, it has to be a linear function over the entire domain, independent of the number of elements.

Similarly, from the condition of continuity between elements 1 and 5, we find

$$\frac{\partial u_y^{(1)}}{\partial x} = \frac{\partial u_y^{(5)}}{\partial x} \quad \text{and} \quad \frac{\partial u_y^{(1)}}{\partial y} = \frac{\partial u_y^{(5)}}{\partial y}$$

and hence both derivatives of u_y are continuous on the interfaces of elements 1 and 5, 5 and 9 and so on. Therefore u_y is also a linear function over the entire domain, independent of the number of elements. Consequently, for the entire domain we have

$$u_x = a_1 + a_2 x + a_3 y \tag{4.66}$$
$$u_y = a_4 + a_5 x - a_2 y \tag{4.67}$$

that is, the dimension of the finite element space is 5, no matter how many elements are used. All other degrees of freedom are needed for satisfying the volumetric and continuity constraints.

Locking does not occur in the p-version because the number of elements and hence the number of constraint conditions is fixed, independent of the number of degrees of freedom. The elements can deform while preserving constant volume. However, the computation of normal stresses from the finite element solution requires special attention. This will be discussed with reference to the following example, see Example 4.6.

Example 4.5   In this example we consider a classical problem of elasticity, that of a rigid circular inclusion in an infinite plate, subjected to unidirectional tension at infinity. Plane strain conditions are assumed. The domain and notation are shown in Fig. 4.18(a). We will use $a = 1$, $b/a = 5$ and thickness $t_z = 1$. The finite element mesh, consisting of four elements, is shown in Fig. 4.18(b).

The boundary conditions are as follows: Along the circular arc AD both displacement components are zero. Along the symmetry lines AB and CD the normal displacement and the shearing stress are zero. Along the circular arc BC the normal and shearing tractions are given by the strong form of the classical solution [60]. The displacement components in polar coordinates are:

$$u_r = \frac{\sigma_\infty}{8Gr} \left\{ (\kappa - 1)\, r^2 + 2\,\gamma\, a^2 + \left[\beta(\kappa + 1)\, a^2 + 2\, r^2 + \frac{2\,\delta\, a^4}{r^2} \right] \cos 2\,\theta \right\} \tag{4.68}$$

$$u_\theta = -\frac{\sigma_\infty}{8Gr} \left[\beta\,(\kappa - 1)\, a^2 + 2\, r^2 - \frac{2\,\delta\, a^4}{r^2} \right] \sin 2\,\theta \tag{4.69}$$

and the stress components are:

$$\sigma_r = \frac{\sigma_\infty}{2} \left[1 - \frac{\gamma a^2}{r^2} + \left(1 - \frac{2\beta a^2}{r^2} - \frac{3\delta a^4}{r^4} \right) \cos 2\theta \right] \tag{4.70}$$

$$\sigma_\theta = \frac{\sigma_\infty}{2} \left[1 + \frac{\gamma a^2}{r^2} - \left(1 - \frac{3\delta a^4}{r^4} \right) \cos 2\theta \right] \tag{4.71}$$

$$\tau_{r\theta} = -\frac{\sigma_\infty}{2} \left(1 + \frac{\beta a^2}{r^2} + \frac{3\delta a^4}{r^4} \right) \sin 2\theta \tag{4.72}$$

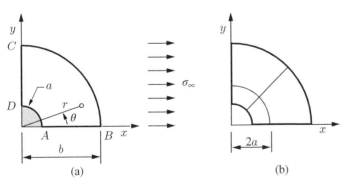

(a)   (b)

Figure 4.18   (a) Rigid circular inclusion in an infinite plate under tension. Notation. (b) Four-element mesh.

where κ, β, γ, δ are constants that depend on Poisson's ratio v only. For plane strain:

$$\kappa = 3 - 4v; \quad \beta = -\frac{2}{3-4v}; \quad \gamma = -(1-2v); \quad \delta = \frac{1}{3-4v} \tag{4.73}$$

and for plane stress:

$$\kappa = \frac{3-v}{1+v}; \quad \beta = -\frac{2(1+v)}{3-v}; \quad \gamma = -\frac{1-v}{1+v}; \quad \delta = \frac{1+v}{3-v}. \tag{4.74}$$

We will use $E = 1$, $v = 0.49999$. The results demonstrate strong p-convergence in energy norm.

Finite element solutions were obtained by means of StressCheck. The load vector corresponding to the normal and shearing tractions acting on the circular arc BC was computed by numerical quadrature using twelve Gauss points per element side. The exact value of the strain energy was computed for $b/a = 5$ as follows:

$$U(\mathbf{u}) = \frac{1}{2} \int_0^{\pi/2} (u_r \sigma_r + u_\theta \tau_{r\theta}) b \, d\theta = 7.31883865 \frac{\sigma_\infty^2 a^2 t_z}{E}. \tag{4.75}$$

The computed values of the strain energy for p ranging from 3 to 8 (product space) and the estimated and exact relative errors in energy norm are listed in Table 4.8. It is seen that the slope of the convergence curve β is increasing with p. This is because the exact solution is an analytic function and hence the rate of p-convergence is exponential. Nevertheless, our a posteriori error estimator, based on the assumption that the rate of convergence is algebraic (see eq. (1.92)), provides very reasonable estimates.

Example 4.6 We consider the problem of a rigid circular inclusion in an infinite plate described in Example 4.5. Plane strain is assumed. The finite element mesh, consisting of four elements, is shown in Fig. 4.18(b). We are interested in the relative error in the sum of normal stresses $\sigma_x + \sigma_y$ measured in maximum norm. Referring to equations (4.70), (4.71) and noting that $\sigma_x + \sigma_y$ is a stress invariant, we define

$$\Sigma \overset{\text{def}}{=} \sigma_r + \sigma_\theta = \sigma_x + \sigma_y = \sigma_\infty \left(1 + \frac{2}{3-4v} \left(\frac{a}{r} \right)^2 \cos 2\theta \right)$$

and we define the percent relative error in maximum norm as follows:

$$(e_r)_\Sigma \overset{\text{def}}{=} 100 \frac{\max |\Sigma - (\sigma_x^{(FE)} + \sigma_y^{(FE)})|}{\sigma_\infty (1 + 2/(3-4v))}. \tag{4.76}$$

Table 4.8 Convergence of the strain energy ($v = 0.49999$). Rigid circular inclusion in an infinite plate under tension. Four-element mesh, product space, $b/a = 5$.

p	N	$\dfrac{U(u)E}{\sigma_\infty^2 a^2 t_z}$	β	Rel. error (%) Est.'d	Rel. error (%) Exact
3	72	6.53074253	–	32.81	32.81
4	128	7.29188870	2.93	6.07	6.07
5	200	7.31760970	3.46	1.30	1.30
6	288	7.31873361	3.37	0.38	0.38
7	392	7.31882930	3.86	0.12	0.11
8	512	7.31883783	3.86	0.04	0.03
Exact		7.31883865	–	–	0

Table 4.9 Example 4.6. Rigid circular inclusion in an infinite plate under tension. Plane strain. Dependence of the maximum relative error $(e_r)_\Sigma$ on Poisson's ratio. Four-element mesh, $p = 8$.

Space	Poisson's ratio			
	0.49	0.499	0.4999	0.49999
Product	0.486	2.725	4.988	32.84
Trunk	2.070	8.981	39.68	233.1

The results of computation for $p = 8$ are listed in Table 4.9. These results show that $(e_r)_\Sigma$ rapidly increases as Poisson's ratio approaches $1/2$, and it increases more rapidly for the trunk space than for the product space. This is an indication that product space is more robust than the trunk space.

The exact value of the sum of normal stresses and the computed values using the product and trunk spaces, $p = 8$, are shown along boundary segment AD in Fig. 4.19 for $v = 0.49999$. For the trunk space the largest error is in the vicinity of 2 degrees, whereas for the product space the largest error is at 45 degrees. At 45 degrees the size of the jump is the same for the product and trunk spaces.

This example illustrates that the product space has significantly better convergence properties than the trunk space for the range of parameters considered. In general, the p-version is more robust than the h-version.

Exercise 4.14 Solve the rigid inclusion problem described in Example 4.5 for $v = 0.3, \ 0.49, \ 0.499$ using $b/a = 5$, $p = 8$, product space. Explain why the error in $\tau_{r\theta}$, measured in maximum norm, is virtually independent of v.

Exercise 4.15 Repeat the computations in Example 4.6 for plane stress. Explain why the values of $(e_r)_\Sigma$ are virtually independent of v.

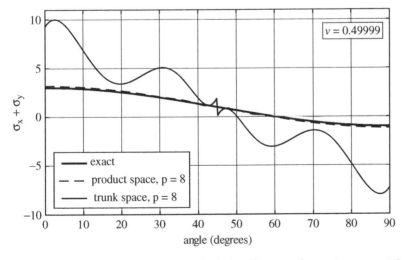

Figure 4.19 Example 4.5. Rigid circular inclusion. The sum of normal stresses at the boundary of the inclusion. Poisson's ratio: 0.49999.

Exercise 4.16   Solve the problem of Example 4.5 using a sequence of quasiuniform meshes of 32, 64 and 128 elements, $p = 2$ (product space) and, using the exact value of the strain energy given by eq. (4.75), compute the relative error in energy norm for each finite element solution and the rate of convergence β for the sequence of solutions. Is the theoretical rate of convergence ($\beta = 1$) realized? I not, why not?

Exercise 4.17   Subdivide each of the four elements in Example 4.6 uniformly into 100 elements. Solve the rigid inclusion problem described in Example 4.5 for $v = 0.3,\ 0.49,\ 0.499$ using $b/a = 5$, $p = 2$, product space. Report the values of $(e_r)_\Sigma$. Partial solution: For $v = 0.49$ you will find $(e_r)_\Sigma = 25.34\%$.

4.6   Solution verification

It is necessary to determine whether the QoIs computed from a finite element solution are sufficiently accurate for the purposes of the analysis. The accuracy of a QoI depends on the accuracy of the finite element solution, measured in energy norm, and the method by which the QoI is computed. We have seen a simple illustration of this in Example 1.9. Only direct methods of computation are considered in this section.

Solution verification involves steps to ascertain that (a) the correct input data was used in the analysis and (b) the quantities of interest are substantially independent of the parameters that characterize the finite element space. This is based on the idea that while we do not know the exact value of the QoI, we know that it exists, it is unique and it is independent of the discretization. Therefore, as long as the computed value of a QoI is changing significantly when the number of degrees of freedom is increased, it cannot be close to its limit value.

This involves obtaining two or more finite element solutions corresponding to a sequence of finite element spaces and examination of the information generated from the finite element solutions. The recommended steps are as follows:

1. Display the solution graphically and check whether the solution is reasonable. For example, plotting the deformed configuration on a 1:1 scale provides information on whether a large error occurred in specifying the loading and constraint conditions or material properties.
2. Estimate the relative error in energy norm and its rate of convergence. The estimated relative error in energy norm is a useful indicator of the overall quality of the solution, roughly equivalent to estimating the root-mean-square error in stresses, see Section 2.6.2. The estimated rate of convergence is an indicator of whether the rate of change of error is consistent with the asymptotic rate for the problem class. Substantial deviations from the theoretical rates of convergence typically indicate errors in the input data or in the finite element mesh. For example, elements may be highly distorted.
3. Check for the presence of jump discontinuities in stresses and fluxes in regions where the stresses or fluxes are large. The normal flux and the normal and shearing stress components at internal element boundaries must be continuous. In regions where the stresses or fluxes are small, some discontinuity is generally acceptable. Significant jump discontinuities in stresses and fluxes at element boundaries usually indicates that the mesh is not fine enough or the polynomial degree is not high enough or the elements are too distorted.
4. Show that the quantities of interest are substantially independent of the mesh and/or the polynomial degree of elements. Estimate the limit value. Report the estimated limit value and the

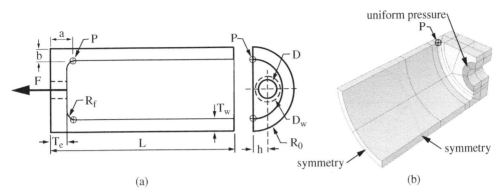

Figure 4.20 Example 4.7. (a) Notation. (b) Isometric view and finite element mesh. Number of elements: 26.

percent difference between the estimated limit value and the computed value corresponding to the highest number of degrees of freedom.

These steps are illustrated by the following example.

Example 4.7 In this example we demonstrate the process of verification. The quantity of interest is maximum von Mises stress in the fillet region of a shear fitting. The geometric configuration, notation and boundary conditions are shown in Fig. 4.20. The fitting is loaded by uniform pressure given by

$$p = \frac{4F}{\pi(D_w^2 - D^2)}$$

where F is the applied force, D is the diameter of the hole, D_w is the outer diameter of the area on which the pressure is acting. The data are as follows: $F = 1000$ lbs, (4448 N), $D = 0.375$ in (9.525 mm), $D_w = 0.650$ in (16.51 mm), outer radius: $R_0 = 1.10$ in (27.94 mm), wall thickness: $T_w = 0.60$ in (15.24 mm), length: $L = 3.0$ in (76.2 mm), fillet radius: $R_f = 0.1$ in (2.54 mm), pad thickness $T_e = 0.50$ in (12.70 mm), offset $h = 0.45$ in (11.43 mm). Modulus of elasticity: $E = 1.05E7$ psi (7.24E4 MPa), Poisson's ratio: $v = 0.30$.

The finite element mesh is shown in Fig. 4.20(b). It consists of 24 hexahedral elements and 2 pentahedral elements. We will perform h- and p-extensions using the trunk space. The initial mesh for h-extension is the mesh shown in Fig. 4.20(b). We denote the diameter of the largest element in the initial mesh by h_1. A sequence of meshes was obtained by uniformly subdividing the elements in the parameter space. This results in $h_k = h_1/k$ ($k = 1, 2, 3, ...$). The corresponding number of elements is given by $M(\Delta_k) = 26k^3$.

Table 4.10 Example 4.7: Estimated error in energy norm.

p	N	π_p	β	$(e_r)_E$
5	2844	−0.41810322	−	3.34
6	4314	−0.41836531	0.99	2.21
7	6263	−0.41848408	1.17	1.43
8	8769	−0.41853071	1.17	0.97
extrapolated:		−0.41856984		

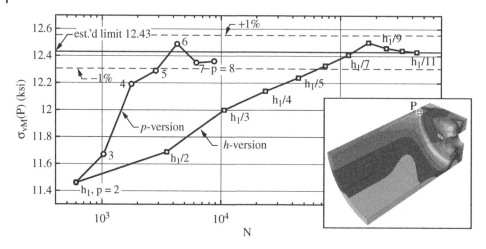

Figure 4.21 Example 4.7. Point convergence.

For *p*-extension the mesh shown in Fig. 4.20(b) was used. The relative error in energy norm at $p = 8$ was estimated to be 0.83%. It was found that the maximum von Mises stress occurs in point P shown in Fig. 4.20. The location parameters are: $a = 0.5881$ in (14.94 mm) and $b = 0.1507$ in (3.83 mm). Point P is located on the intersection between the top surface and the toroidal surface of the fillet. Therefore the stress is uniaxial in point P.

The values of the von Mises stress corresponding to sequences of finite element solutions obtained by *h*- and *p*- extensions are shown in Fig. 4.21.

As the number of degrees of freedom is increased either by *h*- or *p*-extension, the location and magnitude of the maximum stress change. Here we fixed the location where the maximum was found and computed sequences of estimated values of the von Mises stress in that location. We see that convergence is not monotonic either for the *h*- or *p*-version.

Remark 4.7 In the foregoing example we excluded from consideration the neighborhood of load application, based on the following consideration: If the maximum stress would occur in the vicinity of load application then this model would not be suitable for the determination of the maximum stress. This is because the loading is transferred by mechanical contact which is idealized here as a uniform pressure distribution. Should the stress distribution in the contact area be of interest, a different model would have to be formulated[8]. Our region of primary interest was the fillet region. The contact area is a region of secondary interest. □

Remark 4.8 Finite element spaces generated by p-extension are hierarchic whereas finite element spaces generated by h-extension may or may not be hierarchic. In general, sequences of mesh created by mesh generators are not hierarchic. ⊔

Remark 4.9 In many practical problems the QoIs lie in some small subdomain of the solution domain. For example, we may be interested in the stresses in the vicinity of a fastener hole in a large plate. The neighborhood of the fastener hole is the region of primary interest, the rest of the plate is the region of secondary interest. Errors in the computed data may be caused by insufficient

---

8 Mechanical contact is described in Section 9.2.4.

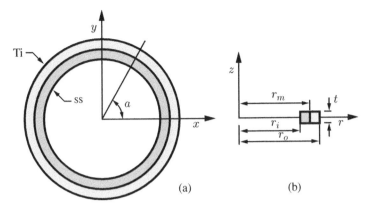

Figure 4.22 Composite ring. Notation.

discretization in the region of primary or secondary interest, or both. Errors caused by insufficient discretization of the region of secondary interest are called pollution errors. □

The following example highlights errors that can and frequently do occur when the analyst does not understand the relationship between input data and the regularity of the exact solution.

Example 4.8   A composite ring was made by joining a stainless steel (ss) and a titanium (Ti) ring.

Two analysts were asked to compute the maximum principal stress $(\sigma_1)_{max}$ when the temperature of the composite ring is increased by 100 C°. They were to assume that the two materials are perfectly bonded and the assumptions of the linear theory of elasticity are applicable. The geometric and material parameters are shown in Table 4.11.

Analyst A formulated this as a problem of plane stress. He used two quadrilateral elements on a 30-degree sector ($\alpha = 30°$) and, having performed p-extension (product space), he reported that the maximum principal stress occurs along the material interface and its value is 22.1 MPa. He

Table 4.11   Geometric and material properties.

Description	Symbol	Value	Units
radius	r_i	100.0	mm
radius	r_o	115.0	mm
radius	r_m	107.5	mm
thickness	t	20	mm
modulus of elasticity, ss	E_s	2.0×10^5	MPa
Poisson's ratio, ss	ν_s	0.295	–
coef. of thermal exp. ss	α_s	1.17×10^{-5}	1/C°
modulus of elasticity, Ti	E_t	1.1×10^5	MPa
Poisson's ratio, Ti	ν_t	0.31	–
coef. of thermal exp. Ti	α_t	8.64×10^{-6}	1/C°

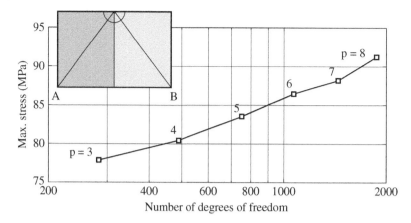

Figure 4.23 Divergence of the maximum principal stress. The results were obtained by *p*-extension on a 16-element geometrically graded mesh. Symmetry boundary conditions were applied on boundary segment AB.

supported this statement by showing that the error in energy norm converged exponentially and the maximum normal stress was substantially constant for $p = 3$, 4, ..., 8.

Analyst B formulated this as an axisymmetric problem. He generated a uniform mesh of 800 nine-node quadrilateral elements (that is, $p = 2$ product space). He reported that the maximum principal stress occurs at $r = r_m, z = t/2$ and its value is 52.8 MPa. Relying on his judgment that the mesh was fine enough, he did not perform solution verification.

Your supervisor asked you to find out why such a large discrepancy exists between the two results. Having checked the input data, you found no errors. You then solved the axisymmetric problem using *p*-extension (product space) on a 16-element, geometrically graded mesh shown in the inset of Fig. 4.23. On plotting the maximum principal stress vs. the number of degrees of freedom on semi-log scale you found that it does not converge to a limit value.

Write a succinct memorandum to your supervisor explaining the reasons for the discrepancy.

Solution
Analyst A solved the problem in the plane of symmetry ($z = 0$) and thus his model does not account for the singularity caused by the abrupt change in material properties in the points $r = r_m, z = \pm t/2$. The exact solution of this problem is an analytic function. That is why *p*-convergence was exponential. Analyst A solved the wrong problem very accurately.

Analyst B solved the correct problem, but with a very large (in fact infinitely large) error in the computed value of $(\sigma_1)_{max}$. The exact value of $(\sigma_1)_{max}$ is not finite in the singular point and hence the absolute error in the reported value of $(\sigma_1)_{max}$ is infinitely large.

You are now faced with the unpleasant problem of having to explain to your supervisor that asking for the maximum stress under the stated modeling assumptions made no sense.

Discussion
Proper formulation of a mathematical model depends on the quantities of interest (QoI). Analyst A did not take into consideration that the plane stress model cannot represent the stress raiser at $r = r_m, z = \pm t/2$ which are singular arcs in 3D, singular points in the axisymmetric formulation. Therefore the dimensionally reduced model chosen by Analyst A is improper, given the QoI. Had the QoI been (say) the maximum radial displacement then both models would have been proper. Analyst A would have reported 0.121 mm, Analyst B would have reported 0.122 mm.

The supervisor was presumably interested in predicting failure (or the probability of failure) and mistakenly assumed that failure can be correlated with the maximum stress corresponding to the exact solution of the problem of linear elasticity. However, in this problem the maximum stress is not a finite number for any temperature change and therefore it cannot be a predictor of failure.

Various theories can be formulated for the prediction of failure events. However, those theories are subject to the restriction that the driver of failure initiation is a finite number that continuously depends on the applied load, in this case the temperature change. For example, one could assume that the generalized stress intensity factor, or the average stress over a volume, is the predictor of failure initiation. Such predictors would have to be calibrated and tested following a process similar to the process described in Chapter 6.

5

Simulation

In Chapter 1 simulation was defined as an imitative representation of the functioning of one system or process by means of the functioning of another. In this chapter we consider the functioning of mechanical systems and their imitative representation by mathematical models. We are interested in predicting certain quantities of interest (QoIs), given the geometry, material properties and loads. The usual QoIs in the simulation of mechanical systems are displacements, stresses, strains, stress intensity factors, ultimate loads, the probability that a system will survive a given number of load cycles, etc. These quantities support engineering decisions concerning design, certification and maintenance.

Simulation begins with a precise statement of an idea of physical reality in the form of a mathematical model. The term "mathematical model" should be understood to collectively refer to all mathematical operations, deterministic or probabilistic, needed for the prediction of the QoIs. The importance of distinguishing between physical reality and an idea of physical reality was emphasized by Wolfgang Pauli[1]:

> *"The layman always means, when he says 'reality' that he is speaking of something self-evidently known; whereas to me it seems the most important and exceedingly difficult task of our time is to work on the construction of a new idea of reality."*

Implied in this statement is that casting ideas of physical reality into mathematical models has subjective aspects: Various ideas can be proposed for the simulation of the functioning of a particular system or process. Therefore it is necessary to have objective criteria by which the relative merit of mathematical models is evaluated. The formulation of mathematical models is described and illustrated in this chapter.

Whereas Pauli had in mind ideas of reality in the realm of quantum mechanics, his statement is universally valid. It is applicable to any engineering or scientific project concerned with the formulation of mathematical models for the purpose to generalize data collected from observations of events. Engineers have the advantage over physicists because it is generally much easier to make observations and test predictions based on mathematical models in the forefront of engineering than in the forefront of physics which include phenomena on scales that range from the subatomic to the astronomical. In either case, the goals are to cast ideas of physical reality into the form of mathematical models, calibrate the mathematical models and test them in validation experiments through comparison of predicted outcomes with the outcomes of physical experiments or observations of physical events.

---

1 Wolfgang Pauli 1900–1958.

Finite Element Analysis: Method, Verification and Validation, Second Edition. Barna Szabó and Ivo Babuška.
© 2021 John Wiley & Sons, Inc. Published 2021 by John Wiley & Sons, Inc.
Companion Website: www.wiley.com/go/szabo/finite_element_analysis

We will view a mathematical model as a transformation of one set of data **D** into another set **F** (the QoI) based on a precisely stated idea of the physical reality **I**. In short hand,

$$(\mathbf{D}, \mathbf{I}) \; \rightarrow \; \mathbf{F} \tag{5.1}$$

where the right arrow represents a mathematical model which (by definition) is the entire process by which (**D**, **I**) is transformed to the quantities of interest.

For example, in the World Wide Failure Exercise [48], investigators sought to predict failure events in test articles made of composite materials. In that case the failure theory was part of the mathematical model and **F** was the set of data needed to predict the outcome of experiments. In addition, the mathematical models included constitutive laws and multiscale procedures that link sub-models of macroscopic and microscopic phenomena. When the goal is to predict the probability of failure then a statistical sub-model is also part of the mathematical model.

A mathematical model formulated for the prediction of the probability of failure events in high cycle fatigue will be discussed in Chapter 6.

Remark 5.1 In view of Pauli's statement, the frequently used term: "physics-based model" does not have a well defined meaning. Mathematical models are precisely formulated ideas relating to specific aspects of physical reality. Usually there are several competing models for the prediction of the same phenomenon. We will discuss objective methods for ranking alternative models in Chapter 6.

5.1 Development of a very useful mathematical model

To illuminate the process by which ideas of physical reality are cast into the form of a mathematical model we trace the history of development of a famous and very useful mathematical model known as the Bernoulli-Euler beam model[2]. Students of mechanical and structural engineering usually learn about this classical model in their second year of study. Interestingly, its development took 188 years and some of the greatest mathematicians of the time contributed to it. We begin with a brief summary of how the formulation is presented today then outline the major milestones in its 188 years of development.

5.1.1 The Bernoulli-Euler beam model

The formulation of the Bernoulli-Euler beam model for prismatic beams of arbitrary cross-section can be found in every textbook on the strength of materials. For the sake of simplicity, the formulation is reviewed here for the special case of beams with rectangular cross-section only. The notation is shown in Fig. 5.1. The origin of the coordinate axes is coincident with the centroid of the cross-section.

The formulation involves three steps: (a) The mode of deformation is assumed and the distribution of strain is inferred. (b) A stress-strain relationship is defined. Here we consider linearly elastic stress-strain relationships only. (c) The equation of equilibrium is written: the bending moment corresponding to the normal stress acting on a cross-section has to be in equilibrium with the applied bending moment. The deformation is assumed to be sufficiently small so that the equilibrium equations can be written with reference to the undeformed configuration.

---

2 Jacob Bernoulli 1655–1705, Leonhard Euler 1707–1783.

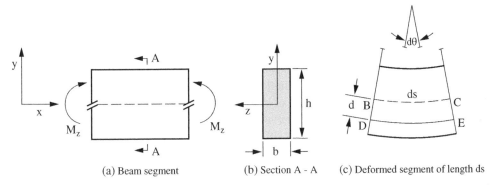

Figure 5.1 Rectangular beam in bending. Notation.

Referring to Fig. 5.1, it is assumed that the bending moment causes the cross-sections to rotate without deformation (plane sections remain plane) hence fibers[3] on the convex side lengthen and fibers on the concave side shorten. Fibers in the xz plane do not deform. The xz plane is called the neutral plane and its intersection with a cross-section is called the neutral axis. In Fig. 5.1 the z axis is coincident with the neutral axis.

We denote the radius of curvature by r and the angle subtended by the arc ds from the center of rotation by $d\theta$. The length of arc DE can be written as $(r + d)d\theta$. By definition, strain is the change in length divided by the original length. Since ds lies in the neutral plane, ds is the original length. Therefore the strain is:

$$\varepsilon = \frac{(r + d)d\theta - rd\theta}{rd\theta} = \frac{d}{r} = -\frac{y}{r}. \tag{5.2}$$

It is assumed that stress is proportional to strain, that is, $\sigma = E\varepsilon$ where E is the modulus of elasticity and the stress must be in equilibrium with the bending moment:

$$M = -\int_{-h/2}^{h/2} \sigma y b \, dy = \frac{E}{r} \int_{-h/2}^{h/2} y^2 b \, dy = \frac{Mbh^3}{12} = \frac{EI}{r} \tag{5.3}$$

where integration is over the cross-section ($dA = bdy$) and I is the moment of inertia of the cross-section about the z axis.

We denote the displacement of the neutral axis by $u = u(x)$. The curvature of u is

$$\frac{1}{r} = \frac{u''}{\left(1 + (u')^2\right)^{3/2}} \approx u'' \tag{5.4}$$

where the approximate equality ($\approx$) is justified when max $(u')^2 << 1$, a condition satisfied in most mechanical and structural engineering applications. If the beam is loaded by a distributed load q (given in force/length units) then, using the convention that q is positive in the positive y direction and M is positive in the sense shown in Fig. 5.1, then the equation of equilibrium is:

$$M'' = q. \tag{5.5}$$

On combining equations (5.3), (5.4) and (5.5) we get the Bernoulli-Euler beam model:

$$\left(EIu''\right)'' = q. \tag{5.6}$$

The usual boundary conditions of interest are: simply supported: $u = M = 0$, fixed or built-in: $u = u' = 0$, free: $M = V = 0$ where V is called shearing force, it is proportional to u'''.

---

3 In this context a fiber is understood to be the set of points $0 < x < \ell$ such that y and z are fixed.

5.1.2 Historical notes on the Bernoulli-Euler beam model

The development of the Bernoulli-Euler beam model began with Galileo[4] who, in a book published in 1638, addressed questions relating to how structures resist loads[5]. He considered a beam of rectangular cross-section, loaded at one end, fixed on the other end and mistakenly assumed that the resistance of the beam to the applied load is uniformly distributed over the fixed cross-section. As a result, Galileo overestimated the strength of beams by a factor of 3. Nevertheless he was able to correctly predict that a beam of rectangular cross-section, bent by moment M about the z-axis (i.e. $M_z = M$ in Fig. 5.1), will be h/b times as strong as the same beam bent by M about the y-axis ($M_y = M$).

Hooke[6] experimented with beams made of wood and observed that the fibers on the convex side are extended whereas the fibers on the concave side are compressed. He also observed that forces and the magnitude of corresponding displacements are in constant ratio. When the components of stress tensor can be written as linear combinations of the components of the strain tensor then that relationship is called Hooke's law.

Jacob Bernoulli investigated the deflected curve of elastic bars. He followed the steps outlined in Section 5.1.1 however he incorrectly assumed that the strain is zero on the innermost fiber on the concave side ($y = h/2$):

$$\varepsilon = \frac{y - h/2}{r_c}, \quad -h/2 < y < h/2$$

where r_c is the radius of the outermost fiber on the convex side. Letting $\eta = h/2 - y$, he arrived at the following formula which is analogous to eq. (5.3) however the moment

$$M = -\int_0^h \sigma \eta \, b d\eta = \frac{Eb}{r_c} \int_0^h \eta^2 \, dy = \frac{Ebh^3}{3r_c}$$

is $4r/r_c$ times the correct moment.

Euler used the methods of variational calculus to arrive at the correct functional form:

$$M = C \frac{u''}{\left(1 + (u')^2\right)^{3/2}}$$

where, according to Euler, C depends on the elastic properties and, for a beam of rectangular cross-section, is proportional to bh^2. The correct statement would have been: is proportional to bh^3, see eq. (5.3). Euler recommended that C should be determined by experimental means. We now know that $C = EI$ where E is a property of the material, determined by experimental means, and I a property of the cross-section, see eq. (5.3).

The formulation of the Bernoulli-Euler beam model was completed by Navier[7] in 1826 when he proved that, under the assumptions that the axial force is zero and the material obeys Hooke's law, the neutral axis passes through the centroid of the cross-section.

The first engineering use of this model on a large scale was in connection with the design of the Eiffel[8] Tower for the 1889 World's Fair. Until then bridges and buildings were designed by precedent. Eiffel demonstrated the practical value of this model.

The foregoing is a greatly abbreviated summary of the history of the development of the Bernoulli-Euler beam model. For additional information we refer to [102, 106]. This summary

---

4 Galileo Galilei 1564–1642.
5 *Two New Sciences*. English translation by H. Crew and A. de Salvio. The Macmillan Company, New York, 1933.
6 Robert Hooke 1635–1703.
7 Claude-Louis Navier 1785–1836.
8 Gustave Eiffel 1832–1923.

is included here to illustrate that casting ideas of physical reality into a mathematical model is a creative process that involves making assumptions based on insight and experience, attempting to validate those assumptions through physical experimentation, and confirming, rejecting or modifying the formulation on the basis of empirical data.

The definition of a mathematical model must always include specification of its range of validity. In the case of the Bernoulli-Euler beam model the assumption that plane sections remain plane is valid when the beam is subjected to pure bending but when shearing forces are present then the shearing deformation may not be negligibly small. This happens when the length to depth ratio is less than about 10 or when the transverse load has a short wavelength (less than a few times the depth of the beam). Another limitation is that the stresses must be in the elastic range. For slender beams the assumption that the deformation is small, and therefore the equilibrium conditions can be stated with respect to the undeformed configuration, may not be realistic. In such cases it is necessary to write the equations of equilibrium with reference to the deformed configuration and a nonlinear mathematical problem has to be solved.

The formulation of mathematical models for slender elastic bodies follows a very different pattern in our time: These models are understood to be dimensionally reduced forms of the general three-dimensional models of nonlinear continua [4]. The Bernoulli-Euler beam model is a special case that, notwithstanding its simplicity and limitations, has great practical value.

The formulation of mathematical models in the field of solid mechanics reached a mature state of development by the mid-twentieth century. Application of these models in engineering practice was severely limited, however, because the solution of mathematical problems by classical methods was feasible only in special and highly simplified cases. This limitation was removed when digital computers became available, opening vast new possibilities in numerical simulation.

5.2 Finite element modeling and numerical simulation

For the reasons that will be discussed in Section 5.2.5, the concept and practice of finite element modeling evolved more than ten years before the theoretical foundations of numerical simulation were established. The terms "finite element modeling" and "numerical simulation" are often used interchangeably. Engineers and engineering analysts often refer to numerical simulations but, more often than not, they have finite element modeling in mind. It is important to clarify the difference between the two.

5.2.1 Numerical simulation

The main elements of numerical simulation are illustrated schematically in Fig. 5.2 which is a restatement of eq. (5.1) with the addition that the quantities of interest ($\mathbf{F}$) are solved numerically and hence we have to rely on a numerical approximation to $\mathbf{F}$ which is denoted by $\mathbf{F}_{num}$. As we have seen in Chapter 1, it is not sufficient to compute $\mathbf{F}_{num}$, it is also necessary to estimate and control the error of approximation $|\mathbf{F} - \mathbf{F}_{num}|$.

The formulation of a mathematical model is a creative process that has room for a wide range of subjective choices. On the other hand, the solution of the numerical problem and control of error in the quantities of interest involve the application of algorithmic procedures based on established theorems of applied mathematics. Once a mathematical model was defined, subjective choices are confined to the selection of discretization schemes and extraction procedures to the extent that the available software tools support such choices.

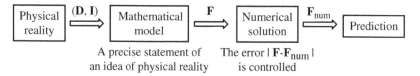

Figure 5.2 The main elements of numerical simulation.

It is essential to recognize the existence of two qualitatively different sources of error: The model form error incurred when making assumptions in the formulation of a mathematical model and the numerical error, coming from the approximation of $\mathbf{F}$ with $\mathbf{F}_{num}$.

The transformation indicated in eq. (5.1) by the right arrow contains coefficients associated with constitutive equations and models of failure initiation. These coefficients have to be calibrated to experimental observations. In a calibration processes the mathematical model is assumed to be correct and its coefficients are determined to match, or nearly match, the outcome of experiments. If $|\mathbf{F} - \mathbf{F}_{num}|$ is not known to be negligibly small in comparison with the experimental errors then the coefficients, and hence the predictions, will be polluted by the numerical errors.

Calibrated models are also validated models within the domain and scope of calibration. Validation experiments performed outside of the domain of calibration provide information about the size of the domain of calibration. If we allow the domain of calibration to be arbitrarily small then just about any model can be validated.

Remark 5.2 It is necessary to ensure that the numerical approximation scheme, the discretization, meets the conditions of consistency[9] and stability[10]. The methods described in this book meet these conditions. For details on consistency and stability we refer to [5].

Remark 5.3 The outcomes of physical experiments are random events, hence calibrated parameters are random numbers. When the statistical dispersion of these numbers is small then randomness is usually neglected and average values are reported. This is common practice in (for example) reporting the modulus of elasticity and Poisson's ratio. However, outcomes can have very large dispersions as (for example) in fatigue testing of materials. In such cases a statistical sub-model has to be incorporated in the mathematical model and the QoIs are the predicted probabilities of outcomes. This is discussed in Chapter 6.

5.2.2 Finite element modeling

Early work on the finite element method was performed entirely by engineers who were familiar with matrix methods of structural analysis. Their idea was to develop computer codes for the solution of problems of elasticity and other field problems similar to those that were already in use at that time for the analysis of structural trusses and frames. They viewed the finite element method as a "matrix method of structural analysis".

Idealized trusses are assemblies of bar elements connected at frictionless hinges. An idealized bar element is assumed to function as a linear spring, loaded by axial tension or compression forces only. Given the cross-sectional area A, the modulus of elasticity E and the length ℓ, the relationship

---

9 Consistency of a discretization refers to a quantitative measure of the extent to which the exact solution satisfies the discrete problem. This quantitative measure depends on the QoI.

10 Stability of a discretization refers to well-posedness of the discretized problem. A discretization is stable if small changes in the data produce small changes in the QoI.

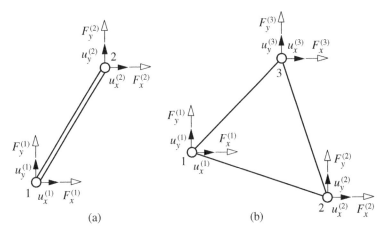

Figure 5.3 Notation: (a) Truss element, (b) 3-node plane stress or plane strain element.

between the axial force F and the corresponding change in length δ is $F = (AE/\ell)\delta$ where the bracketed term is the spring constant of the bar. A bar element is shown in Fig. 5.3(a). The length and orientation of a bar element are determined by the coordinates of the hinges, also called nodes, indicated by the open circles in Fig. 5.3(a).

Letting

$$\{F\} = \{F_x^{(1)} \; F_x^{(2)} \; F_y^{(1)} \; F_y^{(2)}\}^T$$

and

$$\{u\} = \{u_x^{(1)} \; u_x^{(2)} \; u_y^{(1)} \; u_y^{(2)}\}^T$$

the relationship between the nodal force components and the nodal displacement components can be written as

$$\{F\} = [K]\{u\} \tag{5.7}$$

where $[K]$ is a singular matrix, called the element stiffness matrix, see eq. (5.9). The relationship between the nodal forces and the nodal displacements are obtained for the entire truss by writing the equations of force equilibrium for each node and factoring the nodal displacements. Finally, the nodal constraints are enforced and the resulting system of linear equations, the coefficient matrix of which is non-singular, is solved for the nodal displacements.

To extend matrix methods of structural analysis to continuum problems it was necessary to establish an analogous relationship to eq. (5.7). For truss elements the nodal forces are the stress resultants. No such interpretation exists for continuum problems, however. For continuum problems nodal force was interpreted to mean a force-like entity that is energy-equivalent to the strain energy of the element [113]. For example, for the plane stress or plane strain triangle shown in Fig. 5.3(b), we have:

$$\{u\}^T\{F\} = \{u\}^T[K]\{u\}$$

where $\{u\}$ is the 6×1 nodal displacement vector and $[K]$ is a symmetric positive semidefinite 6×6 matrix. The strain energy of the element was related to the nodal displacements through the strains corresponding to the shape functions.

The goal was to formulate element stiffness matrices and load vectors for various applications. This element-oriented view strongly influenced the development of finite element computer codes and the intuitive understanding of the finite element method by the engineering community.

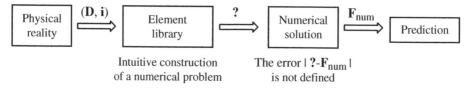

Figure 5.4 The main elements of finite element modeling.

The practice of finite element modeling, which is illustrated schematically in Fig. 5.4, is based on this view.

On comparing Fig. 5.4 with Fig. 5.2 the difference between finite element modeling and numerical simulation should become evident. In finite element modeling a transformation similar to eq. (5.1) is performed. However, in the place of a precisely formulated idea of physical reality $\mathbf{I}$ in the form of a mathematical model, intuition $(\mathbf{i})$ is employed to construct a numerical problem through assembling elements from the element library of a finite element software product:

$$(\mathbf{D}, \mathbf{i}) \rightarrow \mathbf{F}_{num}. \tag{5.8}$$

Practitioners of finite element modeling tacitly transfer the responsibility for defining the mathematical model to the developers of finite element software products. However, the question of whether a numerical problem corresponds to a properly formulated mathematical model, and the quantities of interest $\mathbf{F}$ are properly defined, cannot be answered without knowing which elements were chosen and how those elements were formulated and implemented. In general, the details of implementation are not readily available to users.

The element libraries contain elements identified by the technical theory (e.g. beam, plate, plane stress, shallow-shell, solid, etc.), the number of nodes, the variational principle employed in the formulation and the integration rule applied (full or reduced). For example, a hexahedral element for the solution of thermoelastic problems may be designated as a *"20-node triquadratic displacement, trilinear temperature, hybrid, linear pressure, reduced integration"*[11].

Reduced integration means that fewer than the necessary minimum number of integration points are employed when computing the terms of the stiffness matrix. This topic was addressed in many papers. The general idea is that elements of low polynomial degree are "too stiff" and they become less stiff when reduced integration is employed. See, for example, [114]. This is a typical finite modeling argument in that element properties are modified without consideration of the underlying variational principle. Strang and Fix referred to such practices as "variational crimes" [89].

Subsequently it was realized that reduced integration results in "hourglassing", a term that refers to the appearance of spurious deformation modes that have zero strain energy. Additional numerical schemes had to be invented to counter the effects of hourglassing. See, for example, [27].

Given that errors associated with finite element modeling cannot be estimated, one can reasonably ask whether finite element modeling has any value at all. This question has to be answered in the affirmative, with certain qualifications, however: Importantly, one must distinguish between structural and material strength models. At present the only practical way to analyze complicated engineering structures, such as airframes subjected to static and dynamic loads, automobile bodies under crash conditions and major civil engineering and marine structures, is through finite element modeling. A finite element model is constructed by an intuitive process guided by experience. Finite element models of structural components and assemblies of structural components

---

11 DS Simulia Abaqus/CAE User's Guide, Abaqus 6.14.

are constructed so as to approximate the stiffness of the components and sub-assemblies. Minor details such as fillets, rivets and bolts are omitted. The stiffness of the finite element representation of components and subassemblies are checked against experimental data and the finite element model is modified (tuned) to match the experimental observations. The goal is to obtain reasonable estimates of the force distribution among the structural components.

Interestingly, successful applications of finite element models are made possible by an approximate cancelation of two types of large error: Conceptual errors in the formulation (also known as variational crimes), and the errors of discretization. The construction of a finite element model is not a scientific activity but rather an intuition-based, experience-guided artistic undertaking.

Exercise 5.1 Show that for an ideal truss element

$$[K] = \frac{AE}{\ell} \begin{bmatrix} \cos^2\alpha & -\cos^2\alpha & \sin\alpha\cos\alpha & -\sin\alpha\cos\alpha \\ -\cos^2\alpha & \cos^2\alpha & -\sin\alpha\cos\alpha & \sin\alpha\cos\alpha \\ \sin\alpha\cos\alpha & -\sin\alpha\cos\alpha & \sin^2\alpha & -\sin^2\alpha \\ -\sin\alpha\cos\alpha & \sin\alpha\cos\alpha & -\sin^2\alpha & \sin^2\alpha \end{bmatrix} \tag{5.9}$$

where ℓ is the length of the truss element and, denoting the coordinates of the nodes by (x_i, y_i), $i = 1, 2$,

$$\sin\alpha = (y_2 - y_1)/\ell, \quad \cos\alpha = (x_2 - x_1)/\ell.$$

Exercise 5.2 Refer to Exercise 5.1. Show that the nodal forces satisfy the equations of equilibrium independent of the nodal displacements, no matter what nodal displacements are prescribed.

$$(a)\ F_x^{(1)} + F_x^{(2)} = 0, \quad (b)\ F_y^{(1)} + F_y^{(2)} = 0, \quad (c)\ F_y^{(2)}\cos\alpha - F_x^{(2)}\sin\alpha = 0.$$

5.2.3 Calibration versus tuning

Mathematical models are calibrated. Finite element models are tuned. The differences between the two are addressed in this section.

Calibration

In numerical simulation the functional form of a mathematical model is assumed to be correct for the purpose of calibration. The parameters that represent material properties and boundary conditions are estimated by matching observed data with what would have been predicted if the parameters of interest were known. One very simple example is finding the modulus of elasticity for a material by applying a known force to an elastic bar made of the same material and measuring the corresponding displacement. Knowing the cross-sectional area and the length of the bar, the modulus of elasticity can be readily calculated from this information. An example of a far more challenging calibration problem will be presented in Chapter 6.

Calibrated data have an interval of validity. For example, for elastic moduli that interval is bounded by zero and the proportional limits in tension and compression. The set of intervals on which the coefficients are defined, together with the restrictions associated with the assumptions incorporated in the mathematical model, define the domain of calibration.

In most cases the mathematical problem has to be solved numerically. In such cases it is necessary to ensure that the errors in the computed quantities of interest are negligibly small in comparison with experimental errors and uncertainties.

Tuning

In finite element modeling the term "tuning" refers to adjusting the finite element mesh in such a way as to match experimental observations. Therefore finite element models are interpolators within a small neighborhood of tuned parameters.

Tuning is widely used in analyzing large structural systems such as automobiles under crash conditions and structural analysis of airframes. In fact, finite element modeling is the only practical means available for modeling such large systems at present.

Since validation refers to testing the predictive performance of a mathematical model while ensuring that the errors of discretization are small, and in finite element modeling the model form errors are not separated from the errors of discretization, finite element models cannot be validated.

It is possible to match certain observable quantities reasonably well with data obtained from finite element models and yet have very large errors in other quantities of interest. This will be demonstrated in Section 5.2.8.

5.2.4 Simulation governance

There are many challenging engineering problems that require the continued development and improvement of the methods of numerical simulation. For example, as new materials and material systems are developed, it is necessary to formulate, calibrate and test new constitutive laws and failure criteria. Taking examples from the field of aerospace engineering, one area of great interest is the development of design rules for fiber-reinforced composite laminates. Rule making for condition-based maintenance is another important area: Given information concerning the service history of a high value asset (such as an airframe), what is the optimal interval for inspection and repair? What steps should be taken if damage is detected in a principal structural element[12]? To answer these and similar questions it is necessary to formulate structural and strength models for the prediction of crack formation and crack propagation rates based on the concepts and methods of numerical simulation.

Simulation governance is a managerial function concerned with the exercise of command and control over all aspects of numerical simulation through the establishment of processes for the systematic improvement of the tools of engineering decision making. This includes: (a) proper formulation of mathematical models, (b) selection and adoption of the best available numerical simulation technology, (c) coordination of experimental work with numerical simulation, (d) documentation and archival of experimental data, (e) application of data and solution verification procedures, (f) revision of mathematical models in the light of new information collected from physical experiments and field observations and (g) standardization of design, analysis and certification workflows whenever appropriate.

A plan for simulation governance has to be tailored to fit the mission of each organization or department within an organization: If that mission is to apply established rules of design and certification then emphasis is on solution verification and standardization. If the mission is to formulate design rules, or make condition-based maintenance decisions, then verification, validation and uncertainty quantification must be part of the plan.

---

12 A principal structural element is an element of an aircraft structure that contributes significantly to the carrying of flight ground, or pressurization loads and whose integrity is essential in maintaining the overall structural integrity of the aircraft. See, for example, FAA Advisory Circular No. 25.571-1D.

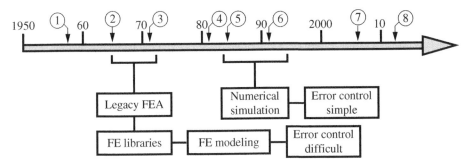

Figure 5.5 FEA timeline.

5.2.5 Milestones in numerical simulation

In order to make a clear distinction between finite element modeling and numerical simulation, and to help understand why and how finite element modeling evolved before the foundations of numerical simulation were established, a brief review of some of the major milestones in the development of the finite element method is presented in the following. These milestones are indicated by numbered arrows on the finite element analysis (FEA) timeline shown in Fig. 5.5.

1. The first engineering paper on the finite element method was published in 1956 [107]. In the following year the Soviet Union launched the first satellite, Sputnik, and the space race began. This brought about substantial investments into engineering and scientific projects in support of the US space program which funded rapid development of various finite element computer codes by academic institutions and aerospace companies.

2. NASA issued a request for proposal in 1965 that eventually led to the development of the finite element software NASTRAN [56], versions of which are still in use today. This marked the beginning of the development of legacy finite element codes. As indicated in the FEA timeline in Fig. 5.5, the infrastructure of legacy codes was substantially established in the course of the following 5 to 7 years. Its development was guided by the understanding of the finite element method that existed in the 1960s. This infrastructure imposes limitations that prevent legacy FEA codes from keeping pace with the evolution of the finite element method.

3. Systematic exploration of the mathematical foundations of the finite element method began around 1972 [12]. This is a fundamentally important milestone because mathematicians view the finite element method very differently from engineers: engineers (generally speaking) think of the finite element method as a modeling tool that permits assembling various elements selected from the library of a finite element software product. They believe that the solution of the corresponding numerical problem approximates the physical response of their objects of interest such as airframes, turbine disks, pressure vessels, etc. to various loading conditions.

Mathematicians, on the other hand, view FEM as a method by which approximate solutions can be obtained to well-defined mathematical problems. For example, the equations of linear elasticity, given a solution domain, material properties, loading and constraint conditions, define a mathematical problem that has a unique exact solution $\mathbf{u}_{EX}$. The finite element solution, denoted by $\mathbf{u}_{FE}$, is an approximation to $\mathbf{u}_{EX}$.

Suppose that we are interested in a quantity $\Phi(\mathbf{u}_{EX})$. A key question is this: How close is $\Phi(\mathbf{u}_{FE})$ to $\Phi(\mathbf{u}_{EX})$? Relying on $\Phi(\mathbf{u}_{FE})$ without having some estimate of the size of the error of approximation in our quantity of interest (QoI) can and often does lead to mistaken conclusions. Finite element models often do not correspond to a well-defined mathematical problem and therefore it is not possible to determine what that error is. It is not uncommon in finite element modeling practice that $\Phi(\mathbf{u}_{EX})$ is not a finite number. In such cases reporting $\Phi(\mathbf{u}_{FE})$ is a serious conceptual error. See, for instance, Example 4.8.

The accuracy of approximation depends on the finite element mesh and the polynomial degree of the elements. In the early implementations of the finite element method the polynomial degree of the elements (denoted by p) was fixed at a low value, typically $p = 1$ or $p = 2$, and the error of approximation was controlled by mesh refinement such that the size of the largest element in the mesh, denoted by h, was reduced. This is known today as the h-version of the finite element method.

In the mid-1970s research indicated that keeping the finite element mesh fixed and increasing p, has important advantages [100]. This is known today as the p-version of the finite element method.

4. In 1981 the p-version was analyzed in [24, 25]. It was proven and demonstrated that for a large class of problems, which includes two-dimensional linear elasticity, the asymptotic rate of convergence of the p-version in L_2 and energy norms is at least twice that of the h-version with respect to the number of degrees of freedom.

5. It was proven and demonstrated in 1984 that in most problems of engineering interest the smoothness of the solution is such that when the finite element mesh is properly graded then the finite element solution converges (in energy norm or L_2 norm) exponentially as p is increased [14, 90]. The h- and p-versions are special cases of the finite element method where both the mesh and the polynomial degree are important in controlling the error of approximation. The distinction between the h- and p-versions is rooted in the history of the development of finite element method rather than in its theoretical foundations. The most important practical advantage of the p-version is that it makes estimation and control of the errors of approximation in the quantities of interest simple and efficient.

6. Any mathematical model can be viewed as a special case of a more comprehensive model. Therefore any mathematical model is a member of a hierarchic sequence. For example, a model based on the assumptions of the linear theory of elasticity is a special case of a model that accounts for plastic deformation. Once the solution of a problem of linear elasticity is available, it is possible to check whether the assumptions incorporated in the model are satisfied. If they are not satisfied then the analyst needs to employ a more comprehensive model. The implementation of a hierarchic modeling framework began in the late 1980s [101]. It provides for seamless transition from lower to higher models. As indicated in Fig. 5.5, the development and documentation of a concept demonstrator for hierarchic models was substantially completed by the mid-1990s.

7. The American Society of Mechanical Engineers (ASME) published its first guideline on verification and validation (V&V) in computational solid mechanics in 2006. The main point is this: Whenever engineering decisions are based on the results of numerical simulation, there is an implied expectation of reliability. Without such expectation it would not be possible to justify the cost of a simulation project. If simulation produces misleading information then it has a negative economic value with possibly severe consequences. There are many well documented instances of expensive repairs, retrofits, project delays and serious safety issues arising from lack of quality assurance in numerical simulation. Therefore assurance of the quality of the results of numerical simulation is essential. Specifically, the following quality control steps are recommended by

ASME: (a) code verification, (b) estimation of the errors of approximation in terms of the quantities of interest and (c) validation of the mathematical model through comparison of predicted outcomes with experimental observations.

8. The concept of simulation governance from the perspective of mechanical and structural engineering was introduced in 2012 [92]. At the 2017 NAFEMS World Congress[13] simulation governance was identified as the first of eight "Big Issues" in the field of numerical simulation.

5.2.6 Example: The Girkmann problem

In order to explore the differences between finite element modeling and numerical simulation in a specific setting, readers of the bulletin of the International Association for Computational Mechanics, called IACM Expressions, were invited to solve the following problem, known as the Girkmann problem[14] [39], using any finite element analysis software available to them.

A spherical shell of thickness $h = 0.06$ m, crown radius $R_c = 15.00$ m is connected to a stiffening ring at the meridional angle $\alpha = 2\pi/9$ (40°). The dimensions of the ring are: $a = 0.60$ m, $b = 0.50$ m. The radius of the mid-surface of the spherical shell is $R_m = R_c/\sin\alpha$. The notation is shown in Fig. 5.6. The z axis is the axis of rotational symmetry. The shell, made of reinforced concrete, is assumed to be homogeneous, isotropic and linearly elastic with Young's modulus $E = 20.59$ GPa and Poisson's ratio $\nu = 0$.

Consider gravity loading only. The equivalent (homogenized) unit weight of the material is 32.69 kN/m^3. Assume that uniform normal pressure p_{AB} is acting at the base AB of the stiffening ring. The resultant of p_{AB} equals the weight of the structure. Assume that the stiffening ring is weightless. The goals of computation are as follows:

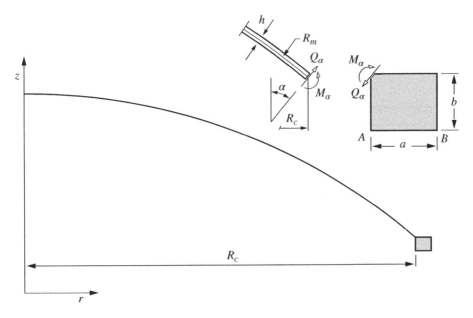

Figure 5.6 The Girkmann problem. Notation.

---

13 NAFEMS is an International Association for Engineering Modeling, Analysis and Simulation. The acronym refers to its parent organization which was formed in 1983 in the United Kingdom: National Agency for Finite Element Methods and Standards.
14 Karl Girkmann 1890–1959.

1. Find the shearing force Q_α in N/m units and the bending moment M_α acting at the junction between the spherical shell and the stiffening ring in Nm/m units.
2. Determine the location (meridional angle) and the magnitude of the maximum bending moment in the shell.
3. Verify that the results are accurate to within 5%.

This problem statement appeared in the January 2008 issue if IACM Expressions [76]. This problem statement differed from the original problem statement by Girkmann in two respects: In [39] kgf (resp. cm) was used for the unit of force (resp. length), in [76] SI units were used. Also, in [39] one of the goals of computation was to determine the radial force per unit length between the ring and the shell. In [76] the goal was to determine the shear force per unit length between the ring and the shell. The responses received were summarized in a follow-on article published in the January 2009 issue [77] and in greater detail in [99].

Of the 16 solutions received 11 were obtained by finite element modeling methods implemented in legacy finite element codes, that is, commercial finite element codes that have infrastructures designed to support finite element modeling. Only two of the respondents who used legacy codes attempted to perform solution verification. However, the quantities of interest either failed to converge or appeared to have converged to the wrong result. The results are summarized in Table 5.1.

One respondent attempted to demonstrate h-convergence for a shell-solid model on one quarter of the stiffened shell using six successive uniform mesh refinements. In the sixth refinement 120 million degrees of freedom were used. The sequence of moments corresponding to the six refinements still had not converged but appeared to tend to approximately −205 Nm/m and the shear force appeared to have converged to approximately 1140 N/m.

Another respondent wrote: *"Regarding verification tasks for structural analysis software that has adequate quality for use in our safety critical profession of structural engineering, the solution of problems such as the Girkmann problem represents a minuscule fraction of what is necessary to assure quality."* This statement is obviously true. That is why it is surprising that the answers obtained with legacy finite element codes had such a large dispersion. As shown in Table 5.1, the reported

Table 5.1 Legacy codes: Summary of results.

Element	Q_α N/m	M_α Nm/m	φ_{max} degrees	M_{max} Nm/m
axisymmetric solid – 4-node elements	953.7	−10.57	–	–
axisymmetric solid – 8-node elements	953.7	−19.67	–	–
axisymmetric shell – solid	593.8	−140.12	–	–
axisymmetric shell – solid	–	−78.63	–	–
shell – solid	1140.0	−205.00	37.70	215.00
shell – solid	16660.0	17976.6	–	–
axisymmetric solid	963.2	−33.73	–	–
shell – solid	1015.7	86.30	–	231.09
axisymmetric shell – solid	949.2	−36.62	–	–
shell – solid	951.3	−38.35	–	–
axisymmetric shell - solid	989.1	−89.11	38.00	238.63

values of the moment at the shell-ring interface ranged between −205 and 17,977 Nm/m. Solution of the Girkmann problem should be a very short exercise for persons having expertise in finite element analysis, yet many of the answers were wildly off. Analysts who cannot find a reasonably accurate solution for the Girkmann problem are not in a position to claim that they can solve much more complicated problems reliably.

Several respondents were not sure how the shell-ring interface should be treated. One respondent stated the problem in this way: *"Since the (axisymmetric) shell elements have three degrees of freedom per node, while the (axisymmetric) ring elements have only two, it is not clear how to reconcile this difference, yet getting this step wrong will give incorrect results that may not be obvious."* This is a typical finite element modeling dilemma one encounters when different types of elements have to be joined.

The results obtained by means of numerical simulation codes are listed in Table 5.2. These codes provide a posteriori error estimation, an essential technical requirement in numerical simulation.

The note "Extraction" in the second entry in Table 5.2 refers to the use of extraction functions for the computation of the shear force and bending moment at the shell-ring interface [99] from the finite element solution. This procedure is analogous to the indirect computation of QoI discussed in Section 1.4.1. The last two entries in Table 5.2 are based on work performed after the completion of the Girkmann round robin study [66].

An interesting and surprising result of this study was that the bending moment M_α obtained by Girkmann, which also appeared in translation in [104], differed by a factor of nearly 3 from the verified solutions obtained with the numerical simulation codes, see Tables 5.2 and 5.3. The reasons for this were investigated by Pitkäranta et al. [78]. It was found that assumptions made by Girkmann, namely that the resultant of the distributed membrane force and the resultant of the distributed axial reaction force pass through the centroid of the footring, were primarily responsible for this discrepancy. The designation: M-B-RE model[15] in Table 5.3 refers a particular adaptation of the classical membrane theory, with bending theory used for accounting for the boundary layer effects at the shell-ring interface, and an engineering theory (minimal-energy model) used for modeling the ring.

Table 5.2 Numerical simulation codes: Summary of results.

Model	Q_α N/m	M_α Nm/m	φ_{max} degrees	M_{max} Nm/m
axisymmetric solid	934.5	−34.81	−	−
axisymmetric solid – Extraction [99]	943.6	−36.81	38.15	255.10
axisymmetric solid	940.9	−36.63	38.20	254.92
thin[15]	948.4	−37.31	38.20	254.50
axisymmetric solid	940.9	−36.80	38.15	254.80
axisymmetric solid *hp* version [66]	943.7	−36.79	38.14	254.90
axisymmetric shell-ring *h* version [66]	942.4	−37.36	38.14	254.10

---

15 Membrane theory (M), bending theory (B) and energy-based ring theory (RE).

Table 5.3 Solution by classical methods: Summary of results.

Method	Q_α N/m	M_α Nm/m
Girkmann [39, 104]	1007.4	−110.5
Pitkäranta (M-B-RE model) [78]	944.9	−36.67

Remark 5.4 According to the problem statement, the shell is made of reinforced concrete, yet it is assumed to be homogeneous and isotropic. While the two statements are contradictory, this is a commonly used idealization in engineering practice. The inhomogeneity and anisotropy caused by the reinforcement are neglected in structural analysis but taken into consideration in strength analysis. In structural analysis problems the QoI are stress resultants, displacements and natural frequencies. These QoI can be well approximated by homogenized models.

Remark 5.5 In this example we were not concerned with the question of whether the mathematical problem is a realistic model of a spherical shell made of reinforced concrete and supported by a footring, given the quantities of interest. We were concerned only with the verification of the numerical solution of this mathematical problem in terms of the QoI.

5.2.7 Example: Fastened structural connection

In this example we analyze the fastened structural connection (lug) shown in Fig. 5.7 with the goal to determine the shear force imposed by each fastener on the lug and the maximum tensile stress in the lug in the vicinity of the fastener holes. We will be interested in evaluating the effects of various modeling assumptions on the quantities of interest. We will consider (a) a model based on the strength of materials, requiring only simple hand calculations, (b) models based on two assumptions concerning the boundary conditions at the fastener holes and (c) a finite element model.

The dimensions (mm) are as follows: $a = 75$, $b = 55$, $s_1 = 32.5$, $s_2 = 40$, $d_1 = 17.5$, $d_2 = 17.5$, $r_h = 3.2$, $r_1 = 25$, $r_2 = 12.5$, the thickness of the lug is $t = 6.4$ (constant). The lug is made of 2014-T6 aluminum. The elastic properties are: $E = 7.52 \times 10^4$ MPa, $v = 0.397$. Let $F = 12.0$ kN, $\alpha = 35°$.

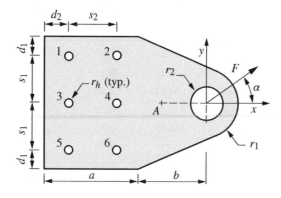

Figure 5.7 Lug problem. Notation.

Model 1: Strength of materials

We estimate the direction and magnitude of the force acting on the plate at each fastener hole by hand calculation using the model described in Appendix J.

In this model the mathematical problem is simplified by assuming that the lug is a perfectly rigid body which undergoes displacement and rotation allowed by the elasticity of the fasteners which is simulated by linear springs. Assuming that the spring rate of each fastener is the same, and using equations (J.14) and (J.15) (details are given in Appendix J), we find the components of the fastener forces and determine their direction and magnitude. The results are shown in Table 5.4.

The errors in these data are caused entirely by the modeling assumption that the lug is perfectly rigid. There is no approximation error in the magnitude and direction of the forces in the fasteners. The errors are entirely model form errors.

To estimate the maximum stress at the most highly loaded fastener, one can make the modeling assumption that the fasteners are sufficiently far apart so that the maximum stress is not significantly affected by the neighboring fasteners. Therefore a surrogate problem can be solved in which the maximum stress is computed for two fasteners separated by a large distance and loaded by smoothly distributed tractions the resultants of which are equal but opposite forces. This approach is representative of how engineers use simplified models to obtain rough estimates of the QoI.

Specifically, consider the problem shown in Fig. 5.8. This problem is concerned with two fastener holes of radius r_h, the centers of which are located at a distance 2ℓ. The perimeter of the holes is loaded by sinusoidal normal tractions T_n corresponding to unit forces:

$$T_n = \begin{cases} (-2/(r_h t \pi)) \cos \theta & \text{for } |\theta| \leq \pi/2 \\ 0 & \text{for } |\theta| > \pi/2 \end{cases} \tag{5.10}$$

where t is the thickness of the plate. Symmetry boundary conditions are prescribed on line segments AB, CD, DE. The line segments EF and FA are traction free.

Table 5.4 Fastener forces estimated by the method described in Appendix J

	Fastener					
	1	2	3	4	5	6
Force (N)	1675.0	3409.5	1812.3	3479.1	4824.2	5665.0
Angle (deg)	27.6	−64.2	154.7	−118.1	170.8	−147.2

Figure 5.8 Surrogate problem. Notation.

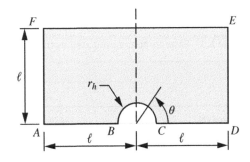

Letting $r_h = 3.2$ mm, $t = 6.4$ mms and solving the problem for ℓ/r_h in the range of 10 to 1000, we find that the maximum value of the principal stress converges to 0.0197 MPa. Therefore the estimated value of the maximum normal stress in the vicinity of the fastener holes is 0.0197 MPa/N. Hence the estimated maximum stress is $0.0197 \times 5665.0 = 111.6$ MPa which occurs on the perimeter of fastener hole #6.

The use of the surrogate problem introduced a second modeling error. This error can be eliminated by applying sinusoidal normal tractions to the fastener holes which are statically equivalent to the bearing forces in Table 5.4. The formula for T_n is the same as in eq. (5.10); however, θ is measured from the line of action of the resultant force.

The maximum tensile stress in the vicinity of the fasteners is found to be 206.7 MPa, located on the perimeter of fastener hole #6. On comparing this result with the estimate from the surrogate problem (111.6 MPa) it is clear that the assumption that the local stress maximum depends only on the force acting on the fastener hole introduces a substantial modeling error.

Remark 5.6 The surrogate problem is analogous to the dipole problem in electrostatics where the electrostatic potential field corresponding to a pair of electric charges of equal magnitude but opposite sign, separated by some distance, is of interest.

Model 2: The fasteners are modeled by linear springs
It is assumed that the normal and shearing tractions at the boundary of each fastener, denoted by T_n and T_t respectively, are

$$T_n = -k_n u_n, \quad T_t = 0 \tag{5.11}$$

where u_n is the normal displacement and k_n is the spring rate, estimated to account for the propping effect[16] of the fastener shank. Assuming plane strain conditions, the spring coefficient, representing the radial stress to displacement ratio of the fastener shank, is estimated from:

$$k_n = \frac{2E}{d(1+v)(1-2v)} \tag{5.12}$$

where E is the modulus of elasticity, d is the diameter and v is Poisson's ratio of the fastener. We assume that the fasteners have the same material properties as the lug. The fastener forces and their directions are shown in Table 5.5.

Since fasteners interact with the lug by mechanical contact, and therefore only compressive tractions can exist between the fasteners and the lug, this model cannot be expected to approximate the fastener forces well. It serves, however, as a necessary first step toward solving the problem discussed next.

Table 5.5 Fastener forces estimated by the linear spring model

	Fastener					
	1	2	3	4	5	6
Force (N)	961.9	3426.5	884.8	3538.2	2363.2	8603.4
Angle (deg)	32.8	−39.0	139.0	−137.6	170.0	−153.4

---

16 The term "propping effect" refers to the resistance of the fastener shank to ovalization of the fastener hole.

Table 5.6 Fastener forces estimated by the nonlinear spring model

	Fastener					
	1	2	3	4	5	6
Force (N)	1228.3	3557.0	1261.0	3593.5	3062.6	7471.1
Angle (deg)	33.1	−49.5	148.2	−132.3	169.3	−152.7

Model 3: The fasteners are modeled by nonlinear springs

This model is the same as Model 2; however, the contact between the fasteners and the lug is represented by springs that act in compression only:

$$T_n = \begin{cases} -k_n u_n & \text{for } u_n > 0 \\ 0 & \text{for } u_n \leq 0 \end{cases} \tag{5.13}$$

therefore the problem is nonlinear and must be solved by iteration. In the iterative solution the first step is to solve Model 2, then continue with cancelation of the tensile spring tractions until a prescribed tolerance is reached. In solving this problem the tolerance τ was set at 0.001. This means that iteration continued until the condition

$$\|\mathbf{a}_k - \mathbf{a}_{k-1}\|_2 \leq \tau \|\mathbf{a}_k\|_2 \tag{5.14}$$

where $\|\mathbf{a}_k\|_2$ is the Euclidean norm[17] of the solution vector at the kth iteration, was reached.

Five iterations were needed. The computed fastener forces are shown in Table 5.6. The maximum stress in the lug occurs on the perimeter of fastener #6. Its value is 254.9 MPa. These results are not sensitive to the spring rate assigned to the fasteners. For example, we find that a 10% reduction in the spring rate causes the maximum principal stress to change by 0.1 N (0.04 %) and the maximum fastener force by 27 N (0.4 %) only.

Model 4: The three-dimensional contact problem

The previously discussed three models were designed to approximate the solution of the three-dimensional multibody contact problem discussed in this section. In this model the lug and fastener shanks are treated as three-dimensional elastic bodies. The fastener shanks are idealized as cylindrical bodies of radius r_h and length ℓ centered on each fastener hole. The boundary conditions at $\ell = \pm t/2$ are $u_x = u_y = T_z = 0$. Frictionless contact condition is prescribed on the cylindrical surface:

$$T_n g = 0 \tag{5.15}$$

where T_n is the normal traction and g is the gap function between the fastener shanks and the lug. The tangential traction, denoted by T_t, is zero. The initial gap function is assumed to be zero. The contact problem is solved by iteratively adjusting T_n until eq. (5.15) is satisfied to within a prescribed tolerance. Denoting the kth increment of T_n by $(\Delta T_n)_k$, the stopping criterion is

$$\max_S (\Delta T_n)_k < \tau \max_S T_n \tag{5.16}$$

where S is the cylindrical surface and τ is the tolerance. For example, letting $\ell = t$ for all fasteners, $\tau = 0.01$ and using the finite element mesh shown in Fig. 5.9(a) with $p = 8$, we get the results shown in Table 5.7.

---

17 See Definition A.3 in the appendix.

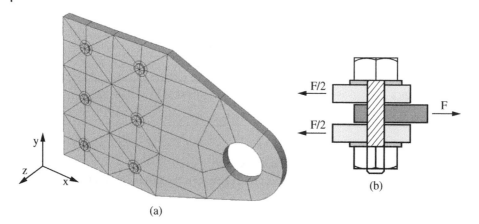

(a)

(b)

Figure 5.9 (a) Three-dimensional contact problem. 50 pentahedral and 108 hexahedral elements. Hand-mesh. (b) Typical bolted connection detail. Double shear.

Table 5.7 Fastener forces estimated by the multibody contact model

	Fastener					
	1	2	3	4	5	6
Force (N)	1102.8	3395.4	1348.6	3511.7	3117.7	7063.5
Angle (deg)	30.1	−49.4	156.0	−132.6	172.9	−153.0

The fastener forces in Table 5.7 were obtained by numerical integration. The error in equilibrium was found to be 0.03%. The estimated maximum stress in the lug is $(\sigma_1)_{max} = 287.4$ MPa which occurs on the perimeter of fastener hole #6. It was verified by p-extension that the relative error in the reported results is less than 1%.

Remark 5.7 In the formulation of models 1 through 4 it was assumed that there is no friction between the fasteners and the lug and between the lug and the plates to which the lug is connected, as shown in Fig. 5.9(b). This assumption is used in the design of shear joints. To test the predictive performance of this model in a validation experiment, the structural connection has to be lubricated. Connections can be designed such that forces are transmitted through friction between the connected plates rather than the shear forces in the fasteners. Such connections are called tension joints.

Discussion

Mathematical models are formulated with the objective to simulate some aspect of physical reality, in the present instance the functioning of a fastened structural connection from the perspective of allowable stress design. The foregoing discussion focused on the definition of four mathematical models. In the case of Model 1 numerical approximation was not involved. In the cases of Models 2, 3 and 4 the numerical approximation errors were verified to be under 1% in the QoI. Therefore the differences in the fastener forces shown in Tables 5.4 through 5.7, as well as the differences in the computed maximum normal stress, are caused by the differences in modeling assumptions.

The credibility of data generated by numerical simulation increases when the number of simplifying assumptions incorporated in a model is decreased. Therefore the results obtained by Model

4 are more credible than the results from the other three models. Note that the complexity of the model increases as the number of simplifying assumptions is decreased. An important question associated with the formulation of any mathematical model is: What is the minimum level of complexity that will make the model good enough to serve its intended purpose? This question[18] cannot be answered without testing the effects of modeling assumptions on the QoI. For example, the maximum stress in the lug, predicted by Model 3 (resp. Model 4), was 254.9 MPa (resp. 288.9 MPa), 11% difference. If this difference is acceptable then Model 3 is good enough for its intended purpose. We could not have assessed this without actually solving the problem using Model 4, however.

Up to this point the model of the lug was treated as a deterministic problem assuming that there is a perfect fit: The holes are exactly the same size as the fastener shanks, there is no gap or interference between the fastener shanks and the lug. In reality there is a substantial uncertainty concerning the quality of the fit: Even under laboratory conditions, care having been taken to achieve a sliding fit by carefully reaming the holes, it was not possible to eliminate the effects of gaps [57]. It is therefore necessary to account for the effects of gaps and interferences. This can be done by modifying the boundary condition given by eq. (5.13) to:

$$T_n = \begin{cases} -k_n(u_n + \delta_n) & \text{for } u_n + \delta_n > 0 \\ 0 & \text{for } u_n + \delta_n \leq 0 \end{cases} \qquad (5.17)$$

where δ_n is the interference or gap, that is, $\delta_n = (d_f - d_h)/2$ where d_f is the diameter of the fastener and d_h is the diameter of the hole. Uncertainties in the values of δ_n are the dominant uncertainties in predicting the distribution of fastener forces and hence the maximum stress in the lug.

Example 5.1 In this example we assume that fasteners 1 through 5 are tightly fitted, that is $\delta_n^{(i)} = 0$ for $i = 1, 2, \ldots 5$, but fastener 6 has a 0.025 mm gap; $\delta_n^{(6)} = -0.025$ mm. We are interested in the distribution of the fastener forces and the maximum principal stress computed by the nonlinear spring model. The results are shown in Table 5.8

On comparing Table 5.8 with Table 5.6, it is seen that the gap at fastener #6 causes a substantial redistribution of forces. The maximum principal stress in the vicinity of the fasteners is 189.2 MPa. It occurs on the perimeter of fastener #6. Therefore there is a substantial change in the maximum stress as well.

The quantities of interest are very sensitive to the quality of fit characterized by $\delta_n^{(i)}$. It would be possible to simulate the response of the structural connection to loads by Monte Carlo simulation. However, the results will be sensitive to the assumptions concerning the statistical distribution of $\delta_n^{(i)}$ which would be difficult to validate.

Table 5.8 Example 5.1. Fastener forces estimated by the nonlinear spring model

	Fastener					
	1	2	3	4	5	6
Force (N)	1560.1	4629.5	2099.2	5848.0	5784.9	2732.4
Angle (deg)	36.6	−54.6	150.7	−131.9	172.0	−146.0

---

18 This question is related to the law of parsimony which is the philosophical principle that if a problem can be solved by more than one method then one should select the method that involves the fewest assumptions. This principle is attributed to William of Ockham (c. 1287–1347), an English philosopher.

5.2.8 Finite element model

In finite element modeling practice it is not uncommon to ignore the diameter of the fasteners and impose displacement constraints on nodes located where the center of each fastener would be. The fastener forces are then estimated by computing the nodal forces from the finite element solution.

The solution of this problem does not lie in the energy space: If the mesh were progressively refined then the strain energy, as well as the maximum displacement, would increase indefinitely. The fastener forces are estimated by summing those nodal force components that correspond to a constrained vertex over all elements that share that vertex. It will be shown that these fastener forces satisfy the equations of equilibrium exactly, independent of the finite element solution.

The fastener forces for the 2227-element mesh of six node triangles, shown in Fig. 5.10, are listed in Table 5.10. Plane stress is assumed. On comparing the entries in Table 5.6 and Table 5.9 it is seen that the fastener forces are not far apart. This result is a consequence of two large errors nearly canceling one another. One error is conceptual; point constraints are not allowed in two- and three-dimensional elasticity. The other error is numerical; the error of approximation in the fastener forces is large. It is a reasonable conjecture that the fastener forces will converge to the fastener forces computed for Model 1, see Table 5.4.

In order to demonstrate that the displacements diverge, we compute the average displacement of the perimeter of the 25 mm diameter hole (radius r_2) in the direction of the applied force. The average displacement is defined by

$$d_{\text{ave}} = \frac{1}{2\pi} \int_0^{2\pi} (u_x \cos \alpha + u_y \sin \alpha) \, d\theta. \tag{5.18}$$

where α is the angle shown in Fig. 5.7.

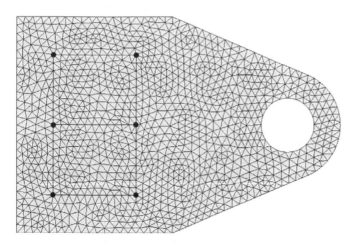

Figure 5.10 Finite element mesh consisting of 2227 triangles. The constrained nodes are indicated by heavy dots.

Table 5.9 Fastener forces estimated by finite element modeling

	Fastener					
	1	2	3	4	5	6
Force (N)	1220.5	3190.9	1175.6	3304.9	3222.8	7760.6
Angle (deg)	29.7	−46.7	146.1	−130.0	171.7	−151.1

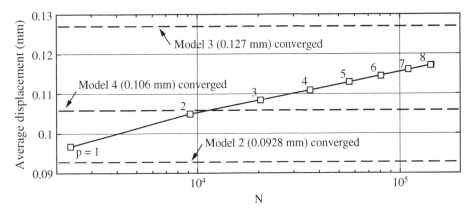

Figure 5.11 Average displacement of the perimeter of the 25 mm diameter hole in the direction of the applied force vs. degrees of freedom.

A sequence of solutions were obtained by increasing the polynomial degree of the elements uniformly from 1 to 8 on the 2227 element mesh shown in Fig. 5.10. The computed values of d_{ave} are plotted against the number of degrees of freedom on a semi-logarithmic graph in Fig. 5.11. Slow (logarithmic) divergence is observed. The exact value of d_{ave} is infinity. In this example the computed value of d_{ave} corresponding to $p = 2$ is 0.105 mm, hence the absolute error is infinity and the relative error is 100%. However, because the displacements diverge very slowly, credible results were obtained with this finite element model: The computed values of d_{ave} fall between d_{ave} predicted by Model 2 and Model 3, see Fig. 5.11. This is an example of near cancelation of two very large errors; the conceptual and the numerical errors.

Remark 5.8 A typical engineering quantity of interest is the spring rate of a structural connection. One possible definition of spring rate is $k = F/d_{ave}$ where F is the applied force indicated in Fig. 5.7. Another definition of spring rate would be $k = F^2/(2U)$ where U is the strain energy. Since $d_{ave} \to \infty$ and $U \to \infty$ as the number of degrees of freedom is increased, the spring rate predicted by the finite element model would converge to zero in both cases if the number of degrees of freedom were progressively increased.

Whereas the displacement and the spring rate of a structural connection could be estimated reasonably well by finite element modeling, the estimated maximum stress in the vicinity of the fasteners, computed from the finite element solution, is very sensitive to the discretization and therefore the probability that a particular discretization would result in near cancelation of the conceptual and numerical errors is very low.

In the specific instance of the finite element mesh shown in Fig. 5.10, using 3-node and 6-node plane stress triangles ($p = 1$ and $p = 2$ respectively), the computed maximum stress values are located at fastener #6 in both cases. However, they differ greatly, as shown in Table 5.10. On the other hand, if we were interested in the strain reading in a point, such as in point A located at $x = -37.0$ mm, $y = 0$ (see Fig. 5.7), denoted by $(\epsilon_x)_A$, then the result from the finite element model would be reasonably close to the result from Model 4 and would in fact converge if the size of the elements were progressively reduced or the polynomial degree increased [20].

The results in Table 5.10 demonstrate that, depending on the quantities of interest, near cancelation of the conceptual and numerical errors may or may not occur in finite element modeling.

Table 5.10 Quantities of interest from finite element modeling and Model 3 (ksi).

Element	N	$(\sigma_1)_{max}$	$(\epsilon_x)_A$
3-node ($p = 1$)	2356	184.0	$2.72E - 4$
6-node ($p = 2$)	9178	468.6	$2.67E - 4$
Model 3		254.7	$2.87E - 4$

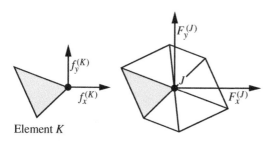

Element K

Figure 5.12 Nodal forces. Notation.

Equilibrium of nodal forces

In this section we show that the nodal forces satisfy the equations of equilibrium exactly, independent of the finite element solution. Therefore equilibrium of nodal forces should not be interpreted to mean that the finite element solution is accurate.

We refer to the example of the point-constrained lug shown in Fig. 5.10 and consider the group of elements that share a constrained node, as shown in Fig. 5.12.

The set of constrained nodes is denoted by $\mathcal{N}$. The set of element numbers that share node $J \in \mathcal{N}$ is denoted by $\mathcal{I}_J$. Uppercase indices refer to global numbering, lowercase indices to local numbering. The polynomial degree assigned to element K is denoted by p_K. Without loss of generality, we assume that the local numbering of vertices is such that vertex 1 is coincident with the constrained node. The inverse of the mapping functions is denoted by

$$\xi = Q_\xi^{(K)}(x, y), \quad \eta = Q_\eta^{(K)}(x, y)$$

and define on element $K \in \mathcal{I}_J$

$$\mathbf{V}_x^{(K)}(x, y) = \{N_1(Q_\xi^{(K)}, Q_\eta^{(K)}) \ 0\}^T, \quad \mathbf{V}_y^{(K)}(x, y) = \{0 \ N_1(Q_\xi^{(K)}, Q_\eta^{(K)})\}^T$$

where $N_1(\xi, \eta)$ is the shape function associated with vertex 1. By definition, the nodal force component $f_x^{(K)}$ is

$$f_x^{(K)} = \sum_{j=1}^{2n_K} k_{1j}^{(K)} a_j^{(K)} = B(\mathbf{u}_{FE}^{(K)}, \mathbf{V}_x^{(K)}), \quad K \in \mathcal{I}_J \tag{5.19}$$

where $k_{ij}^{(K)}$ is an element of the stiffness matrix of finite element K, $a_j^{(K)}$ is the coefficient of shape function j in the local numbering convention,

$$n_K = (p_K + 1)(p_K + 2)/2$$

is the number of shape functions.

By definition, the nodal force component $F_x^{(J)}$ is

$$F_x^{(J)} = \sum_{k \in \mathcal{I}_J} f_x^{(K)} = B(\mathbf{u}_{FE}^{(K)}, \mathbf{v}_x^{(J)}) \tag{5.20}$$

where $\mathbf{v}_x^{(J)}$ defined by

$$\mathbf{v}_x^{(J)} = \sum_{K \in \mathcal{I}_J} \mathbf{v}_x^{(K)} \tag{5.21}$$

is an extraction function for $F_x^{(J)}$. It is a function that lies in the finite element space and is zero on all elements with the exception of elements numbered $K \in \mathcal{I}_J$. Letting $\mathbf{w}_x = \{1\ 0\}^T$ we have

$$B\left(\mathbf{u}_{FE}, \mathbf{w}_x - \sum_{J \in \mathcal{N}} \mathbf{v}_x^{(J)}\right) = r_2 t \int_0^{2\pi} T_x(\theta)d\theta = F\cos\alpha. \tag{5.22}$$

Since $\mathbf{w}_x$ is rigid body displacement we have $B(\mathbf{u}_{FE}, \mathbf{w}_x) = 0$ and, using eq. (5.20), we get the equilibrium equation

$$-\sum_{J \in \mathcal{N}} F_x^{(J)} = F\cos\alpha. \tag{5.23}$$

Analogously letting $\mathbf{w}_y = \{0\ 1\}^T$ we get

$$F_y^{(J)} = \sum_{k \in \mathcal{I}_J} f_y^{(K)} = B(\mathbf{u}_{FE}^{(K)}, \mathbf{v}_y^{(J)}) \tag{5.24}$$

and hence

$$-\sum_{J \in \mathcal{N}} F_y^{(J)} = F\sin\alpha. \tag{5.25}$$

To show that the nodal force components satisfy moment equilibrium, we define the rigid body rotation vector $\mathbf{w}_{\text{rot}} = \{-y\ x\}^T$ and write

$$B\left(\mathbf{u}_{FE}, \mathbf{w}_{\text{rot}} - \sum_{J \in \mathcal{N}}(-y_J \mathbf{v}_x^{(J)} + x_J \mathbf{v}_y^{(J)})\right) = r_2 t \int_0^{2\pi}(-yT_x + xT_y)d\theta \tag{5.26}$$

where x_J, y_J are the coordinates of the Jth node. Since $\mathbf{w}_{\text{rot}}$ is a rigid body rotation, we have $B(\mathbf{u}_{FE}, \mathbf{w}_{\text{rot}}) = 0$ and, using equations (5.20) and (5.24), we get

$$-\sum_{J \in \mathcal{N}}(-y_J F_x^{(J)} + x_J F_y^{(J)}) = r_2 t \int_0^{2\pi}(-yT_x + xT_y)\,d\theta. \tag{5.27}$$

This is the equation of moment equilibrium of the nodal forces and the applied tractions.

Remark 5.9 Whereas the nodal forces depend on the finite element solution, satisfaction of the three equations of equilibrium (5.23), (5.25) and (5.27) does not. Therefore satisfaction of the equations of equilibrium by the nodal forces is not an indication of the quality of the finite element solution.

Remark 5.10 It is possible to show that the nodal forces acting on any subset of elements satisfy the equations of equilibrium. This is used in engineering practice to isolate parts a structure for detailed analysis. An isolated part is visualized as a free body subjected to the nodal forces.

Discussion
The point-constrained lug problem does not have an exact solution, nevertheless predictions of certain data based on finite element approximations of the nonexistent solution can be close to what would be observed in a physical experiment. One of the reasons for this is that observable quantities, for example the displacement of any point, strains measured by strain gages, will give a reasonably close reading to the reading predicted by the finite element model [20].

The second reason is that in some quantities of interest a cancelation of two large errors occur. This is illustrated by the average displacement of the 25 mm diameter hole in Fig. 5.11. This QoI cannot be verified because its exact value is infinity. However, divergence with respect to increasing the number of degrees of freedom, whether by h- or p-extension, is very slow and not easily visible. In the practice of finite element modeling verification of the QoI is rarely performed and the observed displacement can be close to the displacement predicted by the finite element model. This will appear to support the (false) conclusion that the finite element model passed a validation test. Such a conclusion would be false because validation refers to testing the predictive performance of a properly formulated mathematical model. This is possible only if it is first shown that the errors of numerical approximation of the QoI are small. If the QoI corresponding to the exact solution is not finite then the model was not properly formulated and hence cannot be validated.

Since the fastener holes are omitted from the finite element model, this model cannot be used for estimating the maximum tensile stress in the lug in the vicinity of the fastener holes which, according to the problem statement considered here, is a QoI.

Exercise 5.3 Explain why $B(\mathbf{u}_{FE}, \mathbf{w}_{rot}) = 0$ where $\mathbf{w}_{rot} = \{-y\ x\}^T$. Consider two-dimensional elasticity only.

Exercise 5.4 Write down the three displacement vectors that represent infinitesimal rigid body rotations about the x, y and z axes in three-dimensional elasticity.

5.2.9 Example: Coil spring with displacement boundary conditions

The centerline of the coil spring shown in Fig. 5.13 is given by

$$x = r_c \cos\theta \quad -\pi < \theta < 11\pi$$
$$y = r_c \sin\theta \quad -\pi < \theta < 11\pi$$

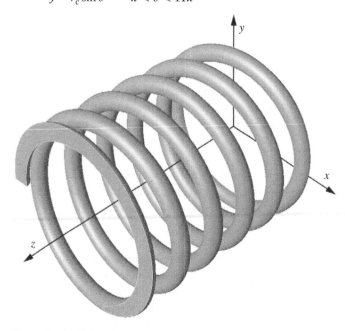

Figure 5.13 Coil spring.

$$z = \begin{cases} 0 & -\pi < \theta \le 0 \\ \theta\, d/(2\pi) & 0 < \theta \le 10\pi \\ 5d & 10\pi < \theta < 11\pi \end{cases}$$

where $r_c = 50.0$ mm is the coil radius, $d = 25.0$ mm is the pitch. The solution domain is such that any section perpendicular to the centerline is circular with radius $r_w = 5.0$ mm (the wire radius). However, the wire is truncated at the ends by the cutting planes $z = 0$, $z = 5d$.

The spring is made of AISI 5160 alloy steel, modulus of elasticity: 200 GPa, Poisson's ratio: 0.285, yield strength: 285 MPa. Assume that the axial displacement u_z at $z = 0$ is zero and at $z = 5d$ it is $u_z = \Delta$, $\Delta < 0$.

The objectives are to determine the spring rates in N/mm units for $\Delta = 0$ and $\Delta = -25$ mm and to verify that the errors of approximation in the reported values are not greater than 3%.

The solution is presented in two parts: First, the solution of the linear model is described. By linear model we understand a problem of linear elasticity: The stress-strain relationship follows Hooke's law and the displacement is sufficiently small so that the equilibrium equations can be written with reference to the undeformed configuration. Second, we solve a geometrically nonlinear model. In this formulation it is assumed that a linear relationship exists between the Cauchy stress and the Almansi strain tensors. For details we refer to Section 9.2.1. In this formulation equilibrium is satisfied in the deformed configuration and the effects of stress stiffening (or stress softening) are taken into account. Solution of the linear problem is the necessary first step toward solving the nonlinear problem.

Solution of the linear model

We compute the strain energy $U(\Delta)$ for an arbitrary fixed value of the displacement Δ imposed on the spring and use $U = k\Delta^2/2$ to determine the spring rate k. Using an automatically generated mesh of 9,691 tetrahedral isoparametric (10-node) elements and letting the polynomial degree p range from 2 to 5 (while keeping the mapping functions fixed), we get the results shown in Table 5.11 where p is the polynomial degree, N is the number of degrees of freedom and the percent error was estimated by extrapolation using the method described in Section 1.5.3.

Using the extrapolated value of U we get

$$k = 2U/\Delta^2 = 20.83 \text{ N/mm}.$$

The estimated relative error in k is less than 0.05%. Note that this error does not account for the curved surfaces being approximated by piecewise quadratic polynomials. However, given the large number of 10-node tetrahedra (see Fig. 5.14), the errors of approximation in the surface representation can be neglected.

Table 5.11 Linear model. Computed values of the strain energy ($\Delta = -25$ mm).

p	N	U (Nmm)	% error
2	58,67	6558.910	0.78
3	173,989	6515.587	0.12
4	385,200	6512.737	0.07
5	721,473	6511.252	0.05
	est. limit	6507.872	–

More generally, the stiffness of the spring can be computed by energy methods. These methods are superconvergent. We assume that the displacements imposed on the boundary at $z = 5d$ can be written as a linear combination of the displacement in the z direction (Δ) and the rotation about the x and y axes, denoted by θ_x and θ_y respectively. The corresponding stress resultants are the axial force F_z and the moments M_x, M_y. Their relationship, determined by energy methods[19], is:

$$\begin{Bmatrix} F_z \\ M_x \\ M_y \end{Bmatrix} = \begin{bmatrix} 20.83 & 0.00 & 69.40 \\ 0.00 & 58073 & 0.00 \\ 69.40 & 0.00 & 58928 \end{bmatrix} \begin{Bmatrix} \Delta_z \\ \theta_x \\ \theta_y \end{Bmatrix}$$

where the units of the numerical data are N/mm, N or Nmm as appropriate for dimensional consistency. It is seen that for the load case $\Delta = -25$ mm, $\theta_x = \theta_y = 0$ we have $F_z = -520.8$ N, $M_x = 0$ and $M_y = -1735$ Nmm. Owing to the end conditions, the line of action of the axial force is not coincident with the z–axis but passes through the point $(x_0,\ y_0)$ where $x_0 = -69.40/20.83 = -3.3$ mm and $y_0 = 0$.

Another method for estimating the spring rate is to compute F_z by integrating the normal stress on the plane surface at $z = 0$ or $z = 5d$ and divide F_z by Δ. This method would converge very slowly, however, because at the edges of the planar surface the stress field is perturbed by singularities. It is much better to "cut" the spring at an arbitrary value of θ by a plane perpendicular to its centerline and compute the resultants of the stress components acting on that plane by numerical integration.

For example, cutting at the mid-point of the spring ($\theta = 5\pi$), we define a local coordinate axis such that the x', y' and z' axes are aligned respectively with the tangent, normal and binormal of the centerline, as shown in Fig. 5.14. The solution at this cut (as well as at any cut away from the ends) has sufficient smoothness to permit accurate determination of the stress resultants by numerical integration.

The computed values of the stress resultants are shown in Table 5.12 for polynomial degrees ranging from 2 to 5. The results show that the dominant stress resultants are the shear force $F_{z'}$ and the twisting moment $M_{x'}$.

The estimated limit values (corresponding to $p \to \infty$) are the stress resultants computed by the superconvergent extraction procedure. The small differences between the estimated limit values and the numerically computed values serve as evidence that the numerically computed values at

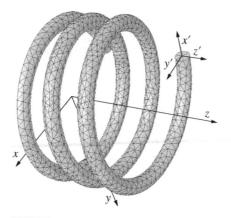

Figure 5.14 Solution domain and finite element mesh in the interval $0 < \theta < 5\pi$.

---

19 The diagonal terms are computed first from the strain energies corresponding to unit values of Δ_z, θ_x, θ_y. Then the off-diagonal terms are computed from the strain energy values corresponding to pairwise imposition of the unit displacement components.

Table 5.12 Linear model. Computed stress resultants in the local coordinate system at $\theta = 5\pi$

p	N	$F_{x'}$	$F_{y'}$	$F_{z'}$	$M_{x'}$	$M_{y'}$	$M_{z'}$
2	58,785	53.1	34.9	−510.7	−24002.0	253.0	1847.6
3	174,046	−43.5	1.0	−521.7	−24220.8	18.8	1919.7
4	385,332	−40.3	1.7	−519.64	−24217.8	10.9	1923.3
5	721,728	−41.9	0.8	−520.0	−24221.2	11.7	1928.7
est. limit		−41.3	0	−519.1	−24228.4	0	1928.0

$p = 5$ are sufficiently accurate and the errors of approximation are well below the specified tolerance of 3%. The development of such evidence, needed for the interpretation of results for the nonlinear problem, is discussed next. For nonlinear problems the resultants have to be computed by integration.

On transforming the force vector in Table 5.12 to the global coordinate system we get $F_x = F_y = 0$, $F_z = -520.7$ N from which $k \approx 20.83$ N/mm. The quantity of interest computed by two different methods has the same value to four significant digits.

Solution of the nonlinear model

In the linear model the equations of equilibrium refer to the undeformed configuration. In the case of this spring the deformed configuration may differ from the original configuration to such an extent that the difference between the two configurations cannot be neglected.

Since the deformed configuration is not known at the outset, we start with the linear model. We update the mapping of each element by adding the linear displacement vector to the position vector corresponding to the original configuration and take into account the nonlinear part of the strain tensor[20]. These steps are repeated until the stopping criterion is satisfied. The stopping criterion is based on the change in value of the apparent potential energy Π_i defined by

$$\Pi_i = \frac{1}{2}\mathbf{x}_i^T[K_i]\mathbf{x}_i - \mathbf{x}_i^T\mathbf{r}_i$$

where the index i refers to the ith iteration, $\mathbf{x}_i$ is the solution vector, $[K_i]$ is the stiffness matrix and $\mathbf{r}_i$ is the load vector. Iteration is stopped when the normalized change in Π_i, denoted by τ and defined by

$$\tau = \sqrt{|\Pi_i - \Pi_{i-1}|/|\Pi_i|} \tag{5.28}$$

is less than a specified value τ_{stop}. In this example $\tau_{\text{stop}} = 0.005$ was used.

The computed stress resultants for $\Delta = -25$ mm in the global coordinate system are listed in Table 2. The first group of entries with p ranging from 2 to 5 were computed using the same 9,695 element mesh on which the linear solution was based. The last entry in Table 5.13, set in italic, was computed using a finer mesh, consisting of 20,907 10-node tetrahedral elements. It is seen that the quantity of interest (F_z) converges to three digits accuracy, yielding the estimate for the spring rate: $k = 20.8$ at $\Delta = -25$ mm.

---

20 Depending on the problem, an incremental procedure may have to be used. In the present case it was not necessary to apply the imposed displacement incrementally.

Table 5.13 Nonlinear model, $\Delta = -25$ mm. Computed stress resultants in the global coordinate system at $\theta = 5\pi$

p	N	F_x	F_y	F_z	M_x	M_y	M_z
2	58,785	33.4	−91.5	−505.0	5070.0	502.0	4488.0
3	174,046	1.1	2.0	−521.5	178.9	−1719.1	−108.0
4	385,332	1.7	−1.0	−519.3	325.7	−1578.8	48.3
5	721,728	0.8	0.5	−519.8	253.6	−1646.0	−21.3
5	1,521,598	0.1	−0.1	−520.4	276.0	−1674.4	4.1

Since we have considered geometric nonlinearities only, it is necessary to ascertain that the material remains elastic in the range $-25 < \Delta < 0$. On computing the maximum von Mises stress, we find that its estimated value is 278.8 MPa which is below the yield point of 285 MPa.

In estimating the maximum von Mises stress the boundaries at $z = 0$ and $z = 5d$ were excluded. This is because the idealized boundary conditions induce singularities along the curved edges. In a physical experiment the displacement would have to be imposed by mechanical contact and therefore the displacement on the boundaries would not be perfectly constant. These singularities are artifacts of the modeling assumptions. Solving the contact problem would have introduced other kinds of local perturbation.

Discussion

Imposing constant normal displacements is one of many possible idealizations of the displacement boundary condition. Such boundary conditions are called "hard" boundary condition. The "soft" displacement boundary condition would be to set the integral of the normal displacements to be a fixed value:

$$\int_{\partial\Omega^+} u_z \, dxdy = \Delta, \quad \int_{\partial\Omega^-} u_z \, dxdy = 0$$

and the integral of rotations about the x and y axes to zero:

$$\int_{\partial\Omega^+} xu_z \, dxdy = \int_{\partial\Omega^-} xu_z \, dxdy = \int_{\partial\Omega^+} yu_z \, dxdy = \int_{\partial\Omega^-} yu_z \, dxdy = 0$$

where $\partial\Omega^+$ and $\partial\Omega^-$ refer to the planar surfaces at $z = 5d$ and $z = 0$ respectively. Between the hard and soft boundary conditions are the "semisoft" boundary conditions corresponding to the nth moment of the displacement u_z ($n = 2, 3 \ldots$) being set to zero.

The choice of boundary conditions is a modeling decision. Assessment of the sensitivity of the quantities of interest to the choice of boundary conditions is one of the tasks in formulating a mathematical model.

5.2.10 Example: Coil spring segment

In this section we analyze a segment of the coil spring described in Example 5.2.9 assuming that the spring is compressed by two equal and opposite forces acting along the z axis. The goals are to estimate the spring rate and the distribution of the von Mises stress.

The spring segment $0 < \theta < \pi/3$ is shown in Fig. K.4 (in Appendix K) together with the 18-element mesh used in this example. The imposition of statically equivalent forces and moments on sections A and B in the local coordinates shown in Fig. K.4 is described in Section K.6. The distribution of tractions corresponding to the technical formulas for rods is described in Section K.6.1 and in Example K.4. These tractions satisfy the equations of equilibrium, hence only rigid body constraints have to be applied.

Solution
The contours of the von Mises stress (MPa) on a 60 degree segment of the coil spring corresponding to axial force $F = 1.0$ N are shown in Fig. 5.15. As explained under Remark K.4, the imposition of tractions through technical formulas induce local perturbations of the solution that decay in accordance with Saint-Venant's principle. This is visible in Fig. 5.15.

In order to minimize the error caused by local perturbations of the solution, we estimate the strain energy of the coil from the strain energy of the set of 6 elements labeled B in Fig. 5.15 that cover a 20-degree segment of the coil. The results are presented in Table 5.14.

Let us compare the spring rate computed in Section 5.2.9 with the spring rate predicted by the current model. In this case we use $k = F^2/(2U)$ where $F = 1.0$ N and $U = 5 \times 18 \times 2.945412E - 4$ Nmm to get $k = 18.86$ N/mm. The difference in spring rates is 9.45%. Since the numerical error in the computed strain energy is negligibly small in both cases, the difference is caused by the differences in modeling assumptions.

The solution of this problem by classical methods, based on work by Wahl [109], is available in engineering handbooks, such as in [111]. Using the notation introduced in Section 5.2.9, the displacement of a coil spring with n active turns is estimated to be

$$\Delta_W = \frac{4Fr_c^3 n}{Gr_w^4}\left(1 - \frac{3}{16}\left(\frac{r_w}{r_c}\right)^2 + \frac{3+v}{2(1+v)}\left(\frac{d}{2\pi r_c}\right)^2\right) \tag{5.29}$$

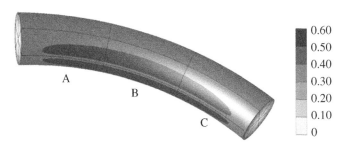

Figure 5.15 Contours of the von Mises stress (MPa) on a 60 degree segment of the coil spring corresponding to axial force $F = 1.0$ N.

Table 5.14 Linear model of the coil segment. Computed values of the strain energy on a 20 degree segment ($F = 1.0$ N)

p	N	U (Nmm)	% error
5	1152	2.945385E-4	0.00
6	1674	2.945397E-4	0.00
7	2340	2.945407E-4	0.00
8	3168	2.945410E-4	0.00
∞	∞	2.945412E-4	–

where F is the axial force and G is the modulus of rigidity. On substituting the parameter values defined in Section 5.2.9, we find the estimated spring rate to be $k_w = 19.33$ N/mm.

The spring rate predicted by the classical formula is greater than what is predicted by solving the three-dimensional problem of elasticity. This can be expected because the curved bar model, on which the classical formula is based, imposes restrictions on the modes of deformation, i.e. the space $E(\Omega)$. Therefore the minimum value of the potential energy for the bar model has to be larger than for the 3D elasticity model. Consequently the bar model underestimates the strain energy of the 3D model and therefore overestimates the spring rate, in this case by 2.5%.

The estimated value of the maximum shearing stress is given by the formula [111]:

$$\tau_{max} = \tau_{nom} \left(1 + \frac{5}{4} \frac{r_w}{r_c} + \frac{7}{8} \left(\frac{r_w}{r_c} \right)^2 \right) \tag{5.30}$$

where

$$\tau_{nom} \equiv \frac{2 F r_c}{\pi r_w^3}. \tag{5.31}$$

On substituting the parameter values defined in Section 5.2.9 we find $\tau_{max}/\tau_{nom} = 1.134$. The three-dimensional finite element solution converges to $\tau_{max}/\tau_{nom} = 1.164$ a difference of 2.6%. We see that the classical formulas give very reasonable estimates for the spring stiffness and the maximum shear stress.

Discussion

The formulation described in this section has the advantage that it is suitable for casting it into the form of a "smart application". Smart applications, also called "smart apps", are expert-designed, user-friendly software products that allow users, who do not need to be trained analysts, to explore design options within a parameter space. The formulation described in this section would allow users to estimate the spring constant, given the parameters of the spring and the elastic properties of the material, without consideration of the effects of constraints imposed on the ends of the spring.

6

Calibration, validation and ranking

This chapter is concerned with the formulation, calibration, validation and ranking of mathematical models with reference to a classical problem of mechanical engineering; that of predicting fatigue failure in metallic machine components and structural elements subjected to alternating loads in the high cycle regime ($> 10^4$ cycles). The mathematical model considered here comprises three sub-models: (a) a problem of linear elasticity, (b) a predictor of fatigue failure defined on elastic stress fields and (c) a statistical model.

In fatigue experiments performed on test coupons (that may be notch-free or notched) special stress fields are imposed on the material. Axial tension-tension or tension-compression tests or bending tests of rotating round bars are the most common types of fatigue tests. There are various ways to generalize those special stress fields to arbitrary stress fields. Predictors of fatigue life are formulated for that purpose. The formulation of predictors is based on intuition and experience. Various predictors have been and can yet be proposed. The question of how and why one would select a predictor from among competing predictors is addressed in this chapter.

The following discussion is focused on the calibration and validation of predictors. This involves the use of data analysis procedures. For readers who are not familiar with those procedures, a brief overview is provided in Appendix I. A description of the experimental data used in this chapter and statistical characterization of data collected from fatigue tests of notch-free coupons are also available in Appendix I.

All numerically computed quantities of interest (QoI) have been verified to ensure that numerical errors in the QoI are negligible in comparison with the uncertainties associated with the physical experiments, such as uncertainties in loading conditions, dimensioning tolerances and residual stresses arising from metal forming operations (such as rolling, drawing and forging) and machining operations (such as milling, boring and turning).

6.1 Fatigue data

Fatigue data are collected from force- or displacement-controlled tests of smooth and notched test specimens. Fatigue testing is governed by standards such as ASTM Standards E466 and E606 [7], [8].

The specimens are subjected to a stress field the amplitude of which is a periodic function. The ratio of the stress components is fixed. Let $\sigma_{\max}$ (resp. $\sigma_{\min}$) be the maximum (resp. minimum)

Finite Element Analysis: Method, Verification and Validation, Second Edition. Barna Szabó and Ivo Babuška.
© 2021 John Wiley & Sons, Inc. Published 2021 by John Wiley & Sons, Inc.
Companion Website: www.wiley.com/go/szabo/finite_element_analysis

principal stress. The ratio $R = (\sigma_{min}/\sigma_{max})$ is called the cycle ratio. The stress amplitude is denoted by σ_a. By definition:

$$\sigma_a = \frac{\sigma_{max} - \sigma_{min}}{2} \equiv \sigma_{max} \frac{1 - R}{2}. \tag{6.1}$$

The mean stress is denoted by σ_m. By definition:

$$\sigma_m = \frac{\sigma_{max} + \sigma_{min}}{2} \equiv \sigma_{max} \frac{1 + R}{2}. \tag{6.2}$$

Under sinusoidal loading the stress field in a test coupon is

$$\sigma_{ij}(\mathbf{x}, t) = \sigma_{max}(\mathbf{x}_0) \left(\frac{1 + R}{2} + \frac{1 - R}{2} \sin(2\pi ft) \right) \Sigma_{ij}(\mathbf{x}) \tag{6.3}$$

where $\mathbf{x}$ is the position vector, $\sigma_{max}(\mathbf{x}_0)$ is the maximum value of a reference stress in point $\mathbf{x}_0$, t is time (s), f is the frequency (Hz), $\Sigma_{ij}(\mathbf{x})$ is a dimensionless symmetric matrix the elements of which are the ratios between the stress components and $\sigma_{max}(\mathbf{x}_0)$. For example, in specimens where the stress is uniaxial and substantially constant within the test section the elements of $\Sigma_{ij}(\mathbf{x})$ are zero except for $\Sigma_{11}(\mathbf{x}) = 1$. Such a specimen is shown in Fig. 6.1. This type of specimen was used in experiments reported in [40].

In specimens subjected to pure shear $\tau(t)$ we have $\sigma_{max} = \tau_{max}$ the elements of $\Sigma_{ij}(\mathbf{x})$ are zero, except for $\Sigma_{12}(\mathbf{x}) = \Sigma_{21}(\mathbf{x}) = 1$.

The test records contain the following information: (a) specimen label, (b) specimen geometry, (c) the maximum test stress in the test section S_{max}, (d) the cycle ratio R, (e) the number of cycles at the end of the test n and (f) notation indicating whether failure occurred outside of the test section or the test was stopped prior to failure (runout). Typical fatigue test data are shown in Appendix I, Fig. 1.2.

Remark 6.1 More than one type of load such as shear, bending and axial load may be applied simultaneously. Cyclic loads may be acting in-phase or out of phase.

6.1.1 Equivalent stress

In many important applications one stress component is so dominant that all other stress components are negligible in comparison. The family of predictors

$$\sigma_{eq} = \sigma_{max}^{1-c} \, \sigma_a^c, \quad 0 < c < 1 \tag{6.4}$$

where σ_{eq} is called equivalent stress, is widely used in engineering practice for the special case of uniaxial stress. Substituting eq. (6.1) we get:

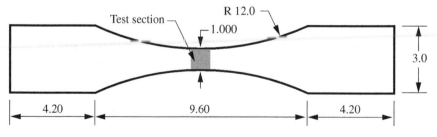

Figure 6.1 Notch-free test coupon. The dimensions are in inches. Thickness: 0.090 in. The test section is within $\pm 1/2$ inches from the minimum cross-section.

$$\sigma_{eq} = \sigma_{max} \left(\frac{1-R}{2} \right)^c . \tag{6.5}$$

The predictor with $c = 1/2$ is one of the predictors proposed in [88]. We will also use $c = 1/2$ in the following. In fatigue tests σ_{eq} is fixed and the number of cycles (n) at which failure occurs is recorded. The set of data $(n_i, \sigma_{eq}^{(i)})$, $i = 1, 2, \ldots$ recorded in fatigue tests where the first principal stress in the test section is either constant or varies with a small gradient is called the S-N data. For example, S-N data may be collected from rotating round bars, the test section of which is loaded by a constant bending moment. In that case the principal stress varies linearly between its maximum and minimum value and, since the test article is subjected to full stress reversal ($R = -1$), $\sigma_{eq} = \sigma_{max}$.

Remark 6.2   The definition of equivalent stress is not unique. Consider, for example, the following definition of equivalent stress:

$$\sigma_{eq}^{(\alpha)} = \left(\alpha \sigma_{max} + (1 - \alpha)\overline{\sigma}_{max} \right) \left(\frac{1-R}{2} \right)^c , \quad 0 \leq \alpha \leq 1 \tag{6.6}$$

where $\overline{\sigma}_{max}$ is the maximum von Mises stress. If the stress is uniaxial then the definitions given by equations (6.5) and (6.6) are identical for any alpha. However, the two definitions differ for any other stress condition.

6.1.2   Statistical models

Plotting the values of σ_{eq} at which fatigue failure occurs in notch-free stress coupons subjected to cyclic loading against $\log_{10} n$ results in a cluster of points that can be well approximated by a curve, called the S-N curve, see for example Fig. 1.2 in Appendix I.

The S-N curve will be understood to be the median of the statistical distribution of $\log_{10} n$, given σ_{eq}, and will be denoted by $\mu(\sigma_{eq})$. Many plausible statistical models can be formulated to represent the population from which the S-N data were supposedly sampled. We will use one statistical model for this purpose. This model belongs to the family of random fatigue limit models [73] and is based on the following assumptions:

1. The statistical distribution of $\log_{10} n$ is normal with mean $\mu(\sigma_{eq})$ and constant standard deviation s.
2. The functional form of the mean is:

$$\mu(\sigma_{eq}) = A_1 - A_2 \log_{10}(\sigma_{eq} - A_3), \quad \sigma_{eq} - A_3 > 0 \tag{6.7}$$

   where the parameter A_3 is called the fatigue limit or endurance limit.
3. The fatigue limit is a random variable and $\log_{10} A_3$ has normal distribution with mean μ_f and standard deviation s_f.

Statistical models differ by the assumptions concerning the probability density function of $\log_{10} n$, given σ_{eq}, the choice of the functional form $\mu(\sigma_{eq})$, the definition of σ_{eq}, the functional form of the standard deviation and, when A_3 is treated as a random variable, the functional form of the probability density function representing the dispersion of A_3 and so on.

The parameters of statistical models are estimated by the maximum likelihood method which is described and illustrated by examples for two functional forms of $\mu(\sigma_{eq})$ in Appendix I (Section I.3). Estimation of parameters for statistical models where c is one of the model parameters and $\log_{10} A_3$ has either normal or smallest extreme value (sev) distribution is described in [17].

Since countless plausible statistical models can be formulated, it is necessary to have a procedure by which the relative merit of those models can be evaluated with respect to the available data. Several such methods are described and illustrated by examples in [17]. We will use the Bayes factor defined in Appendix I for that purpose.

The five parameters corresponding to the random fatigue limit model defined above are displayed in Table I.5 in the appendix. These parameters maximize the likelihood function for the S-N data shown in Fig. I.2.

Remark 6.3   It was assumed that the crack initiation process consumed substantially all of the fatigue lives of the test coupons and once a small crack formed, it propagated very quickly, so that the number of cycles to failure was substantially the same as the number of cycles at the end of the initiation process. This assumption was justified by the results of an investigation (not detailed here) of crack growth rates in a 2024-T3 aluminum sheet assuming that a 0.05 inch (1.3 mm) crack had formed by an initiation process.

6.1.3   The effect of notches

The effect of notches, fillets, keyways, oil holes and other stress raisers on the fatigue life of mechanical components had been extensively investigated well before computers came to be used for the solution of stress concentration problems. We mention only the seminal work of Neuber[1] [65] and Peterson[2] [74] in this regard and note that a very rich technical literature exists on this important subject. For a survey we refer to [80].

Designers of machine elements relied on handbooks and design manuals where they could look up the stress concentration factors associated with various stress raisers. By definition, the stress concentration factor is the ratio of the maximum stress σ_{max} to a reference stress σ_{ref}:

$$K_t = \sigma_{max}/\sigma_{ref}. \tag{6.8}$$

Knowing σ_{max} and R, the equivalent stress can be calculated from eq. (6.5) and, using σ_{eq}, the expected value of the fatigue life can be estimated from the S-N curve under constant cycle loading.

It was found that this method would significantly underestimate the fatigue limit for notched specimens when the stress gradient is steep, as in the neighborhood of notch roots. This is illustrated in Fig. 6.2 where the results of fatigue experiments performed on nine notched specimen types are displayed. The curve $\mu(\sigma_{eq})$ is the mean of the random fatigue limit model[3] calibrated on the basis of S-N data obtained for notch-free coupons.

6.1.4   Formulation of predictors of fatigue life

The physical processes that result in the formation of cracks caused by cyclic loading are highly complex, irreversible processes on the microscopic scale. These processes violate the assumptions of small strain continuum mechanics which include the assumption that a body can be subdivided into arbitrarily small volumes, the properties of which remain the same as those of the bulk material. Nevertheless, the predictors used for correlating the formation of cracks with the number of load cycles in high cycle fatigue are usually determined from the solutions of problems of linear elasticity.

---

1 Heinz August Paul Neuber 1906–1989.
2 Rudolph Earl Peterson 1901–1982.
3 Since by assumption $\log_{10} n$ is normal, the mean and the median of $\log_{10} n$ are coincident.

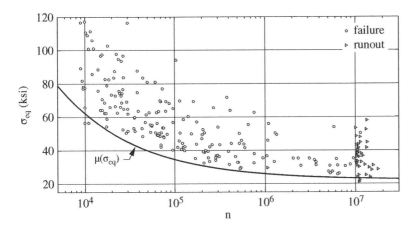

Figure 6.2 The results of fatigue experiments performed on nine notched specimen types. See Table I.2 in Appendix I for details.

This apparent contradiction is resolved through the assumptions that (a) the stresses (or strains) computed from the solutions of problems based on the linear theory of elasticity represent average stresses or strains over representative volume elements[4] (RVEs) rather than pointwise stresses, (b) those average stresses can be correlated with crack initiation events through suitably chosen predictors of fatigue failure and (c) the possibility that small amounts of plastic deformation may occur is not ruled out. However, it is assumed that any plastic zone is surrounded by elastic material and therefore the elastic stress field drives the process that causes fatigue failure to occur. Plastic strains are necessarily small.

Predictors of fatigue life and statistical models have to be formulated for the generalization of S-N data to multiaxial stresses and conditions where the stresses have steep gradients, such as in the vicinity of notches. These are phenomenological models that have to be calibrated for each material.

6.2 The predictors of Peterson and Neuber

The fatigue stress concentration factor K_f was introduced in order to correct for the difference between predictions based on the maximum equivalent stress and observations of the number of cycles to failure in notched specimens. By its original definition, K_f is the fatigue limit of a notch-free specimen divided by the fatigue limit of a notched specimen. However, it will be shown that K_f can be also used for predicting fatigue life at stress levels higher than the fatigue limit in the high cycle range:

$$\sigma'_{\max} = K_f \sigma_{\text{ref}}. \tag{6.9}$$

The relationship between K_f and K_t was established by the formula

$$K_f = 1 + q(K_t - 1) \tag{6.10}$$

---

4 By definition, the representative volume element is the smallest volume over which the averaged material properties are substantially the same as the averaged material properties over the entire volume.

where q is called notch sensitivity factor. Peterson proposed the following empirical formula for q:

$$q = \frac{1}{1 + a/r} \tag{6.11}$$

were a is a material constant and r is the notch radius. An alternative empirical formula, based on Neuber's work on stress concentrations, was employed by Kuhn and Hardrath in [54]:

$$q_N = \frac{1}{1 + f(\omega)\sqrt{A/r}}, \quad f(\omega) = \frac{\pi}{\pi - \omega} \tag{6.12}$$

where ω is the flank angle of the notch and A is a material parameter.

Equation (6.9) is an example of an intuitively constructed predictor of fatigue failure caused by cyclic loading. Neuber and Peterson defined K_f differently. This raises the question: Whose definition is better? It is possible to show that for a sufficiently small interval of notch radii both work well but for a large interval neither does. These predictors are based on the following assumptions:

1. The notch sensitivity factor is characterized by a single geometric parameter, the notch radius. This assumption is applicable to notches and fillets in machine components but is not applicable to corrosion pits, scratches and other surface defects caused by wear, the kind of damage that has to be considered when making condition-based maintenance decisions.
2. The parameters a and A are material parameters that must be determined by calibration.
3. The parameters a and A are strongly correlated with the ultimate tensile strength of the material.

Remark 6.4 Reduction of the peak stress using equations (6.10) to (6.12) is closely related to averaging the normal stress over a material-dependent distance a.

Let us consider, for example, a circular hole of radius r_0 in an infinite plate subjected to constant stress σ_∞ acting in the x-direction. Let (r, θ) be polar coordinates with the origin in the center of the circle with θ measured from the positive x-axis. Then at $\theta = \pm\pi/2$;

$$\sigma_\theta = \sigma_x = \sigma_\infty \left(1 + \frac{1}{2}\frac{r_0^2}{r^2} + \frac{3}{2}\frac{r_0^4}{r^4} \right) \tag{6.13}$$

see, for example, [105]. The maximum stress is at $r = r_0$: $\sigma_{max} = 3\sigma_\infty$, that is, $K_t = 3$. Therefore, using equations (6.9) through (6.11), we get

$$\sigma'_{max} = K_f \sigma_\infty = \left(1 + \frac{K_t - 1}{1 + a/r_0} \right) \sigma_\infty = \left(1 + \frac{2}{1 + a/r_0} \right) \sigma_\infty. \tag{6.14}$$

The average stress over the interval defined by the endpoints $(r_0, \pi/2)$ and $(r_0 + a, \pi/2)$ is

$$\begin{aligned}
\left(\sigma_x \right)_{ave} &= \frac{1}{a} \int_{r_0}^{r_0+a} \sigma_x(r, \pi/2)\, dr = \frac{\sigma_\infty}{a} \int_{r_0}^{r_0+a} \left(1 + \frac{1}{2}\frac{r_0^2}{r^2} + \frac{3}{2}\frac{r_0^4}{r^4} \right) dr \\
&= \left(1 + \frac{1}{2}\frac{1}{1 + a/r_0} + \frac{3}{2}\frac{1 + a/r_0 + a^2/(3r_0^2)}{(1 + a/r_0)^3} \right) \sigma_\infty \\
&= \left(1 + \frac{2}{1 + a/r_0} - \frac{3}{2}\frac{a/r_0 + 2a^2/(3r_0^2)}{(1 + a/r_0)^3} \right) \sigma_\infty \\
&= \left(1 + \frac{2}{1 + a/r_0} - O(a/r_0) \right) \sigma_\infty.
\end{aligned} \tag{6.15}$$

On comparing eq. (6.14) with eq. (6.15) it is seen that the notch sensitivity factors differ by a term which is of order a/r_0. When a is much smaller than r_0 then this term can be neglected.

6.2.1 The effect of notches – calibration

Calibration of Peterson's notch sensitivity factor is demonstrated in the following with the aid of the fatigue test data summarized in Table I.2 in the appendix. We will use test records for six notched specimen types for estimating parameter a in eq. (6.11). These test records correspond to line items 2 through 7 in Table I.2 and will be called the calibration set. The remaining test records, called the validation set, will be used for demonstrating the validation process. This will emulate the process by which test data are collected over time and the calibration and ranking of models are updated as new data become available. For example, we use data from reports [40] to [44] collected over an eight-year period. Suppose that we would have formulated a predictor when only half of the data were available. We would have had to update and possibly revise the predictor in the light of new information. In industrial and research organizations the management of activities of this kind fall under the purview of simulation governance.

Not all test records summarized in Table I.2 conform with the assumptions on which Peterson's formula is based. For example, one test record for the edge-notched specimen with $r = 0.0313$ inch indicates that $\sigma_{max} = 247.8$ ksi, $\sigma_{min} = 102.1$ ksi and failure occurred at 4500 cycles. The yield strength of this material is approximately 54 ksi hence the plastic strains are not small (of the order of 5%). This test is clearly outside of the scope of Peterson's formula.

The available S-N data is in the range of $8500 < n_f < 10^7$ where n_f is the number of cycles at failure. Since we are interested in generalizing these data to notched specimens, we exclude from consideration notched specimen records that failed below 8500 cycles. The total number of records available for the six specimen types in the calibration set are shown under the heading N in Table 6.1. The number of failed specimens and the number of runouts in the retained test records, called qualified test records, are shown under the headings N_f and N_r respectively.

Using equations (6.8) through (6.10) for each specimen type we have

$$\frac{\sigma'_{max}}{\sigma_{max}} = \frac{\sigma'_{eq}}{\sigma_{eq}} = \frac{1 + q_k(K_t - 1)}{K_t}, \qquad q_k = q(r_k) \tag{6.16}$$

where σ'_{eq} is defined analogously to σ_{eq}:

$$\sigma'_{eq} = \sigma'_{max}\left(\frac{1-R}{2}\right)^{1/2}. \tag{6.17}$$

We denote the ratio σ'_{eq}/σ_{eq} by x_k and observe that x_k depends on K_t and q_k only. Therefore x_k is a characterizing parameter for the kth specimen type made of the same material. Given a set of independent test records $(n_i, \sigma_{eq}^{(i)})$, $(i = 1, 2, \dots, m_k)$ for the kth notched specimen type we seek x_k

Table 6.1 Calibration of Peterson's material parameter a based on the random fatigue limit model.

k	Specimen	r_k (in)	K_t	N	N_f	N_r	x_k	q_k	a_k (in)
2	Open hole	1.5000	2.11	39	25	5	0.7943	0.6091	0.9628
3	Edge notch	0.3175	2.17	42	27	4	0.7601	0.5550	0.2546
4	Fillet notch	0.1736	2.19	32	23	5	0.7159	0.4772	0.1902
5	Edge notch	0.0570	4.43	34	20	5	0.6664	0.5691	0.0431
6	Fillet notch	0.0195	4.83	36	23	4	0.5198	0.3945	0.0299
7	Edge notch	0.0313	5.83	46	22	6	0.5490	0.4557	0.0374

such that the set of data $(n_i, x_k \sigma_{eq}^{(i)})$ maximizes the log likelihood function of the random fatigue limit model defined in Section I.3 in the appendix:

$$
LL_k(x_k | \hat{\theta}_3) = \sum_{i=1}^{m_k} \left[(1 - \delta_i) \ln \left(\phi_M(w_i, x_k \sigma_{eq}^{(i)}) \right) \right.
$$
$$
\left. + \delta_i \ln \left(1 - \Phi_M(w_i, x_k \sigma_{eq}^{(i)}) \right) \right]
\tag{6.18}
$$

where $\hat{\theta}_3$ is the vector of the five statistical parameters that characterize the random fatigue limit model (see Table I.5) and $w_i = \log_{10} n_i$. The functions ϕ_M and Φ_M are the marginal probability density and marginal cumulative distribution functions defined by equations (I.18) and (I.19) respectively, $\delta_i = 0$ if the test resulted in failure, $\delta_i = 1$ indicates a runout, that is the specimen did not fail when the test was stopped at the recorded number of cycles.

Considering qualified test records only, x_k is determined such that $LL_k(x_k | \hat{\theta}_3)$ is maximum. The values of x_k are listed in Table 6.1 along with the computed values of q_k and a_k. The formulae for computing q_k and a_k are:

$$
q_k = \frac{x_k K_t - 1}{K_t - 1} \quad \text{and} \quad a_k = r \left(\frac{1}{q_k} - 1 \right).
\tag{6.19}
$$

Substituting q_k for q in eq. (6.10) we find that for the kth specimen type $K_f = x_k K_t$. Therefore K_f can be used for predicting fatigue life at any stress level in the high cycle range, not just at the fatigue limit.

Peterson assumed that a is a material constant. However, the results displayed in Table 6.1 indicate otherwise. In fact, the coefficient of variation[5] of a_k is 141%. Therefore the assumption that the parameter a is a material constant, independent of the notch radius, does not hold for this material. Consequently this assumption, and therefore Peterson's formula, has to be rejected on the basis of the evidence in Table 6.1. The same conclusion applies to Neuber's definition of the notch sensitivity factor given by eq. (6.12).

This should be understood to mean that Peterson's predictor is not applicable in the interval of notch radii $0.0195 \leq r_k \leq 1.50$ (inches) used in calibration (see Table 6.1). However, it may be applicable in a narrow interval of notch radii, say $0.2 \leq r_k \leq 0.4$ (inches) which are commonly used in machine design. Note, however, that if the domain of parameters is sufficiently small then any predictor can be calibrated. A calibrated predictor is a validated predictor within the domain of calibration.

The relationship between $\log_{10} a$ and $\log_{10} r$ is shown in Fig. 6.3 where the open squares represent the data in Table 6.1. Observe that this relationship is nearly linear within the interval of notch radii considered here. Using the least squares method we find

$$
\log_{10} a \approx -0.16641 + 0.83899 \log_{10} r, \quad R^2 = 0.979
\tag{6.20}
$$

where R^2 is the coefficient of determination.

On substituting the eq. (6.20) into eq. (6.11) we get the revised form of Peterson's definition of the notch sensitivity factor for the 24S-T3 aluminum alloy:

$$
q_{rev} = \frac{1}{1 + 0.6817 r^{-0.1610}}, \quad 0.02 < r < 1.5
\tag{6.21}
$$

where $0.02 < r < 1.5$ is the domain of calibration. Note that whereas Peterson's notch sensitivity factor has one parameter, the revised formula has two.

---

5 The coefficient of variation is defined as the ratio of the standard deviation to the absolute value of the mean. It is a measure of dispersion.

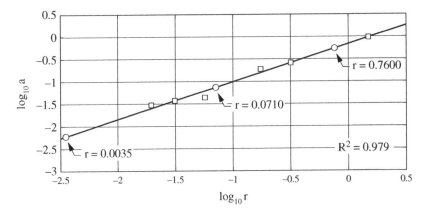

Figure 6.3 Empirical relationship between the parameters a and r for 24S-T3 (7024-T3) aluminum alloy calibrated on the basis of the six data points represented by the open squares.

Remark 6.5 The definition of x_k is not restricted to the ratio σ'_{eq}/σ_{eq}. It can be understood as a scaling factor for the applied load and hence the entire stress field $\sigma_{ij} = \sigma_{ij}(\mathbf{x})$. Therefore it is applicable to any predictor $\mathcal{P} = \mathcal{P}(\sigma_{ij})$ that has the property $\mathcal{P}(x_k\sigma_{ij}) = x_k\mathcal{P}(\sigma_{ij})$. Another predictor that has this property will be discussed in Section 6.3.

Remark 6.6 Fitting the data points in Fig. 6.3 by the least squares method is equivalent to assuming that the mean of $\log_{10} a$ is a function of $\log_{10} r$ and the probability density of $\log_{10} a$ is normal with variance s^2. We can then find the unknown coefficients by maximizing the likelihood function. To show this let $x = \log_{10} r$ and $y = \log_{10} a$ and write

$$y = c_1 + c_2 x + \epsilon, \quad \epsilon \sim \mathcal{N}(0, s^2).$$

By assumption, the probability density of ϵ is

$$f = \frac{1}{\sqrt{2\pi}s} \exp\left(-\epsilon^2/(2s^2)\right).$$

Let $\epsilon_i = y_i - c_1 - c_2 x_i$ ($i = 1, 2, \dots, n$). Then the likelihood function is

$$L = \prod_{i=1}^{n} \frac{1}{\sqrt{2\pi}s} \exp\left(-\epsilon_i^2/(2s^2)\right)$$

and the log likelihood function is

$$LL = \ln\left(\frac{1}{\sqrt{2\pi}s}\right)^n - \frac{1}{2s^2}\sum_{i=1}^{n}(y_i - c_1 - c_2 x_i)^2.$$

To find the maximum of LL we differentiate it with respect to c_1 and c_2 and let the derivatives equal to zero. We then get the same set of linear equations for c_1, c_2 as with the least squares method.

6.2.2 The effect of notches – validation

An ideal predictor would generalize the statistical model calibrated for the notch-free specimens to the notched specimens. This can never be fully ascertained, however. The best we can do is to test the hypothesis that the statistical distribution of the data obtained from testing the notched specimens is the same as the statistical distribution of the S-N data. Specifically, our hypothesis is that the probability distribution of the fatigue data $(n_i, \bar{x}_k \sigma_{eq}^{(i)})$, $k = 1, 2, \ldots, m_k$, collected from records of fatigue tests of notched specimens, is consistent with the random fatigue limit model.

The following discussion is concerned with testing Peterson's revised predictor against the three sets of records corresponding to line item 8 through 10 in Table I.2 in Appendix I, the validation set. The open circles in Fig. 6.3 correspond to those test records.

In order to emulate an ideal validation scenario, we assume that only a test plan is available at this point and this test plan provides details concerning the type and number of specimens to be tested and the loading conditions to be applied. Based on this information, the maximum equivalent stress $(\sigma_{eq})_i$ can be computed for each of the planned tests.

For each specimen type in the validation set we compute $\bar{x}_k$:

$$\bar{x}_k = \frac{\bar{q}_k(K_t - 1) + 1}{K_t} \quad \text{where } \bar{q}_k = q_{rev}(r_k) \tag{6.22}$$

and for each specimen type in the validation set we determine the predicted median from the inverse of the cumulative distribution function. Specifically, we find $\bar{w}_i \equiv \log_{10} \bar{n}_i$ such that

$$\Phi_M(\bar{w}_i, \bar{x}_k \sigma_{eq}^{(i)}) - 0.5 = 0 \tag{6.23}$$

where Φ_M is the marginal cumulative distribution function of the random fatigue limit model. The predicted median is the cumulative distribution function of $\bar{n}_i = 10^{\bar{w}_i}$. When no real root is found or $\bar{n}_i > 10^8$ then a runout is predicted. Additional discussion in available in Appendix I under Remark I.3. The 0.05 and 0.95 quantiles are computed analogously. We expect the cumulative distribution function of the outcome of the validation experiments to lie substantially within these quantiles.

Once the outcome of validation experiments is available, we plot and compare the empirical cumulative distribution function with the predicted median. An example is shown in Fig. 6.4.

Of the three sets of validation experiments the edge notched specimen corresponding to $k = 9$ is the most important one because the notch radius $r = 0.0035$ inch falls well outside of the interval of notch radii in the calibration set. There are 12 qualified test records for this specimen type. One runout was predicted, four runouts were recorded.

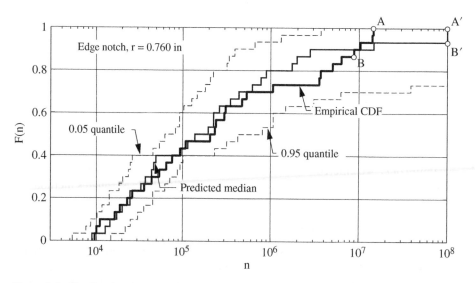

Figure 6.4 Predicted and empirical cumulative distribution functions for the edge notched specimens with $r = 0.760$ inch.

In the calibration process the statistical model and the functional form of the predictor are assumed to be correct and the parameters are determined such that the predicted and experimental outcomes are identical. In other words, the predictor is trained to match the calibration data. Therefore a calibrated predictor interpolates or approximates the calibration data within the domain of calibration in some sense.

The value of new data outside of the domain of calibration is that the predictive performance of the model can be tested against that data and the domain of the calibration set can be enlarged. The value of new data within the domain of calibration is that the dataset on which calibration is based is enriched.

Edge notched specimen with r = 0.760 inch

For the edge notched specimens with $r = 0.760$ inch the results of the validation experiments are presented in the form of the empirical cumulative distribution function shown in Fig. 6.4. The predicted median and the 0.05 and 0.95 quantiles are also shown. The segments labeled $A' - B'$ represent the predicted runouts, whereas the segments labeled $A - B$ represent the recorded runouts. Four runouts were recorded, two runouts were predicted. The empirical CDF lies within the 0.05 and 0.95 quantiles predicted on the basis of the random fatigue limit model calibrated to the S-N data of the notch-free specimens, as described in Appendix I.

Remark 6.7 The number of runouts cannot be reliably predicted. A runout is recorded when a test is stopped prior to the occurrence of failure for any reason. A runout is predicted when eq. (6.23) does not have a real root or the estimated number of cycles for a given σ'_{eq} exceeds 100 million. For additional discussion see Remark I.3 in the appendix.

Edge notched specimen with r = 0.0035 inch

The predicted and empirical cumulative distribution functions for the edge notched specimens with $r = 0.0035$ inch are shown in Fig. 6.5. Four runouts were recorded, one runout was predicted.

Conclusion

The realized cumulative distribution functions do not contradict the hypothesis that the notched fatigue data came from the same population as the S-N data. Therefore we find no reason to reject Peterson's revised predictor.

6.2.3 Updated calibration

Having completed the validation experiments, the validation set is merged with the calibration set and the calibration process is repeated. This is to take into account all of the available information concerning the notched fatigue properties of this material.

The data in Table 6.1 is augmented with the calibration data in Table 6.2 and, using the full set of nine data points, a new empirical relationship between the parameters a and r is obtained. This relationship is

$$\log_{10} a = -0.16763 + 0.82291 \log_{10} r, \quad R^2 = 0.979. \tag{6.24}$$

Using this relationship, Peterson's revised formula is updated to

$$q_{upd} = \frac{1}{1 + 0.6798 r^{-0.1791}}, \quad 0.003 < r < 1.5. \tag{6.25}$$

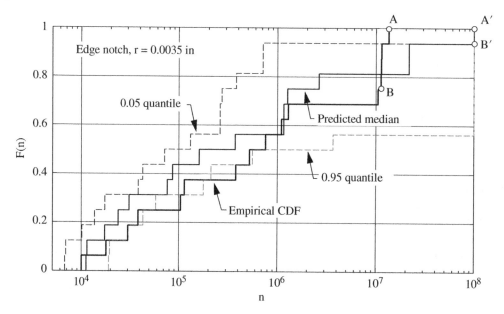

Figure 6.5 Predicted and empirical cumulative distribution functions for the edge notched specimens with $r = 0.0035$ inch.

Table 6.2 Additional calibration data to augment Table 6.1.

k	Specimen	r_k (in)	K_t	N	N_f	N_r	x_k	q_k	a_k (in)
8	Edge notch	0.7600	1.62	31	25	4	0.8164	0.5203	0.7007
9	Edge notch	0.0035	4.48	17	12	4	0.4553	0.2987	0.0082
10	Edge notch	0.0710	4.41	19	12	4	0.6560	0.5551	0.0569

Merging the validation set with the calibration set has resulted in a substantial increase in the size of the domain of calibration, see eq. (6.21). Using q_{upd} in eq. (6.16), we compute the updated $\bar{x}_k = q_{\text{upd}}(r_k)$ for the qualified test records of nine notched specimen types used for the updated calibration. The results are shown in Fig. 6.6 where $\sigma'_{\text{eq}} = \bar{x}_k \sigma_{\text{eq}}$. Compare Fig. 6.6 with Fig. 6.2. The median and quantile functions were computed from the random fatigue limit model. On comparing the two figures it is obvious that σ'_{eq} is a far better predictor than σ_{eq} was.

Remark 6.8 When comparing Fig. 6.6 to Fig. 6.2 it is obvious that σ'_{eq} is a much better predictor than σ_{eq}. However, if we ask the question how much better or worse is σ'_{eq} based on q_{upd} than σ'_{eq} based on q_{rev}, simply plotting the two distributions would not provide an answer. A quantitative measure is needed.

The relative performance of two models is evaluated by the likelihood ratio, in this case $L_{\text{upd}}/L_{\text{rev}}$, where L is the likelihood function defined by eq. (I.20) for the random fatigue limit model with the parameters $\hat{\theta}_3$ that can be found in Table I.5. On computing the log likelihood function LL defined by eq. (6.18) we find $LL_{\text{upd}} = -2534.440$ and $LL_{\text{rev}} = -2537.068$. Therefore

$$L_{\text{upd}}/L_{\text{rev}} = \exp(LL_{\text{upd}} - LL_{\text{rev}}) = 13.8$$

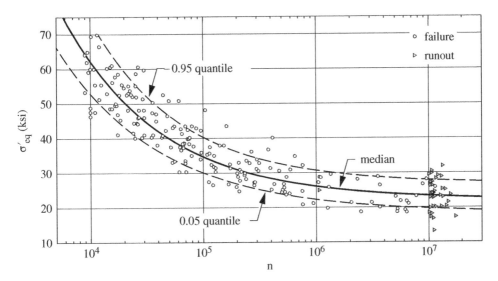

Figure 6.6 The results of fatigue experiments performed on nine notched specimen types. Compare with Fig. 6.2.

which is a strong indication that the updated model is better than the revised model, given the data for the nine notched specimen types.

The parameter $a = 0.02$ is given in [75] for aluminum alloy sheets and bars for use in Peterson's formula, see eq. (6.11). The hypothesis that a is a material constant was rejected on the basis of the results of calibration summarized in Table 6.1, having observed that the coefficient of variation was high. We are now in a position to obtain a quantitative measure of the difference between the predictors based on Peterson's original definition of q, denoted by q_{Pet}, and q_{upd}. On computing the log likelihood function corresponding to q_{Pet} we find $LL_{Pet} = -3463.283$ hence $L_{upd}/L_{Pet} = \exp(LL_{upd} - LL_{Pet})$ is a very large number indicating that rejection of this model was fully justified.

In reference [37] estimates are given for the parameter A in Neuber's formula for aluminum alloys, see eq. (6.12). By interpolation for the ultimate tensile strength of 24S-T3 aluminum (73 ksi), we find $A = 0.018$ inch. On computing the log likelihood function corresponding to Neuber's definition of q, denoted by q_{Neu} we find $LL_{Neu} = -2971.704$. Therefore the likelihood ratio L_{upd}/L_{Neu} is a very large number indicating that Neuber's model should be rejected also.

The q vs. r curves are shown in Fig. 6.7. On comparing Peterson's and Neuber's formulas with the revised and updated formulas for q, substantial differences are evident, whereas the difference between the revised and updated formulas is small. Therefore confidence in the updated formula within the indicated domain of calibration is justified. Observe that the notch sensitivity factor approaches unity very differently for the Peterson, Neuber and updated Peterson formulas as the notch radius increases.

6.2.4 The fatigue limit

The notch sensitivity factors of Peterson and Neuber were proposed for estimating the fatigue limit of notched specimens. In the following we compare the fatigue limit values of the random fatigue limit model with those computed from the original formulas of Peterson and Neuber. For reference we will use the mean of the random fatigue limit of the notch-free specimens which

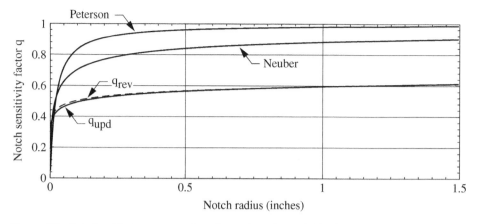

Figure 6.7 Relationship between notch radius and notch sensitivity factor for 24S-T3 aluminum alloy sheets.

Table 6.3 24S-T3 alloy: Fatigue limit values predicted by the random fatigue limit (RFL), Peterson and Neuber models.

k	Specimen	r_k	K_t	$\bar{x}_k$	RFL	Peterson	Neuber
2	Open hole	1.5000	2.11	0.7962	17.58	21.93	20.93
3	Edge notch	0.3175	2.17	0.7547	16.66	21.37	19.79
4	Fillet notch	0.1736	2.19	0.7381	16.30	20.84	19.16
5	Edge notch	0.0570	4.43	0.5883	12.99	17.64	15.93
6	Fillet notch	0.0195	4.83	0.5408	11.94	13.21	13.50
7	Edge notch	0.0313	5.83	0.5374	11.87	14.95	14.19
8	Edge notch	0.7600	1.62	0.8406	18.56	21.86	20.95
9	Edge notch	0.0035	4.48	0.4937	10.90	7.48	10.18
10	Edge notch	0.0710	4.41	0.5964	13.17	18.33	16.36

is $10^{\mu_f} = 10^{1.344} = 22.08$ ksi, see Table I.5. For the notched specimens this number is scaled by $\bar{x}_k$ defined by eq. (6.22). The results, shown in Table 6.3, indicate that the Peterson and Neuber formulas overestimate the fatigue limit values by a significant margin.

The random fatigue limit model estimates the probability of survival at the mean fatigue limit to be 50%, see Appendix A, whereas the fatigue limit values computed by the Peterson and Neuber formulas are deterministic numbers, generally understood to mean that the probability of survival at stress levels below the fatigue limit is 100%. Therefore those numbers should be smaller, not larger.

Example 6.1 The fatigue strength of 2024-T3 aluminum alloy is estimated to be 20 ksi (138 MPa) by the Aluminum Association[6] (AA). By this definition fatigue strength is the maximum completely reversed normal stress that the material is expected to resist at 500 million cycles. This definition is not consistent with random fatigue limit models which treat fatigue limit and fatigue strength as

---

6 Source: ASM Material Data Sheet.

random variables. Using the random fatigue limit model characterized by the parameters in Table I.5, the estimated probability that fatigue failure will occur below 500 million cycles is

$$\Pr\left(\text{failure} | n < 5.0 \times 10^8, \ \sigma_{eq} = 20\right) = \Phi_M(w, \sigma_{eq}) = 0.1697 \ (17\%)$$

where Φ_M is the marginal cumulative distribution function defined in Appendix I by eq. (I.19). In other words, the probability of survival at 500 million cycles of fully reversed 20 ksi maximum stress loading is 83%. Note, however, that 500 million cycles is well outside of the domain of calibration of the random fatigue limit model described in this chapter and the validity of the model has not been established for such high number of cycles.

The random fatigue limit model described in this chapter has been calibrated to 10 million cycles and thus it is possible to answer the following question: "What is the maximum stress at fully reversed loading such that the probability of survival is 99% at 10 million cycles?" For this we have to solve the following problem: Find σ_{eq} such that

$$1 - \Pr\left(\text{failure} | n < 10^7, \ \sigma_{eq}\right) = 0.99.$$

The solution is $\sigma_{eq} = 18.2$ ksi. One should bear in mind, however, that prediction of events that have a low probability of occurrence involves extrapolation even when the event is within the domain of calibration. This is because statistical models are calibrated against data that have a high probability of occurrence. At low probability values $\Phi_M(w, \sigma_{eq})$ is sensitive to the choice of the statistical model and therefore large model form uncertainties exist. Consequently, validation of statistical models for the prediction of fatigue limit is not feasible.

6.2.5 Discussion

At this point the reader would be fully justified in expecting a yes or no answer to the question of whether or not Peterson's updated predictor passed the validation test. Disappointing as it may sound, it is not possible to answer the question in that way. To answer the question yes or no it would have been necessary to define a validation metric and state a priori the acceptable magnitude of disagreement between the predicted and realized outcomes in terms of the chosen validation metric. We have not set such a tolerance because any number would have been arbitrary and difficult to justify.

The purpose of validation is to assess the predictive performance of a mathematical model. The outcome of a validation experiment is one of the following: (a) the prediction is not consistent with the observations, hence the model has to be rejected, or (b) the predictions are consistent with the observations therefore there is no reason to reject the model. A model is validated only within its domain of calibration. Here we can say with a high degree of confidence that Peterson's predictor with the updated notch sensitivity factor q_{upd} has been validated in the domain of calibration indicated in eq. (6.25).

This claim is not based on some arbitrarily chosen tolerance. Rather, it is in the spirit of Hume[7] who wrote: "*A wise man proportions his belief to the evidence.*" The evidence here is that (a) the empirical cumulative distribution functions lie substantially within the 0.05 and 0.95 predicted quantiles as seen (for example) in Fig. 6.4 and (b) upon introduction of new data, which included data that lie outside of the domain of calibration, predictions based on the formula changed by a small amount only, indicating that predictions based on q_{rev} are substantially the same as predictions based on q_{upd}. This criterion is discussed in greater detail in [16] where the distance between

---

7 David Hume 1711–1776.

the cumulative distribution functions associated with the prior and posterior models is assumed to be an acceptable approximation of the distance between the posterior model and the (unknown) true model.

Another important objective of validation is to develop information that provides a rational basis for choosing from among competing models the one that is best suited for the intended use of the model. Here we used the ratio of likelihoods to show that, given the available data, Peterson's predictor with q_{upd} is better than with q_{rev} and both are far better than Peterson's or Neuber's predictor with the original definitions for q. Ranking Peterson's revised predictor against another kind of predictor is discussed in Section 6.3.3.

Both Peterson's and Neuber's definitions of the notch sensitivity factor are based on the assumption that notch sensitivity is characterized by a material constant. The results of experiments reported in [94] show, however, that, at least for 24S-T3 and 75S-T6 aluminum alloys and SAE 4130 steel, this assumption does not hold. Therefore this assumption cannot be valid for all metallic alloys and may not be valid for any metallic material.

Remark 6.9 Peterson's revised notch sensitivity formula has been calibrated for a very special stress condition only: Since the boundary surface at the notch is stress-free and the stress in the transverse direction is negligibly small in comparison with the first principal stress, the maximum stress at the notch is substantially uniaxial. In other words, the domain of calibration is limited to uniaxial stresses. Generalization to biaxial stresses will be discussed in Section 6.4.

6.3 The predictor G_α

A predictor of fatigue failure, proposed in [93], will be discussed in the following. This predictor is based on the assumption that the onset of fatigue failure can be correlated with the averaged volume integral of a linear combination of two stress invariants. It is defined as follows:

$$G_\alpha(\sigma_{ij}, R) = \frac{1}{V_c} \int_{\Omega_c} \left(\alpha I_1 + (1 - \alpha)\overline{\sigma} \right) \, dV \left(\frac{1-R}{2} \right)^{1/2}, \quad 0 \le \alpha \le 1 \tag{6.26}$$

where σ_{ij} is the stress tensor field, $I_1 = \sigma_{kk}$ is the first stress invariant, $\overline{\sigma}$ is the von Mises stress,

$$\overline{\sigma} = \sqrt{\frac{3}{2} \left(\sigma_{ij} - \frac{1}{3}\delta_{ij}\sigma_{kk} \right) \left(\sigma_{ij} - \frac{1}{3}\delta_{ij}\sigma_{kk} \right)} \tag{6.27}$$

and V_c is the volume of the domain of integration defined by

$$\Omega_c = \{\mathbf{x} | \sigma_1 > \beta\sigma_{max} > 0\} \tag{6.28}$$

where σ_1 is the first principal stress and $\sigma_{max} > 0$ is the maximum macroscopic stress. This is a generalization of the uniaxial stress to triaxial stress in the sense that G_α is defined for triaxial stress and, in the special case when constant uniaxial stress σ_1 is applied, it has the value of the equivalent stress

$$G_\alpha(\sigma_{ij}, R) = \sigma_1 \left(\frac{1-R}{2} \right)^{1/2} = \sigma_{eq}. \tag{6.29}$$

Therefore we will generalize the statistical model defined in Section I.3 by replacing σ_{eq} with G_α in the definition of the mean:

$$\mu(G_\alpha) = A_1 - A_2 \log_{10}(G_\alpha - A_3), \quad G_\alpha - A_3 > 0. \tag{6.30}$$

Many generalizations are possible. Predictors are scalars that must be independent of the choice of the coordinate system. The predictor that has the largest likelihood, given a set of experimental data, is the one to be preferred. We will compare G_α with Peterson's modified predictor using experimental data for nine notched specimen types.

The predictor G_α has two parameters that have to be determined by calibration. Parameter α establishes a convex combination of the first stress invariant I_1 and the von Mises stress $\bar\sigma$. Parameter β defines the domain of integration which depends on the material properties and the stress field in the vicinity of stress raisers and thus it plays a role analogous to Peterson's notch sensitivity factor.

Peterson's notch sensitivity factor depends on the notch radius. There are other kinds of stress raisers, such as scratches and corrosion pits, that cannot be characterized by a notch radius. For this reason we introduce the highly stressed volume V for characterizing stress raisers. By definition,

$$V = \int_{\Omega_\varrho} y(\mathbf{x})\, dV \quad \text{where} \quad y = \begin{cases} 1 & \text{when } \sigma_1(\mathbf{x}) > \gamma\sigma_{\max} \\ 0 & \text{otherwise.} \end{cases} \tag{6.31}$$

The domain of integration Ω_ϱ is the neighborhood of a stress raiser. Denoting the location where σ_1 is maximum at a stress raiser by $\mathbf{x}_0$ we have $\Omega_\varrho = \{\mathbf{x}\,|\,|\mathbf{x} - \mathbf{x}_0| < \varrho\}$ where ϱ is chosen large enough to include all points where $\sigma_1(\mathbf{x}) > \gamma\sigma_{\max}$ and small enough to include only one stress raiser or one group of closely situated stress raisers.

The parameter γ is independent of the material properties. Its sole purpose is to define the highly stressed volume V which depends on the stress distribution, and hence the type of loading, but is independent of the magnitude of the load. The predicted number of fatigue cycles is not sensitive to the choice of γ. As in reference [93], we will use $\gamma = 0.85$.

6.3.1 Calibration of $\beta(V, \alpha)$

The calibration of $\beta = \beta(V, \alpha)$ is analogous to the calibration of q discussed in Section 6.2.1 where $x_k = \sigma'_{\max}/\sigma_{\max} = \sigma'_{eq}/\sigma_{eq}$ was found through maximization of the log likelihood function for each specimen type in the calibration set. Here we used the nine specimen types listed in line items 2 through 10 in Table I.2 for calibration and, for a fixed α and each specimen type, we found β_k by iteration such that

$$G_\alpha^{(k)} = x_k \sigma_{\max} \left(\frac{1-R}{2} \right)^{1/2}, \quad k = 2, 3, \dots, 10 \tag{6.32}$$

where the values of x_k are the same as those in Sections 6.2.1 and 6.2.3. See Remark 6.5. The β_k values were fitted by least squares for each α using the assumed functional form:

$$\overline{\beta}_k = a_1 + a_2 \log_{10} A_k + \epsilon, \quad \epsilon \sim \mathcal{N}(0, \sigma_\epsilon^2) \tag{6.33}$$

where V is the highly stressed volume and ϵ is a random variable which was assumed to have normal distribution with zero mean and standard deviation σ_ϵ. For example, for $\alpha = 0$ we find $a_1 = 0.9422$, $a_2 = 0.08184$. The calibration curve for β is shown in Fig. 6.8. Note that the interval of the highly stressed volume for which calibration data are available is approximately $(1.0E - 8 < V < 1.0E - 2)$ in^3.

Using $\overline{\beta}_k$, we find $\overline{x}_k$ by evaluating

$$\overline{x}_k = \frac{1}{\sigma_{\max} V_c} \int_{\Omega_c} \left(\alpha I_1 + (1 - \alpha)\overline{\sigma} \right)\, dV \tag{6.34}$$

on the domain $\Omega_c = \{\mathbf{x}\,|\,\sigma_1 > \overline{\beta}_k \sigma_{\max} > 0\}$. The numerical results for $\alpha = 0$ are shown in Table 6.4.

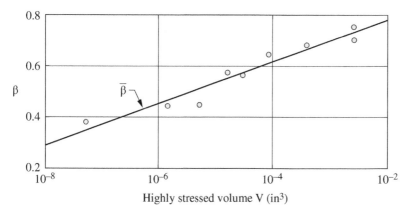

Figure 6.8 The computed β_k values and the $\overline{\beta}$ function corresponding to $\alpha = 0$.

Table 6.4 24S-T3 aluminum alloy: Computed values of V_k, β_k, $\overline{\beta}_k$ and $\overline{x}_k$ for $\alpha = 0$.

k	Specimen	K_t	$V_k \left(\text{in}^3 \right)$	x_k	β_k	$\overline{\beta}_k$	$\overline{x}_k$
2	Open Hole	2.11	2.898E-02	0.7943	0.7021	0.7329	0.8163
3	Edge Notch	2.17	4.325E-03	0.7601	0.6826	0.6652	0.7487
4	Fillet Notch	2.19	9.561E-04	0.7159	0.6455	0.6116	0.6951
5	Edge Notch	4.43	1.827E-04	0.6664	0.5742	0.5528	0.6363
6	Fillet Notch	4.83	1.618E-05	0.5198	0.4448	0.4666	0.5501
7	Edge Notch	5.83	5.888E-05	0.5490	0.4487	0.5125	0.5960
8	Edge Notch	1.62	2.875E-02	0.8164	0.7539	0.7326	0.8161
9	Edge Notch	4.48	5.908E-07	0.4553	0.3809	0.3490	0.4325
10	Edge Notch	4.41	3.324E-04	0.6560	0.5645	0.5740	0.6575

Remark 6.10 The values of x_k, and hence the computed values of β_k, are influenced by uncertainties in loading conditions, residual stress, variations in surface finish, and tolerances in the dimensions of the test articles as well as numerical errors. While the presence of systematic errors cannot be ruled out, the source reports for the experimental data indicate that the investigators exercised a great deal of care in planning and performing the experiments to avoid such errors. The relative errors in all numerically computed data have been verified to be less than 1%.

6.3.2 Ranking

Using $\overline{x}_k$, we evaluate for all specimen types the log likelihood function for a sequence of the model form parameter $\alpha = \alpha_j$. Specifically, the sequence $\alpha_j = 0.2(j - 1), j = 1, 2, \ldots, 6$ will be used.

$$LL_k^{(j)}(\alpha_j | \hat{\theta}_3) = \sum_{i=1}^{m_k} \left[(1 - \delta_i) \ln \left(\phi_M(w_i, \overline{x}_k \sigma_{eq}^{(i)}) \right) \right.$$

$$\left. + \delta_i \ln \left(1 - \Phi_M(w_i, \overline{x}_k \sigma_{eq}^{(i)}) \right) \right], \quad k = 2, 3, \ldots, 10 \tag{6.35}$$

Table 6.5 Computed log likelihood values $LL^{(j)}$.

$\alpha = 0$	$\alpha = 0.2$	$\alpha = 0.4$	$\alpha = 0.6$	$\alpha = 0.8$	$\alpha = 1.0$
-2487.33	-2495.33	-2510.72	-2536.07	-2574.97	-2634.26

where $\hat{\theta}_3$ is the vector of the five statistical parameters that characterize the random fatigue limit model (see Table I.5) and $w_i = \log_{10} n_i$. The functions ϕ_M and Φ_M are the marginal probability density and marginal cumulative distribution functions defined by equations (I.18) and (I.19) respectively, $\delta_i = 0$ if the test resulted in failure, $\delta_i = 1$ indicates runout. The computed values of the log likelihood function

$$LL^{(j)} = \sum_{k=2}^{10} LL_k^{(j)}(\alpha_j|\hat{\theta}_3), \quad j = 1, 2, \dots, 6 \tag{6.36}$$

are shown in Table 6.5. It is seen that $LL^{(j)}$ is maximum for $\alpha = 0$. Therefore the model form parameter α should be set to zero. In the following we will be concerned with this model only.

6.3.3 Comparison of G_α with Peterson's revised predictor

Let us compare the predictive performance of G_α with that of the Peterson's predictor based on the revised and updated notch sensitivity index. The basis for comparison is the value of the log likelihood function which is our validation metric.

The maximum log likelihood for Peterson's revised and updated predictor was $LL_{\mathrm{upd}} = -2534.44$, see Section 6.2.3. This value is substantially lower than the value of the maximum log likelihood function of the present model with $\alpha = 0$ but is slightly higher than the present model with $\alpha = 0.6$, see Table 6.5. Based on this evidence we conclude that G_α with $\alpha = 0$ is a better predictor than Peterson's revised predictor, given the data at our disposal and our choice of the statistical model. Mathematical models (and hence predictors) are ranked on the basis of the log likelihood function, given the available data.

The values of G_α are plotted against n for the nine notched specimen types considered here and for $\alpha = 0$ in Fig. 6.9. On comparing Fig. 6.9 with Fig. 6.6 it becomes evident that G_α is preferable to Peterson's revised and updated predictor.

It is not possible to claim that G_α, or any other predictor that may be formulated in the future, is the best predictor. It is possible to say only, given the data, the statistical model and two predictors, which one is better. When new data become available, or when a new predictor or statistical model is proposed, then it is possible to evaluate their relative merit objectively on the basis of the likelihood function following the procedure outlined here.

The formulation of mathematical models is an open-ended problem. There have to be established policies and procedures for the systematic revision and updating of models when new ideas are proposed, or when new data become available. The formulation and management of such policies and procedures are in the domain of simulation governance.

6.4 Biaxial test data

The scope of validation discussed up to this point was confined to substantially uniaxial stress conditions in the high cycle fatigue regime. In this section we examine whether the predictor G_α

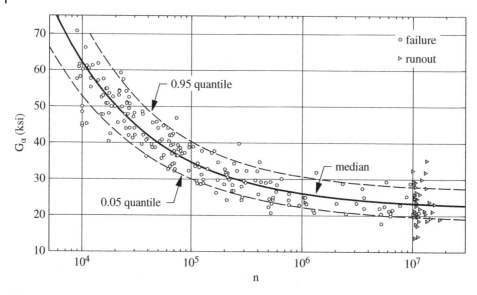

Figure 6.9 24S-T3 aluminum alloy: Combined qualified test records for the nine notched specimen types, $\alpha = 0$. The median and quantile functions are those of the random fatigue limit model. Compare with Fig. 6.6.

($\alpha = 0$), calibrated under uniaxial loading conditions, will pass validation tests under biaxial loading. We will use results of uniaxial and biaxial fatigue tests performed on 2024-T3 aluminum alloy specimens in the intermediate to high cycle fatigue regime published in [38]. The analysis described in this section is based on reference [95].

The specimens were fabricated from drawn tubing so as to conform to ASTM standard E2207[8] . These specimens feature a 30 mm long cylindrical section with outside diameter of 29 mm, inside diameter 25.4 mm, wall thickness 1.8 mm. A 3.2 mm diameter cylindrical hole was cut into the test section by drilling and reaming. The axis of the hole was perpendicular to the test section. A specimen is shown in Fig. 6.10(a). The surfaces were polished to remove machining marks. The tests were performed under in-phase, fully reversed ($R = -1$) axial, torsional and combined loading conditions with load control in the 0.2 to 7.0 Hz frequency range.

Crack initiation and growth were observed using a 2.0 megapixel digital microscopic camera capable of 10X to 230X optical zoom. The fatigue crack initiation event was defined as the first appearance of a 0.2 mm surface crack [38].

6.4.1 Axial, torsional and combined in-phase loading

The test results are shown in Fig. 6.11 where N is the number of cycles at failure and G_α ($\alpha = 0$) is the predictor defined by eq. (6.26). The solid line represents the median of the random fatigue limit model characterized by the five parameters in Table I.5.

As previously stated, the purpose of a validation experiment is to test the predictive performance of a mathematical model. In this case; given the notched specimen described previously, and the experimental procedure described in [38], the prediction being tested is formulated as follows: "The probability is x% (say, 90%) that the number of cycles at which failure will occur (N_f) lies in

---

8 ASTM E2207 – 15 Standard Practice for Strain-Controlled Axial-Torsional Fatigue Testing with Thin-Walled Tubular Specimens.

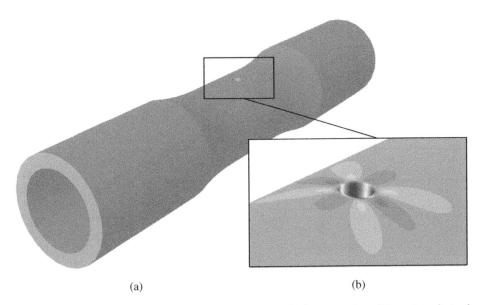

(a)   (b)

Figure 6.10 (a) Specimen used in validation experiments. (b) Contours of von Mises stress in torsion.

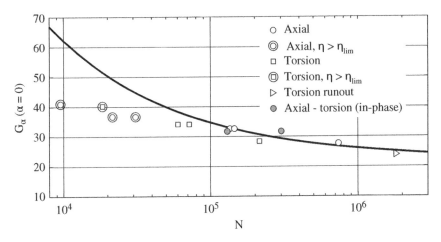

Figure 6.11 Outcomes of axial, torsion and combined in-phase fatigue experiments performed on notched 2024-T3 aluminum specimens. Source: reference [38].

the interval $N_1 \leq N_f \leq N_2$". The limit values N_1 and N_2 are determined from the survival function $S(N)$ which, given G_α, is defined by

$$S(N) = 1 - \Phi_M(N|G_\alpha) \tag{6.37}$$

where Φ_M is the marginal cumulative distribution function (CDF) defined by eq. (I.19).

Example 6.2 Let us consider two biaxial fatigue tests reported in [38] with nominal axial stress of 81 MPa and nominal shear stress of 50 MPa applied in-phase in both tests. The corresponding value of the predictor G_α ($\alpha = 0$) was computed and verified to be $G_\alpha = 31.6$ ksi (218 MPa). The survival function is shown in Fig. 6.12.

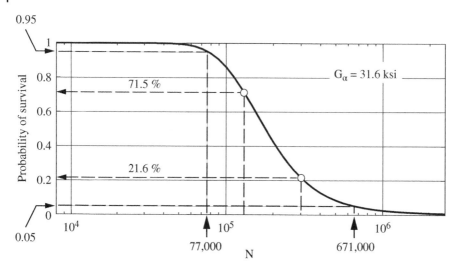

Figure 6.12 Survival function corresponding to $G_\alpha = 31.6$ ksi. Outcomes of combined in-phase axial-torsion fatigue experiments performed on notched 2024-T3 aluminum specimens.

We determine N_1 and N_2 from the condition $S(N_1) = 0.05$ and $S(N_2) = 0.95$ and find $N_1 = 671{,}000$, $N_2 = 77{,}000$. Therefore our model predicts that the probability that failure will occur in the interval of 77,000 to 671,000 cycles is 90%. The reported experimental values (N_{exp}), shown in Fig. 6.12 are 302,000 and 130,000 cycles. Therefore the outcome of experiments is within the predicted interval.

An alternative way of reporting the results of validation experiments is to say that the probabilities of survival, computed from eq. (6.37) by substituting the reported number of cycles at failure N_{exp} for N, are 21.6% and 71.5% respectively. Since these probabilities lie in the interval of 5% to 95%, we consider the predictor G_α ($\alpha = 0$) to have passed this validation test.

6.4.2 The domain of calibration

Mathematical models are defined on sets of admissible parameters, specific to the model. Models are calibrated on a subset of the admissible parameters that define the domain of calibration. The domain of calibration is an essential attribute of a mathematical model.

The predictor G_α was formulated for the purpose of generalization of the S-N data to notched specimens in the high cycle regime. The lower limit of what is considered to be high-cycle fatigue is usually and somewhat arbitrarily set at 10 thousand cycles. In reference [94] fatigue test records for notched specimens that failed at fewer than 8500 cycles, the lower limit of the number of cycles for which S-N data are available, were not included in the calibration set.

We introduce a restriction based on the size of the plastic zone relative to the volume of integration V_c. We define V_{yld} as the volume where the von Mises stress $\bar{\sigma}$, determined from the linear solution, is greater than the yield stress σ_{yld}.

$$\eta \overset{\text{def}}{=} V_{yld}/V_c. \tag{6.38}$$

It would be overly restrictive to limit the predictor to purely elastic stress fields because the peak elastic stress can be very high at notch roots. Therefore such a restriction would severely limit the applicability of the predictor. It is possible to admit small amounts for plastic deformation, provided that the plastic zone is sufficiently small so that plastic deformation is controlled by the surrounding elastic stress field. The effects of plasticity controlled by the elastic stress field are taken into account through calibration of the parameter β. Naturally, this leads us to the question: How large the plastic zone may be? Equivalently, what is the limiting value of η, denoted by $\eta_{\lim}$? This value is a delimiter of the domain of calibration.

Test records that satisfy conditions imposed by the assumptions on which the formulation of the mathematical model is based are called qualified test records. In the present case the size of the plastic zone is limited by the condition $\eta < \eta_{\lim}$. The value of $\eta_{\lim}$ will be inferred from the outcome of validation experiments.

The results of experiments, taken from Table 5 of reference [38], and the values of η are listed in Table 6.6 where σ_{nom} and τ_{nom} represent the maximum nominal normal and shearing stresses in the test section, G_α is the computed value of the predictor, converted to ksi units for consistency with the calibration records, N_{exp} is the number of cycles at which failure occurred.

The estimated probabilities of survival to the number of cycles at which failure occurred, computed from eq. (6.37), are listed under the heading "Prob".

The results in Table 6.6 show that if we limit the domain of the predictor to $N_{exp} > 8500$ then 4 out of 13 outcomes fail the validation experiments in the sense that the predicted probabilities of survival fall outside of the 5% to 95% interval. On closer examination we find, however, that these specimens had larger amounts of plasticity than the others. For example, if we set $\eta_{\lim} = 0.15$ then two records are disqualified for excessive plastic deformation, the number of qualified records is 11 of which 10 passed the test. This is consistent with the prediction. If we set $\eta_{\lim} < 0.08$ then four records are disqualified and the remaining 7 specimens pass the validation test. In either case, there is no reason to reject the model.

Since we have access only to a limited number of records of biaxial fatigue tests in the high cycle regime, we are not in a position to propose a sharp definition for the threshold value of η. A tentative value of 10% would appear to be reasonable, however, and consistent with the available data.

Table 6.6 Results of validation experiments performed on notched 2024-T3 aluminum specimens. Source for the experimental data: Table 5 of Reference [38].

Loading	σ_{nom} MPa	τ_{nom} MPa	G_α ksi	N_{exp} cycles	Prob (%)	Result	η (%)	Remarks
Axial	145	0	41.0	9,500	100.0	Fail	17.1	$\eta > 10\%$
Axial	130	0	36.7	21,670	99.2	Fail	8.0	
Axial	130	0	36.7	31,000	92.1	Pass	8.0	
Axial	115	0	32.5	135,450	26.5	Pass	1.6	
Axial	115	0	32.5	145,600	22.4	Pass	1.6	
Axial	98	0	27.7	735,000	13.5	Pass	0.0	
Torsion	0	108	40.3	18,500	97.4	Fail	34.4	$\eta > 10\%$
Torsion	0	91	34.0	71,890	61.0	Pass	6.9	
Torsion	0	91	34.0	60,140	75.6	Pass	6.0	
Torsion	0	76	28.4	215,000	57.6	Pass	0.0	
Torsion	0	64	23.9	1,800,000	43.8	Pass	0.0	runout
Combined	81	50	31.6	302,000	21.6	Pass	0.7	in-phase
Combined	81	50	31.6	130,000	71.5	Pass	0.7	in-phase

Based on the evidence developed in the calibration and validation processes, it was asserted in Section 6.2.5 that Peterson's updated predictor was validated in the interval of notch radii $0.003 < r < 1.5$ inches. Here we assert that another predictor, based on integral averages, has been validated also and, furthermore, it performs better than Peterson's updated predictor when tested against the experimental data summarized in Table I.2. This is based on the observation that the log likelihood function corresponding to the predictor defined in Section 6.3, evaluated for the entire set of qualified test records, is larger than that corresponding to Peterson's updated predictor and has a larger domain of calibration in the sense that it has been calibrated with reference to the highly stressed volume and therefore stress raisers do not have to be characterized by notch radii.

These claims of validation and ranking are based on the available experimental data. It is of course possible that for a different set of data the ranking would change. It is also possible that other kinds of predictor will be formulated, calibrated and tested that will perform better than the predictor considered here.

6.4.3 Out-of-phase biaxial loading

The test conditions and validation data for combined out-of-phase axial and torsional loading in the high cycle range are summarized in Table 6.7 where σ_{nom} and τ_{nom} are the nominal normal and shearing stresses[9] corresponding to the axial and torsional components of the applied load and N_{exp} is the number of fully reversed cycles at which failure occurred [38].

The nominal first principal stresses for the axial, torsional and combined loadings are shown in Fig. 6.13 where we assumed that

$$(\sigma_1)_{axl} = \sigma_{nom}\sin(2\pi ft) \tag{6.39}$$

$$(\sigma_1)_{tor} = \tau_{nom}\sin(2\pi ft + \phi) \tag{6.40}$$

where f is the frequency (Hz), ϕ is the phase angle, in the present case $\phi = \pi/2$. Therefore the combined value, denoted by $(\sigma_1)_{com}$, is

$$(\sigma_1)_{com} \stackrel{\text{def}}{=} \frac{(\sigma_1)_{axl}}{2} + \left(\left(\frac{(\sigma_1)_{axl}}{2}\right)^2 + (\sigma_1)^2_{tor} \right)^{1/2}. \tag{6.41}$$

Accounting for out-of-phase loading requires a modeling assumption in addition to those made for the simulation of in-phase loading described in [95]. This modeling assumption accounts for the effects of out-of-phase loading on the notch. Various possibilities exist. The only requirement

Table 6.7 Results of experiments for 90° out-of-phase fully reversed axial and torsional loading.

Exp.	σ_{nom} MPa	τ_{nom} MPa	N_{exp} cycles
1	115	71	42760
2	115	71	78100
3	100	58	96800

Source for the data: Table 5 of Reference [38].

9 Nominal stress is the maximum stress in the test section without the notch.

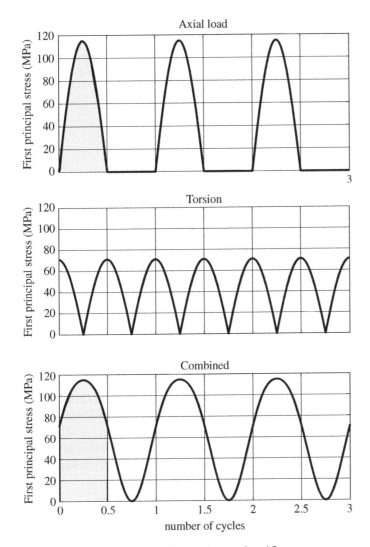

Figure 6.13   Nominal stresses in experiments 1 and 2.

is that when the phase angle is zero then the predictor has to be the same as the predictor defined for in-phase loading. We will discuss two extensions in the following.

Extension A

It is seen in Fig. 6.13 that the maximum value of the axial and combined loading is the same. However, the material is subjected to high nominal stress over a greater fraction of the cycle under combined loading. We will assume that the magnifying effects of the out-of-phase loading are proportional to the ratio of the shaded areas in Fig. 6.13. Specifically, we define

$$A_1 \stackrel{\text{def}}{=} \int_0^{1/2} (\sigma_1)_{\text{axl}} \, dt \tag{6.42}$$

$$A_2 \stackrel{\text{def}}{=} \int_0^{1/2} (\sigma_1)_{\text{com}} \, dt \tag{6.43}$$

Table 6.8 Test records used for the calibration of Model B.

#	$(G_\alpha)_{\text{axl}}$ ksi	$(G_\alpha)_{\text{tor}}$ ksi	N_{exp} cycles	A_1 MPa	A_2 MPa
1	32.5	26.5	42.7E3	36.61	50.52
2			78.1E3		

Table 6.9 The outcome of the validation experiment for 90° out-of-phase axial and torsional loading.

#	$(G_\alpha)_{\text{axl}}$ ksi	$(G_\alpha)_{\text{tor}}$ ksi	$(G_\alpha)_{\text{eff}}$ ksi	N_1 cycles	N_2 cycles	N_{exp} cycles	Prob (%)	Result
3	28.3	21.7	33.2	60.4E3	395.3E3	96.8E3	72.6	Pass

and

$$(G_\alpha)_{\text{eff}} \overset{\text{def}}{=} \kappa (G_\alpha)_{\text{axl}} \frac{A_2}{A_1} \tag{6.44}$$

where κ is a parameter, independent of A_1 and A_2, determined by calibration.

We will use the data in the first two rows in Table 6.7 for calibration, the data in the third row for validation. The computed values of $(G_\alpha)_{\text{axl}}$, $(G_\alpha)_{\text{tor}}$, A_1 and A_2 are shown in Table 6.8. On substituting A_1, A_2 and $(G_\alpha)_{\text{axl}}$ into eq. (6.44), we find $(G_\alpha)_{\text{eff}} = 44.8\kappa$ ksi.

The objective of calibration is to determine κ such that the data in the calibration set maximizes the log likelihood function corresponding to the statistical model. Specifically, letting $w_1 = \log_{10} 42.7E3$ and $w_2 = \log_{10} 78.1E3$, we have

$$LL(x) = \log(\phi_M(w_1, 44.8x)) + \log(\phi_M(w_2, 44.8x)) \tag{6.45}$$

where ϕ_M is the marginal pdf of the random fatigue limit model given by eq. (I.18) in the appendix. From this we find

$$\kappa = \arg\max LL(x) = 0.8680. \tag{6.46}$$

This completes the calibration process. We now examine the predictive performance of $(G_\alpha)_{\text{eff}}$. Specifically, we test the following prediction: "The probability that the number of cycles N_f at which failure will occur lies in the interval $N_1 < N_f < N_2$ is 90%". Another way of saying this is that 9 out of 10 specimens are predicted to fail in the given (load-dependent) interval.

For the test record used in validation, we find $A_1 = 31.83$ MPa, $A_2 = 42.95$ MPa, hence

$$(G_\alpha)_{\text{eff}} = \kappa (G_\alpha)_{\text{axl}} \frac{A_2}{A_1} = 33.2 \text{ ksi}.$$

Given $(G_\alpha)_{\text{eff}}$, the limits N_1, N_2 and the probability of survival to N_{exp} cycles are computed. The results are shown in Table 6.9, where the probability of survival to the experimentally observed number of cycles is shown under the heading "Prob". Based on these results, we conclude that the model passes the validation test.

Finally we update the calibration records and recompute κ: The two records used for calibration are augmented by the record used for validation and eq. (6.45) becomes

$$LL_1(x) = \log(\phi_M(w_1, 44.8x)) + \log(\phi_M(w_2, 44.8x)) + \log(\phi_M(w_3, 38.2x)) \tag{6.47}$$

where $w_3 = \log_{10} 96.8E3$ and find the updated value of κ, denoted by κ_{upd};

$$\kappa_{upd} = \arg\ \max\ LL_1(x) = 0.8840. \tag{6.48}$$

Extension B

We next consider the following extension of in-phase multiaxial loading:

$$(G_{\alpha,p})_{eff} \overset{def}{=} \left((G_\alpha)_{axl}^p + (G_\alpha)_{tor}^p\right)^{1/p}, \quad p \geq 1 \tag{6.49}$$

where $(G_\alpha)_{axl}$ and $(G_\alpha)_{tor}$ are the values of the predictor G_α corresponding to the axial and torsional loading, respectively. It is possible to select p in such a way that, for the available out-of-phase biaxial test records, the likelihood function is maximum. This would calibrate $(G_{\alpha,p})_{eff}$. Since we have very few test records at our disposal, and we are primarily interested in the process of validation and ranking, we forego that process and let $p = 2$ and define $(G_\alpha)_{eff} \equiv (G_{\alpha,2})_{eff}$.

The limit values N_1 and N_2 are determined from the survival function of the statistical model, see eq. (6.37). The computed values of $(G_\alpha)_{eff}$ and the limits of the predicted interval N_1 and N_2, corresponding to 5% and 95% probabilities of survival respectively, are listed in Table 6.10. The column with the heading N_{exp} shows the probability of survival to the realized number of cycles calculated from eq. (6.37). Observe that all failures occurred in the predicted intervals, in other words, there were three successful predictions in three trials.

Remark 6.11 We could have set the survival probabilities to (say) the 98% range. However, the tail segments of survival functions are very sensitive to the choice of the statistical model and should not be relied upon in validation experiments. Generally speaking, validation deals with events that have high probability of occurrence.

Exercise 6.1 Verify the survival probabilities for the $((G_\alpha)_{eff}, N_{exp})$, entries in Table 6.10.

Ranking

We have formulated two possible extensions for the predictor validated for in-phase loading. Both predictors passed the validation tests. We now ask: Which one is better? Should we prefer one over another? Questions such as this have to be addressed in numerical simulation where the intuitive process of formulation has to be moderated by objective methods of ranking. To that end, we evaluate the log likelihood (LL) function for each predictor with reference to the outcome of the

Table 6.10 The outcome of validation experiments for 90° out-of-phase axial and torsional loading.

#	$(G_\alpha)_{axl}$ ksi	$(G_\alpha)_{tor}$ ksi	$(G_\alpha)_{eff}$ ksi	N_1 cycles	N_2 cycles	N_{exp} cycles	Prob (%)	Result
1	32.5	26.5	42.0	22.4E3	80.5E3	42.7E3	45.3	Pass
2	32.5	26.5	42.0	22.4E3	80.5E3	78.1E3	5.7	Pass
3	28.3	21.7	35.6	43.9E3	219.8E3	96.8E3	42.7	Pass

Table 6.11 The computed log likelihood (*LL*) values for predictors A and B.

Predictor	$(G_\alpha)_{eff}$ (ksi)			N_{exp} (cycles)			*LL*
	1	2	3	1	2	3	
A	39.6	39.6	33.8	42,700	78,100	96,800	−34.33
B	42.0	42.0	35.6				−34.95

validation experiments. The predictor with the greater *LL* value is the better generalization of the S-N data to notched specimens subjected to out-of-phase biaxial loading, given the available data. The updated predictors and the corresponding values of the log likelihood function are listed in Table 6.11.

The evidence for preferring predictor A to predictor B is based on the likelihood ratio

$$L_A/L_B = \exp(LL_A - LL_B) = 1.86 \tag{6.50}$$

where L_A and L_B are the likelihood values corresponding to predictors A and B, respectively. If this ratio is in the interval (1/3, 3), which is the case here, then the difference is barely worth mentioning. In other words, the two predictors are in a virtual tie.

Predictive performance

We test the predictive performance of the mathematical model by calculating the number of cycles N_1 (resp. N_2) corresponding to the 5% (resp. 95%) survival probabilities for each specimen, given the test conditions. Therefore, if the model is correct then 9 out of 10 specimens will fail in the predicted interval. We denote the number of outcomes that fall in the predicted interval divided by the number of tests by θ. Our problem now is to find the most probable value of θ, denoted by θ_0, and the associated confidence interval, given the experimental data, and prior information concerning the probability density function of θ.

Bayes' theorem states:

$$Pr(\theta|D) = \frac{Pr(D|\theta)Pr(\theta)}{Pr(D)} \tag{6.51}$$

where $Pr(\theta|D)$ is the posterior probability density function, $D = (m, n)$ stands for the available experimental data in terms of the number of experiments m and the number of successful outcomes n. The term $Pr(\theta|D)$ is called the posterior probability density function.

$Pr(D|\theta)$ is the probability of obtaining the observed data, given θ. The probability of n successes in m trials is given by the binomial distribution:

$$Pr(D|\theta) = \frac{m!}{n!(m-n)!}\theta^n(1-\theta)^{m-n}. \tag{6.52}$$

$Pr(\theta)$ is our prior information (the prior) concerning the probability density function (pdf) of θ. We will use the Beta distribution to estimate this:

$$Pr(\theta) = \frac{\Gamma(\alpha+\beta)}{\Gamma(\alpha)\Gamma(\beta)}\theta^{\alpha-1}(1-\theta)^{\beta-1} \tag{6.53}$$

where $\Gamma(\cdot)$ is the Gamma function, $\alpha > 0$ and $\beta > 0$ are shape parameters. The shape parameters define the prior probability distribution.

Substituting equations (6.52) and (6.53) into eq. (6.51) we get

$$Pr(\theta|D) = C\theta^{n+\alpha-1}(1-\theta)^{m-n+\beta-1} \tag{6.54}$$

where $C = 1/Pr(D)$ is a normalizing constant that must be chosen such that

$$C \int_0^1 \theta^{n+\alpha-1}(1-\theta)^{m-n+\beta-1}\, d\theta = 1. \tag{6.55}$$

Noting that C is the normalizing constant for the Beta distribution, we have

$$C = \frac{\Gamma(m+\alpha+\beta)}{\Gamma(n+\alpha)\Gamma(m-n+\beta)}. \tag{6.56}$$

When the posterior pdf has the same functional form as the prior, as in the present case, then the prior is said to be conjugate to the posterior pdf. Using conjugate priors has the advantage that it is simple to update the posterior pdf.

Selection of the prior

In Bayesian data analysis the selection of the prior pdf is guided by various considerations that fall into the following broad categories.

1. An uninformative prior, also called objective prior, expresses ignorance concerning the expected probability distribution of the data. In our case, for example, if we think that a successful outcome of an experiment is just as likely as an unsuccessful one then, invoking the principle of indifference, we assume that θ has uniform distribution, that is $Pr(\theta) \sim U(0, 1)$. The principle of indifference states that if we can enumerate a set of basic, mutually exclusive, probabilities and have no reason to believe that any one of these is more likely to be true than another, then we should assign the same probability to all [87]. Uninformative priors may express objective information. For example, in this case we know that θ must lie in the interval $(0, 1)$.
2. An informative prior expresses specific, definite information about a variable. For example, we would have an informative prior if we had access to test records of experiments that existed before we came into possession of new test records. Those existing test records would provide specific information relating to the probability distribution of θ.
3. A weakly informative prior expresses partial information about a variable.

Although we had specific, definite prior information about the pdf of θ, detailed in Section 6.4.2 where validation of G_α for in-phase (IP) biaxial loading conditions was considered, we chose an uninformative prior based on the following consideration: Our goal was to test two extensions of the definition of G_α formulated to account for out-of phase (OP) biaxial loading conditions. The prior information was for in-phase loading only. Using anything other than an uninformative prior would have tended to obscure the relevant information content of the experimental data.

Remark 6.12 Suppose that we would have formulated and tested G_α for out-of phase conditions first and then tested it under in-phase conditions, then it would have been proper to use the posterior pdf developed from OP data because the IP condition is a special case of the OP condition.

Inferential statistics

For both predictors we had $n = 3$ successful outcomes in $m = 3$ experiments. The uniform distribution is equivalent to the Beta distribution with the shape parameters set to $\alpha = \beta = 1$. In this case $C = 4$ and the posterior pdf is

$$Pr(\theta|D) = 4\theta^3. \tag{6.57}$$

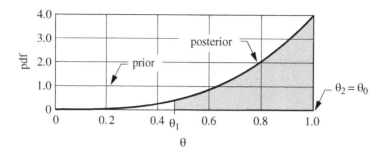

Figure 6.14 Posterior pdf corresponding to three successes in three trials.

The best estimate is given by the maximum of the posterior pdf, in the present case the best estimator of θ, denoted by θ_0, is $\theta_0 = 1$. It is customary to provide the shortest 95% confidence interval, also called credible interval. This means that we have to find θ_1 and θ_2 such that

$$Pr(\theta_1 < \theta_0 < \theta_2 | D) = 4 \int_{\theta_1}^{\theta_2} \theta^3 \, d\theta \approx 0.95 \tag{6.58}$$

subject to the condition that $\theta_2 - \theta_1$ is minimum. In the present case we find $\theta_1 = 0.473$, $\theta_2 = 1$. The shaded area in Fig. 6.14 represents how much we are justified in believing that θ lies in the interval (θ_1, θ_2).

The data tell us that the most likely outcome of the next experiment will be "pass". Put differently, the justifiable degree of belief that the next experiment will result in a pass of the validation experiment is very close to unity, given the available data. This degree of belief (also called plausibility) is modified through the process of conditionalization when new data become available. Conditionalization will be illustrated in Example 6.3.

Remark 6.13 In this example the pdf is highly asymmetric. While θ_0 is the most probable value of θ, the expected value, denoted by θ_M, may be more representative because it takes into account the skewness of the distribution [87]. Its value is: $\theta_M = 0.80$. Note that the predicted value of θ happens to lie in the middle of the interval (θ_M, θ_0).

Validation

Validation criteria are usually stated in terms of a metric, or a similar measure, such as the Kullback-Leibler divergence, that quantifies the difference between the predicted and realized pdfs, and an acceptable tolerance [69]. The problem is that setting any particular tolerance would seem arbitrary and difficult to justify. In validation we are asking the question: "Is there a reason that would justify rejecting the model?" – In the present case all three experiments passed the validation test. Clearly, there is no reason to reject this model. Had θ_0 been outside of the 95% confidence interval, we could not trust predictions based on the model and therefore we would have to consider rejecting it. In such obvious cases the answer is clear. In most cases, however, the problem will be like this: We performed m experiments, predicted n successes and observed n_{obs} successes, $n \neq n_{obs}$.

In general, a mathematical model is not rejected until and unless a better model is found, in which case the justification for rejection is based on the relative merit of competing models, measured by the posterior ratio. When the posterior ratio happens to be close to unity, as in the present instance, the competing models remain under consideration until new data become available that will serve to justify a decision one way or another.

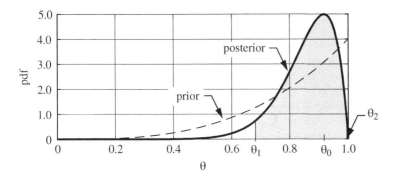

Figure 6.15 Example 2: Prior and posterior pdfs.

Considering that we used an uninformative prior, and only three data points, it is not surprising that the confidence interval is rather wide. In order to reduce the width of the confidence interval, additional experiments would have to be performed. The length of the confidence interval is approximately proportional to the square root of the number of experiments [87].

Example 6.3   Suppose, for the sake of illustration, that additional data became available indicating $n = 8$ successful outcomes in $m = 9$ experiments. Let us find the best estimate of θ and the confidence interval using the previously obtained posterior pdf for the prior:

$$Pr(\theta) \propto \theta^{\alpha-1}(1-\theta)^{\beta-1}$$

where $\alpha = 4$, $\beta = 1$. Therefore the new value of C from eq. (6.56) is:

$$C = \frac{\Gamma(14)}{\Gamma(11)\Gamma(3)} = 156$$

and the posterior pdf is

$$Pr(\theta|D) = 156 \times \theta^{11}(1-\theta). \tag{6.59}$$

The best estimate for θ_0, given by the maximum of the posterior pdf, is $\theta_0 = 0.917$ and the boundary points of the shortest 95% confidence interval are found to be $\theta_1 = 0.681, \theta_2 = 0.995$. In view of these results, our justifiable degree of belief is 95% that the limit value of θ_0 lies in this interval.

Remark 6.14   We used the posterior pdf given by eq. (6.57 for the prior in Example 6.3. The result would have been the same if we used the uniform distribution for the prior and all available records ($n = 11$ successes out of $m = 12$ trials). In other words, the results are independent of the ordering of the outcomes.

The updated domain of calibration
A mathematical model is a validated model within its domain of calibration. On a sufficiently small domain of calibration just about any model can be validated. New tests within a domain of calibration are useful for updating the parameters but cannot be considered as proper validation experiments. For a validation experiments to be proper, at least one of the admissible parameters must be outside of the domain of calibration, or the experimental conditions (such as loading conditions) must be more general than the conditions employed in the calibration experiments. On successful conclusion of a set of validation experiments, the parameters are updated and the domain of calibration is revised.

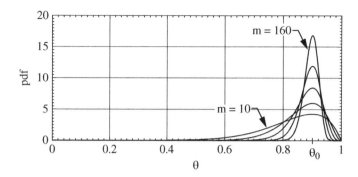

Figure 6.16   Ideal posterior pdfs corresponding to $m = 10, 20, 40, 80, 160$ ideal data points, $\theta_0 = 0.9$.

Table 6.12   Ideal data ($\theta_0 = 0.9$): Minimum 95% confidence intervals (*CI*).

m	10	20	40	80	160
θ_1	0.048	0.044	0.040	0.036	0.033
θ_2	0.998	0.994	0.990	0.986	0.983
CI	0.361	0.259	0.184	0.131	0.093

The extension of G_a to biaxial in-phase loading was validated with reference to one specimen type only. The values of the highly stressed volume were within the domain of calibration. For example, for axial loading: $V = 7.28E - 6$ in^3, for torsional loading: $V = 4.29E - 6$ in^3. Whereas calibration of $G_a(\beta)$ was under uniaxial loading only, the validation experiments included torsional and combined torsional and axial loading conditions.

The number of experiments
For the purpose of illustration, let us assume that we have ideal data such that $n = m\theta_0$, where $\theta_0 = 0.9$ is the ratio of the predicted number of successful outcomes to the number of experiments. The ideal posterior pdfs are illustrated for $m = 10, 20, 40, 80, 160$ in Fig. 6.16.

This figure illustrates that the confidence interval converges very slowly. The minimum 95% confidence intervals are listed in Table 6.12 where θ_1 and θ_2 are the lower and upper limits of the confidence interval (*CI*).

The empirical relationship is

$$CI \approx 1.123 \, m^{-0.4907}. \tag{6.60}$$

which is consistent with the estimate that the confidence interval is approximately proportional to $1/\sqrt{m}$, see, for example, [87]. The results indicate that, in order to reduce the size of the minimum 95% confidence interval to 0.1, well over 100 experiments would be necessary, which is many times more than what is usually available in practice.

6.5   Management of model development

While the procedures illustrated in this chapter were specific to the objective to generalize the results of fatigue tests performed on notch-free and notched coupons under constant cycle uniaxial

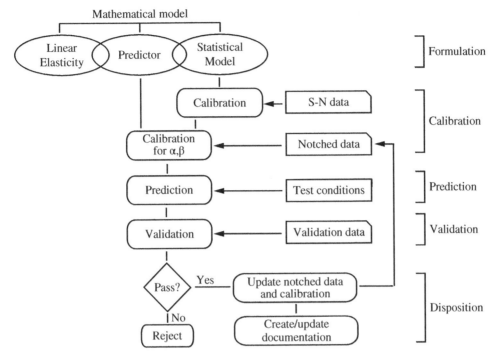

Figure 6.17 Schematic representation of the validation process.

loading conditions to variable cycle triaxial conditions, those procedures fit into the generic tasks of formulation, calibration, prediction, validation and disposition, common to all model development projects. This is illustrated schematically in Fig. 6.17.

The process is open-ended: The formulation of mathematical models involves making assumptions based on intuition, insight and personal preferences. This is a creative process, no boundaries can be set other than the requirement of logical consistency and the need to recognize that every assumption imposes some limitation on the model. Those limitations, together with limitations on the data available for calibration and testing, define the domain of calibration. The domain of calibration is an essential attribute of a model. New data within the domain of calibration present an opportunity to update the model parameters, and new data outside of the domain of calibration present an opportunity to revise the ranking of candidate models and update both the domain of calibration and the model parameters.

The development of mathematical models typically involves multiple disciplines and cuts across traditional departmental boundaries in industrial and research organizations. It is necessary for those organizations to exercise simulation governance, that is, create conditions that facilitate incremental improvement of simulation practices over time, see Section 5.2.4.

The economic stakes associated with the management of numerical simulation resources can be very substantial. Properly managed, numerical simulation can be a strategic corporate asset, poorly managed it can become a major liability. There are many well documented cases of substantial economic loss attributable to poor management of numerical simulation projects.

The justification for employing a particular mathematical model is based on negation, such as "we found no reason to reject the model" or "no one proposed a better model". Conceptually, one could reject a model based on some fixed criterion. In reality, however, it is very difficult to justify a particular criterion for rejection. A mathematical model is not rejected until and unless a better model is found. For example, one would not reject Peterson's predictor until a better one, such as Peterson's revised predictor, was found. The justification for preferring the predictor G_α over Peterson's revised predictor is that G_α has been shown to produce better predictions on a larger domain of calibration. Mathematical models evolve over time through testing, evaluating, comparing, updating and documenting.

6.5.1 Obstacles to progress

We mention two strongly correlated obstacles to the full realization of the potential of numerical simulation technology in engineering. One is that management has not yet grasped that current simulation practices in industrial settings are still very far from what is needed for full realization of the potential of the technology. The second is widespread confusion in the professional community evidenced by the all too frequent references to fuzzy terminology and empty notions.

An example of fuzzy terminology is "finite element modeling". As noted in Section 5.2.5, this term mixes two conceptually and functionally different entities: the mathematical model and its numerical treatment.

The prime example of empty notions is "physics-based model". To explain why it is empty, we quote from a renown theoretical physicist and an influential philosopher of science:

> *"I take the positivist viewpoint that a physical theory is just a mathematical model and that it is meaningless to ask whether it corresponds to reality. All that one can ask is that its predictions should be in agreement with observation."*

Stephen Hawking[10]

> *"The empirical basis of objective science has nothing absolute about it. Science does not rest on solid bedrock. The whole towering structure, the often fantastic and audacious construction of scientific theories, is built over a swamp. Its foundations are pillars driven from above into the swamp – not down to any natural 'given', ground, but driven just as deeply as is necessary to support the structure. The reason why we stop driving the pillars deeper into the ground is not that we have reached solid rock. No, our decision is based on the hope that the pillars will support the structure."*

Karl Popper[11]

They are telling us that physical reality cannot be known. Some phenomenological aspects of reality can be simulated by mathematical means with remarkable precision and success. However, limitations are always present in the mathematical models used in the foundational as well as in the applied sciences. Therefore the notion of a "physics-based model" is empty. A mathematical model is a precise statement of an idea of some aspects of reality and nothing more. The importance of proper terminology stands long recognized:

---

10 Stephen William Hawking 1942–2018. "The nature of space and time". Princeton University Press, 2010 (with Roger Penrose).
11 Karl Raimund Popper 1902–1994. "The Two Fundamental Problems of the Theory of Knowledge". Routledge, 2014.

"If names be not correct, language is not in accordance with the truth of things. If language be not in accordance with the truth of things, affairs cannot be carried on to success."

Confucius, The Analects – 13

Progress is possible through evolution. The best management can do is to create a friendly environment for the evolutionary process to take its course unimpeded. There are many opportunities for improvement in that area.

7

Beams, plates and shells

Dimensional reduction was discussed in Chapter 2 in connection with planar and axisymmetric models. Another very important class of dimensionally reduced models is discussed in this chapter. Our starting point is the generalized formulation of the problem of linear elasticity.

7.1 Beams

In order to present the main points in a simple setting, mathematical models for beams are derived from the generalized formulation of the problem of two-dimensional elasticity. The formulation of models for beams in three dimensions is analogous but, of course, more complicated. Referring to Fig. 7.1, the following assumptions are made:

1. The xy plane is a principal plane, that is, loads applied in the xy plane will not cause displacement in the direction of the z axis.
2. The x-axis passes through the centroid of the cross-section. The domain of the cross-section is denoted by ω.
3. The material is elastic and isotropic.

The displacement vector components are written in the following form:

$$u_x = u_{x|0}(x) + u_{x|1}(x)y + u_{x|2}(x)y^2 + \cdots + u_{x|m}(x)y^m \tag{7.1}$$

$$u_y = u_{y|0}(x) + u_{y|1}(x)y + u_{y|2}(x)y^2 + \cdots + u_{y|n}(x)y^n \tag{7.2}$$

where the functions $u_{x|k}(x)$, $u_{y|k}(x)$ are called field functions; their multipliers (powers of y) are called director functions. Writing the displacement components in this form allows us to consider a hierarchic family of models for beams characterized by the pair of indices (m, n). The highest member of the hierarchy is the mathematical model based on two-dimensional elasticity which corresponds to $m, n \to \infty$.

In the following we will denote the exact solution of a hierarchic beam model characterized by m and n by $\mathbf{u}_{EX}^{(m,n)}$ and the exact solution of the mathematical model based on two-dimensional elasticity by $\mathbf{u}_{EX}^{(2D)}$.

Finite Element Analysis: Method, Verification and Validation, Second Edition. Barna Szabó and Ivo Babuška.
© 2021 John Wiley & Sons, Inc. Published 2021 by John Wiley & Sons, Inc.
Companion Website: www.wiley.com/go/szabo/finite_element_analysis

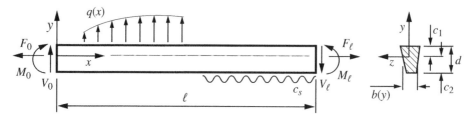

Figure 7.1 Notation.

Any of the boundary conditions described in connection with two-dimensional elasticity may be specified. However, we will be concerned only with the loadings and constraints typically used in the analysis of beams. Specifically, the following types of load will be considered:

1. The traction component $T_n = T_y$ acting on the surface $y = c_1$ (see Fig. 7.1):

$$T_n = \frac{q(x)}{b(c_1)} \tag{7.3}$$

   where q_y is the distributed load (in N/m units).

2. Normal tractions acting on cross-sections written in terms of the axial force F and the bending moment M. The sign conventions are indicated in Fig. 7.1.

$$T_n = \frac{F}{A} - \frac{My}{I} \tag{7.4}$$

   where A is the area of the cross-section and I is the moment of inertia of ω with respect to the z axis:

$$A = \int_\omega dydz, \qquad I = \int_\omega y^2\, dydz. \tag{7.5}$$

3. The shearing tractions $T_t(y, z)$ acting on cross-sections are either constant or defined by a function such that $T_t(c_1, z) = T_t(-c_2, z) = 0$ and the resultant is the shear force V:

$$T_t = -\frac{V}{A} \quad \text{or} \quad T_t = -\frac{VQ(y)}{Ib(y)} \tag{7.6}$$

   where $Q(y)$ is the static moment about the z axis of that portion of the cross-section which extends from y to c_1:

$$Q(y) = \int_y^{c_1} sb(s)\, ds. \tag{7.7}$$

The beam may be supported by an elastic foundation, with foundation modulus $c_s(x) \geq 0$ (N/m² units) and kinematic boundary conditions may be prescribed. We will assume that the elastic foundation reacts in the y direction only, i.e., the foundation generates a distributed transverse load $q_s(x) = -c_s(x)u_y(x, -c_2)$ (N/m units).

We will write the generalized formulations for beam models as applications of the principle of minimum potential energy. The strain energy is:

$$U = \frac{1}{2}\int_\Omega (\sigma_x \epsilon_x + \sigma_y \epsilon_y + \tau_{xy}\gamma_{xy})\, dV + \frac{1}{2}\int_0^\ell c_s u_y^2(x, -c_2)\, dx \tag{7.8}$$

and the potential of external forces is

$$P = \int_0^\ell qu_y(x, c_1)\, dx + \sum_{i=1}^2 \int_{\omega_i} (T_x u_x + T_y u_y)\, dS \tag{7.9}$$

where ω_i $(i = 1, 2)$ are the cross-sections at $x = 0$ and $x = \ell$ respectively. The energy space is defined in the usual way. The exact solution of a particular model is the minimizer of the potential energy on the space of admissible functions:

$$\Pi(\mathbf{u}_{EX}^{(m,n)}) = \min_{\mathbf{u}^{(m,n)} \in \tilde{E}(\Omega)} \Pi(\mathbf{u}^{(m,n)})$$

where $\Pi(\mathbf{u}) = U(\mathbf{u}) - P(\mathbf{u})$ and $\tilde{E}(\Omega)$ is the space of admissible functions.

Remark 7.1 Proper selection of a beam model characterized by the indices (m, n) is problem-dependent. When the field functions are smooth and the thickness is small then m and n can be small numbers. The assumptions incorporated into the commonly used beam models imply that the field functions do not change significantly over distances comparable to the thickness.

Remark 7.2 The distinction between the notions of mathematical model and its discretization is blurred by conventions in terminology: It is customary to refer to the various beam, plate and shell formulations as theories or models. These models are semidiscretizations of the fully three-dimensional model. Therefore errors that can be attributed to the choice of indices, such as the indices (m, n) in equations (7.1) and (7.2), and analogous indices defined for plate and shell models, identify both: a particular semidiscretization of the fully three-dimensional model and the definition of a beam, plate or shell model.

7.1.1 The Timoshenko beam

Let us consider the simplest model, the model corresponding to the indices $m = 1, n = 0$. We introduce the notation:

$$u_{x|0}(x) = u(x), \quad u_{x|1}(x) = -\beta(x), \quad u_{y|0}(x) = w(x).$$

The function β represents a positive (i.e., counterclockwise) angle of rotation (with respect to the z axis). Since a positive angle of rotation multiplied by positive y would result in negative displacement in the x direction, $u_{x|1} = -\beta$. The strain components are as follows:

$$\epsilon_x = \frac{\partial u_x}{\partial x} = u' - \beta'y, \quad \epsilon_y = \frac{\partial u_y}{\partial y} = 0, \quad \gamma_{xy} = \frac{\partial u_x}{\partial y} + \frac{\partial u_y}{\partial x} = -\beta + w' \tag{7.10}$$

where the primes represent differentiation with respect to x. Assuming that the normal stresses σ_y and σ_z are negligibly small in comparison with σ_x, the stress components are:

$$\sigma_x = E\epsilon_x = E(u' - \beta'y), \quad \sigma_y = 0, \quad \tau_{xy} = G\gamma_{xy} = G(-\beta + w').$$

Therefore the strain energy is:

$$U = \frac{1}{2} \int_0^\ell \left(\int_\omega [E(u' - \beta'y)^2 + G(-\beta + w')^2] \, dydz \right) dx + \frac{1}{2} \int_0^\ell c_s w^2 \, dx$$

where the volume integral was decomposed into an area integral over the cross-section and a line integral over the length of the beam. Using the notation introduced in eq. (7.5), and noting that since the y and z axes are centroidal axes, we have

$$\int_\omega y \, dydz = 0,$$

the strain energy can be written as:

$$U = \frac{1}{2} \int_0^\ell [EA(u')^2 + EI(\beta')^2 + GA(-\beta + w')^2]\, dx + \frac{1}{2} \int_0^\ell c_s w^2\, dx. \tag{7.11}$$

For the reasons discussed in the following section, the strain energy expression is usually modified by multiplying the shear term by a factor known as the shear correction factor, denoted by κ. The modified expression for the strain energy is:

$$U_\kappa = \frac{1}{2} \int_0^\ell [EA(u')^2 + EI(\beta')^2 + \kappa GA(-\beta + w')^2]\, dx + \frac{1}{2} \int_0^\ell c_s w^2\, dx. \tag{7.12}$$

The potential of external forces is

$$P = \int_0^\ell qw\, dx + F_\ell u(\ell) + M_\ell \beta(\ell) - V_\ell w(\ell) - F_0 u(0) - M_0 \beta(0) + V_0 w(0) \tag{7.13}$$

where F, M, V are respectively the axial force, bending moment and sher force, collectively called stress resultants, which are defined as follows:

$$F = \int_\omega \sigma_x\, dydz, \quad M = -\int_\omega \sigma_x y\, dydz, \quad V = -\int_\omega \tau_{xy}\, dydz. \tag{7.14}$$

The subscripts 0 and ℓ refer to the location $x = 0$ and $x = \ell$, respectively.

The potential energy is defined by:

$$\Pi_\kappa = U_\kappa - P. \tag{7.15}$$

This formulation is known as the Timoshenko beam model[1]. Particular applications of the principle of minimum potential energy depend on the specified loading and boundary conditions.

Shear correction

In this formulation the shear strain is constant on any cross-section (see eq. (7.10)). This is inconsistent with the assumption that no shear stresses are applied at the top and bottom surfaces of the beam, i.e., at $y = c_1$ and $y = -c_2$, see Fig. 7.1. From equilibrium considerations, it is known that the shear stress distribution is reasonably well approximated for a wide range of practical problems by the technical formula

$$\tau_{xy} = -\frac{VQ}{Ib}$$

where V is the shear force, $Q = Q(y)$ is the function defined by eq. (7.7), I is the moment of inertia about the z axis and $b(y)$ is the width of the cross section, as shown in Fig. 7.1. The derivation of this formula can be found in standard texts on strength of materials.

We will consider a rectangular cross-section of depth d and width b in the following. In this case the shear stress distribution is:

$$\tau_{xy} = \frac{3}{2} \frac{V}{db} \left(1 - 4\frac{y^2}{d^2}\right)$$

and the strain energy corresponding to this shear stress in a beam of length Δx is

$$\Delta U_\tau = \frac{1}{2} \frac{b\Delta x}{G} \int_{-d/2}^{+d/2} \tau_{xy}^2\, dy = \frac{1}{2} \frac{\Delta x}{G} \frac{V^2}{db} \frac{6}{5}.$$

---

1 Stephen P. Timoshenko 1878–1972.

We will adjust the shear modulus G in the $(1,0)$ beam model so that the strain energy will be the same:

$$\Delta U_\tau^{(1,0)} = \frac{1}{2}\frac{b\Delta x}{\kappa G}\int_{-d/2}^{+d/2}\tau_{xy}^2\,dy = \frac{1}{2}\frac{b\Delta x}{\kappa G}\int_{-d/2}^{+d/2}\left(\frac{V}{db}\right)^2\,dy = \frac{1}{2}\frac{\Delta x}{\kappa G}\frac{V^2}{db}$$

where κ is the shear correction factor. Letting $\Delta U_\tau = \Delta U_\tau^{(1,0)}$ we find that $\kappa = 5/6$. For this reason the strain energy expression given by eq. (7.11) is replaced by eq. (7.12). Most commonly $\kappa = 5/6$ is used, independently of the cross-section, although it is clear from the foregoing discussion that κ depends on the cross-section.

In static analyses of slender beams the strain energy is typically dominated by the bending term, hence the solution is not influenced significantly by κ. As the length to depth ratio decreases, the influence of κ increases. In vibrating beams the influence of κ increases with the natural frequency.

Exercise 7.1 Based on the formulation outlined in Section 1.7, write down the generalized formulation of the undamped elastic vibration problem for the Timoshenko beam. Assume $u_{x|0} = 0$. Would you expect the natural frequencies computed for the Timoshenko model to be larger or smaller than the corresponding natural frequencies computed for the beam model $(2,3)$? Why?

Numerical solution

The numerical solution of this problem by the finite element method is very similar to the solution of the one-dimensional model problem described in Section 1.3. The main difference is that here we have three field functions $u(x)$, $\beta(x)$ and $w(x)$, which are approximated by the finite element method. Let us write

$$u = \sum_{i=1}^{M_u} a_i \Phi_i(x), \quad \beta = \sum_{i=1}^{M_\beta} b_i \Phi_i(x), \quad w = \sum_{i=1}^{M_w} c_i \Phi_i(x)$$

and denote

$$a = \{a_1\ a_2\ \dots\ a_{M_u}\}^T, \quad b = \{b_1\ b_2\ \dots\ b_{M_\beta}\}^T, \quad c = \{c_1\ c_2\ \dots\ c_{M_w}\}^T.$$

The structure of the stiffness matrix becomes visible if we write the strain energy in the following form:

$$U_\kappa = \frac{1}{2}\{a^T\ b^T\ c^T\}\begin{bmatrix} K_u & 0 & 0 \\ 0 & K_\beta & K_{\beta w} \\ 0 & K_{\beta w}^T & K_w \end{bmatrix}\begin{Bmatrix} a \\ b \\ c \end{Bmatrix}.$$

Given the assumption that the shearing tractions on the top and bottom surfaces of the beam are zero and the x axis is coincident with the centroidal axis (see Fig. 7.1), the function u, representing axial deformation, is not coupled with the rotation β and the transverse displacement w and therefore can be solved independently of β and w.

At the element level, assuming constant sectional and material properties, K_β can be constructed directly from the matrices given by (1.66) and (1.70) and K_w can be constructed from (1.66).

For example, for $p_k = 2$ matrix $[K_\beta^{(k)}]$ is

$$[K_\beta^{(k)}] = \begin{bmatrix} \dfrac{EI}{\ell_k} + \dfrac{\kappa GA\ell_k}{3} & -\dfrac{EI}{\ell_k} + \dfrac{\kappa GA\ell_k}{6} & -\dfrac{\kappa GA\ell_k}{2\sqrt{6}} \\[3mm] & \dfrac{EI}{\ell_k} + \dfrac{\kappa GA\ell_k}{3} & -\dfrac{\kappa GA\ell_k}{2\sqrt{6}} \\[3mm] \text{(sym)} & & \dfrac{2EI}{\ell_k} + \dfrac{\kappa GA\ell_k}{5} \end{bmatrix}. \tag{7.16}$$

The terms of the coupling matrix $K_{\beta w}^{(k)}$ are:

$$k_{ij}^{(\beta w)} = GA \int_{-1}^{+1} N_i \frac{dN_j}{d\xi} \, d\xi.$$

Remark 7.3 A different polynomial degree may be assigned to each field function on each element. In the following we will assume that all fields have the same polynomial degree p_k on an element but p_k may vary from element to element.

Exercise 7.2 Write down the terms of the element-level matrices K_w and $K_{\beta w}^{(k)}$ for $p_k = 2$.

Example 7.1 Consider a beam of constant cross-section with built-in (fixed) support (i.e., $u_x = u_y = 0$) at $x = 0$ and simple support ($u_y = 0$) at $x = \ell$. There is an intermediate support ($u_y = 0$) at $x = \ell/2$, as shown in Fig. 7.2(a). The beam is uniformly loaded by a constant distributed load $q = -q_0$. The goal is to determine the location and magnitude of the maximum bending moment. In this case the axial displacement u is zero and the potential energy is defined as follows:

$$\Pi = \frac{1}{2} \int_0^\ell [EI(\beta')^2 + \kappa GA(-\beta + w')^2] \, dx + \int_0^\ell q_0 w \, dx. \tag{7.17}$$

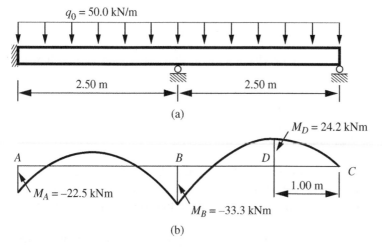

Figure 7.2 (a) Problem definition. (b) Bending moment diagram.

The spaces of admissible functions are defined as follows:

$$\tilde{E}_\beta \overset{\text{def}}{=} \left\{ \beta \mid \int_0^\ell [(\beta')^2 + \beta^2] \, dx \le C < \infty, \ \beta(0) = 0 \right\}$$

$$\tilde{E}_w \overset{\text{def}}{=} \left\{ w \mid \int_0^\ell (w')^2 \, dx \le C < \infty, \ w(0) = 0, \ w(\ell/2) = 0, \ w(\ell) = 0 \right\}.$$

The problem is to find β_{EX} and w_{EX} by minimization of the potential energy:

$$\Pi(\beta_{EX}, w_{EX}) = \min_{\substack{\beta \in \tilde{E}_\beta \\ w \in \tilde{E}_w}} \Pi(\beta, w).$$

Assume that the beam is an S200 × 27 American Standard steel beam[2]. The section properties are: $A = 3490$ mm, $I = 24.0 \times 10^6$ mm^4. Let $\ell = 5.00$ m, $E = 200$ GPa, $v = 0.3$ and $q_0 = 50.0$ kN/m. The weight of the beam ($27 \times 9.81 \times 10^{-3} = 0.265$ kN/m) is negligible in relation to the applied load.

The solution was obtained using two finite elements and the shear correction factor $\kappa = 5/6$. At $p \ge 4$ the exact solution is obtained (up to round-off errors). The results are shown in Fig. 7.2(b).

Shear locking in Timoshenko beams

Let us consider a beam of rectangular cross-section of dimension $b \times d$ subject to a distributed load $q = d^3 f(x)$ and assume that the axial force is zero. In this case the potential energy is:

$$\Pi = \frac{1}{2} \int_0^\ell \left[\frac{Ebd^3}{12}(\beta')^2 + \kappa Gbd(-\beta + w')^2 \right] dx - d^3 \int_0^\ell f(x)w \, dx. \tag{7.18}$$

On factoring d^3 we have:

$$\Pi = \frac{d^3}{2} \int_0^\ell \left[\frac{Eb}{12}(\beta')^2 + \frac{\kappa Gb}{d^2}(-\beta + w')^2 \right] dx - d^3 \int_0^\ell f(x)w \, dx. \tag{7.19}$$

For sufficiently small d values the term $\kappa Gb/d^2$ is much larger than $Eb/12$. Since the minimum of Π is sought, this forces $\beta \to w'$ as $d \to 0$:

$$\lim_{d \to 0} \int_0^\ell (-\beta + w')^2 \, dx = 0. \tag{7.20}$$

The effect of the constraint $\beta = w'$ is that the number of degrees of freedom is reduced. Convergence can be very slow when low polynomial degrees are used. This is called shear locking. As $d \to 0$, the solution of the Timoshenko model converges to the solution of the Bernoulli-Euler model which is described in the following section.

7.1.2 The Bernoulli-Euler beam

The Bernoulli-Euler beam model[3] is the limiting case of the Timoshenko model, and all higher order models, with respect to $d \to 0$. Assuming that $u = 0$ and $c_s = 0$ and letting $\beta = w'$, for homogeneous boundary conditions the potential energy expression (7.15) becomes

$$\Pi = \frac{1}{2} \int_0^\ell EI(w'')^2 \, dx - \int_0^\ell qw \, dx. \tag{7.21}$$

---

2 In this designation S indicates the cross-section; 200 is the nominal depth in mm, and 27 is the mass per unit length (kg/m).
3 Leonhard Euler 1707–1783, James Bernoulli 1654–1705.

This is the generalized formulation corresponding to the familiar fourth order ordinary differential equation that can be found in every introductory text on the strength of materials:

$$(EIw'')'' = q(x) \tag{7.22}$$

which is the Bernoulli-Euler beam model. To show this, let $w \in \tilde{E}(I)$ be the minimizer of Π and let $v \in E^0(I)$ be an arbitrary perturbation of w. For the sake of simplicity let us assume that the prescribed displacements, rotations, moments and shear forces are zero. Then,

$$\Pi(w + \varepsilon v) = \frac{1}{2} \int_0^\ell EI(w'' + \varepsilon v'')^2 \, dx - \int_0^\ell q(w + \varepsilon v) \, dx$$

will be minimum at $\varepsilon = 0$:

$$\left(\frac{\partial \Pi}{\partial \varepsilon} \right)_{\varepsilon=0} = 0.$$

Therefore

$$\int_0^\ell EIw''v'' \, dx - \int_0^\ell qv \, dx = 0.$$

On integrating by parts twice, we get:

$$(\underbrace{EIw'' \, v'}_{M})_0^\ell - \underbrace{((EIw'')'v)_0^\ell}_{V} + \int_0^\ell [(EIw'')'' - q]v \, dx = 0.$$

This must hold for any choice of v that does not perturb the prescribed essential boundary conditions. Since $M = EIw''$ and $V = (EIw'')'$, the boundary terms vanish and we have the strong form of the Bernoulli-Euler beam model given by eq. (7.22).

Exercise 7.3 Show that eq. (7.22) can be obtained without the assumption that moments or shear forces prescribed on the boundaries are zero. For the definition of moments and shear forces refer to eq. (7.14). Hint: If a non-zero moment and/or shear force is prescribed on a boundary then the expression for the potential energy given by eq. (7.21) must be modified to account for this term.

Exercise 7.4 Derive the strong form of the Bernoulli-Euler beam model for the case $c_s \neq 0$.

Exercise 7.5 Assume that $u = 0$ and $c_s = 0$. Show that the strong form of the Timoshenko model is:

$$(EI\beta')' + \kappa GA(-\beta + w') = 0 \tag{7.23}$$
$$[\kappa GA(-\beta + w')]' = -q \tag{7.24}$$

and hence show that in Example 7.1 the exact solution is obtained when $p \geq 4$. Hint: The procedure is analogous to that described in Section 7.1.2.

Numerical solution

For the Bernoulli-Euler beam model the energy space is

$$E(I) \stackrel{\text{def}}{=} \{ w \mid \int_0^\ell (w'')^2 \, dx \leq C < \infty \}, \qquad I = \{ x \mid 0 < x < \ell \}. \tag{7.25}$$

This implies that $w \in E(I)$ is continuous and its first derivative is also continuous. Functions that are continuous up to and including their nth derivatives belong in the space $C^n(\Omega)$, see

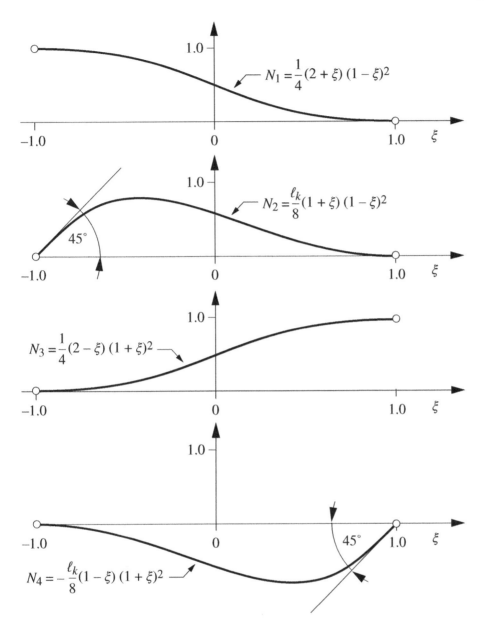

$$N_1 = \frac{1}{4}(2 + \xi)(1 - \xi)^2$$

$$N_2 = \frac{\ell_k}{8}(1 + \xi)(1 - \xi)^2$$

$$N_3 = \frac{1}{4}(2 - \xi)(1 + \xi)^2$$

$$N_4 = -\frac{\ell_k}{8}(1 - \xi)(1 + \xi)^2$$

Figure 7.3 The first four C^1 shape functions in one dimension.

Section A.2.1. Until now we have considered functions in $C^0(\Omega)$ only. Functions that lie in $E(I)$, defined by eq. (7.25), must also lie in $C^1(I)$.

The first four shape functions defined on the standard beam element ($-1 < \xi < +1$) are shown in Fig. 7.3. Note that N_2 and N_4 are scaled by the factor $\ell_k/2$. This is because

$$\theta = \frac{dw}{dx} = \frac{2}{\ell_k}\frac{dw}{d\xi}.$$

To define $N_i(\xi)$, $i \geq 5$, it is convenient to introduce the function $\psi_j(\xi)$ which is analogous to $\phi_j(\xi)$ given by eq. (3.24):

$$\psi_j(\xi) = \sqrt{\frac{2j-3}{2}} \int_{-1}^{\xi} \int_{-1}^{s} P_{j-2}(t) \, dt \, ds, \quad j = 4, 5, \ldots \tag{7.26}$$

where $P_{j-2}(t)$ is the Legendre polynomial of degree $j - 2$. For example,

$$\psi_4(\xi) = \frac{1}{8} \sqrt{\frac{5}{2}} (\xi^2 - 1)^2.$$

Then, for $i \geq 5$;

$$N_i(\xi) = \psi_{i-1}(\xi).$$

The stiffness matrix for a Bernoulli-Euler beam element of length ℓ_k, constant EI and $p = 5$ is:

$$[K] = \frac{EI}{\ell_k^3} \begin{bmatrix} 12 & 6\ell_k & -12 & 6\ell_k & 0 & 0 \\ & 4\ell_k^2 & -6\ell_k & 2\ell_k^2 & 0 & 0 \\ & & 12 & -6\ell_k & 0 & 0 \\ & \text{(sym.)} & & 4\ell_k^2 & 0 & 0 \\ & & & & 8 & 0 \\ & & & & & 8 \end{bmatrix}. \tag{7.27}$$

Remark 7.4 The simple model given by eq. (7.22) has been used successfully for the solution of practical problems for almost 200 years. It is remarkably accurate for the computation of deflections, rotations, moments and shear forces when the solution does not change substantially over distances of size d. This is not the case, however, for vibrating beams when the mode shapes have wavelengths close to d.

Timoshenko proposed high frequency vibrations. Of course, the Timoshenko model has limitations as well. Furthermore, the accuracy of a particular model cannot be ascertained unless it can be shown that the data of interest are substantially independent of the model characterized by the indices (m, n), the boundary conditions and other modeling decisions. For these reasons there is a need for a model hierarchy.

Remark 7.5 The Bernoulli-Euler and Timoshenko beam models are used with the objective to determine reactions, shear force and bending moment diagrams. Having determined the bending moment M, the normal stress is computed from the formula:

$$\sigma = -\frac{My}{I} \tag{7.28}$$

where y is a centroidal axis, see Fig. 7.1. For the Bernoulli-Euler model this formula is derived from

$$\sigma \equiv \sigma_x = -Ew''y \quad \text{and} \quad M = -\int_A \sigma_x y \, dA = Ew'' \int_A y^2 \, dA = EIw''.$$

For the Timoshenko model the derivation is analogous. Stresses computed in this way can be very accurate away from the supports and points where concentrated forces are applied but very inaccurate in the neighborhoods of those points where the assumptions incorporated into these models do not hold. Nevertheless, engineering design is based on the maximum stress computed from the formula (7.28) subject to the requirement that $\|\sigma\|_{\max}$ must be less than an allowable value which is approximately two-thirds of the yield stress. This makes sense only if we understand that the

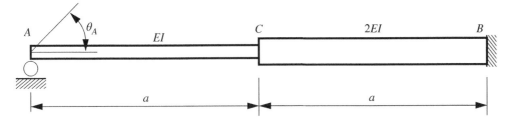

Figure 7.4 Problem definition for Exercise 7.8.

actual goal is to design beams such that $\|M\|_{\max}$ is much less than the moment that would cause extensive plastic deformation.

Exercise 7.6 Determine $\psi_5(\xi)$ and verify k_{66} in eq. (7.27).

Exercise 7.7 Verify the value of k_{34} in eq. (7.27).

Exercise 7.8 The beam shown Fig. 7.4 is simply supported on the left and fixed on the right. A positive rotation θ_A is imposed on the simply supported end.

1. Using the Bernoulli-Euler beam model, determine the displacement and rotation of the centroidal axis at point C in terms of θ_A and a.
2. Given the exact value of the strain energy:

$$U_{EX} = \frac{12}{11}\frac{EI}{a}\theta_A^2$$

   what is the moment M_A that had to be imposed to cause the θ_A rotation of the centroidal axis? Is the moment positive or negative? (Hint: The work done by the moment equals the strain energy.)
3. If the Timoshenko beam model were used, would the exact value of M_A be larger or smaller? Explain.
4. If the number of degrees of freedom is increased by uniform mesh refinement, does the strain energy increase, decrease or remain the same? Explain.

Exercise 7.9 The multi-span beam shown in Fig. 7.5 is fixed at the ends and simply supported in three points. The bending stiffness EI is constant.

1. Taking advantage of the symmetry, state the principle of minimum potential energy and specify the space of admissible functions for the Bernoulli-Euler beam model.

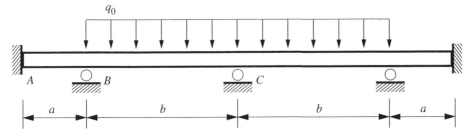

Figure 7.5 Problem definition for Exercise 7.9.

2. Find an expression for the rotation at support B in terms of the parameters q_0, EI, a and b for the Bernoulli-Euler beam model.

3. Is it possible to assign polynomial degrees to the elements such that the finite element solution is the exact solution (up to round-off errors)? Explain.

Exercise 7.10 Based on the formulation outlined in Section 1.7, write down the generalized formulation of the undamped elastic vibration problem for the Bernoulli-Euler beam model.

Exercise 7.11 Model the problem in Example 3.3 as a beam.

(a) Using the Bernoulli-Euler beam model, one element on the interval $0 < x < \ell/2$, find the rotation at $x = \ell/2$ in terms of δ/ℓ. Hint: Write down the constrained stiffness matrix and load vector for $p = 3$. Impose antisymmetry condition at $x = \ell/2$.

(b) Using the result obtained for part (a), compute the bending moment M_0 acting on boundary at $x = 0$ by extraction. Hint: Select $v = -N_2(\xi)$ for the extraction function. For the definition of $N_2(\xi)$ see Fig. 7.3.

7.2 Plates

The formulation of plate models is analogous to the formulation of beam models. The middle surface of the plate is assumed to lie in the $x\,y$ plane. The two-dimensional domain occupied by the middle surface is denoted by Ω and the boundary of Ω is denoted by Γ. The thickness of the plate is denoted by d and the side surface of the plate is denoted by S, that is: $S = \Gamma \times (-d/2,\ d/2)$. The displacement vector components are written in the following form:

$$
\begin{aligned}
u_x &= u_{x|0}(x, y) + u_{x|1}(x, y)z + \cdots + u_{x|m_x}(x, y)z^{m_x} \\
u_y &= u_{y|0}(x, y) + u_{y|1}(x, y)z + \cdots + u_{y|m_y}(x, y)z^{m_y} \\
u_z &= u_{z|0}(x, y) + u_{z|1}(x, y)z + \cdots + u_{z|n}(x, y)z^n
\end{aligned}
\tag{7.29}
$$

where $u_{x|0}, u_{x|1}, u_{y|0}$, etc. are independent field functions. In the following we will refer to a particular plate model by the indices $(m_x,\ m_y,\ n)$ and denote the corresponding exact solution by $\mathbf{u}_{EX}^{(m_x,m_y,n)}$. The exact solution of the corresponding problem of three-dimensional elasticity will be denoted by $\mathbf{u}_{EX}^{(3D)}$.

In analyses of plates the stress resultants rather than the stresses are of interest. The stress resultants are the membrane forces:

$$
F_x = \int_{-d/2}^{+d/2} \sigma_x\, dz \quad F_y = \int_{-d/2}^{+d/2} \sigma_y\, dz \quad F_{xy} = F_{yx} = \int_{-d/2}^{+d/2} \tau_{xy}\, dz
\tag{7.30}
$$

the transverse shear forces:

$$
Q_x = -\int_{-d/2}^{+d/2} \tau_{xz}\, dz \quad Q_y = -\int_{-d/2}^{+d/2} \tau_{yz}\, dz
\tag{7.31}
$$

and the bending and twisting moments:

$$
M_x = -\int_{-d/2}^{+d/2} \sigma_x\, z\, dz \quad M_y = -\int_{-d/2}^{+d/2} \sigma_y\, z\, dz
\tag{7.32}
$$

$$
M_{xy} = -M_{yx} = -\int_{-d/2}^{+d/2} \tau_{xy}\, z\, dz.
\tag{7.33}
$$

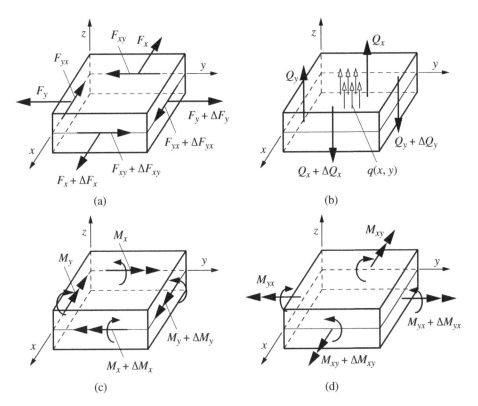

Figure 7.6 Sign convention for stress resultants.

M_x, M_y are called bending moments; M_{xy} is called twisting moment. The negative sign in the expressions for the shear force and bending moment components is made necessary by the conventions adopted for the stress resultants shown in Fig. 7.6 and the convention that tensile stresses are positive. Since $\tau_{xy} = \tau_{yx}$, the convention adopted for the twisting moments results in $M_{yx} = -M_{xy}$.

The starting point for the formulation of plate models is the principle of virtual work or, equivalently, the principle of minimum potential energy written in terms of the field functions with the integration performed in the z direction. We will consider a restricted form of the three-dimensional elasticity problems, using constraints and loads typically used in connection with the analysis of plates.

The boundary conditions are usually given in terms of the normal-tangent (n, t) system. It is left to the reader in the following exercises to derive the transformations from the (x, y) to the (n, t) systems.

Exercise 7.12 Refer to Fig. 7.7(a) and show that

$$Q_n = -\int_{-d/2}^{+d/2} \tau_{nz}\, dz = Q_x \cos\alpha + Q_y \sin\alpha. \tag{7.34}$$

Exercise 7.13 For the infinitesimal plate element shown in Fig. 7.7(b), subjected to bending and twisting moments only, show that

$$M_n = M_x\cos^2\alpha + M_y\sin^2\alpha + 2M_{xy} \sin\alpha \cos\alpha \tag{7.35}$$

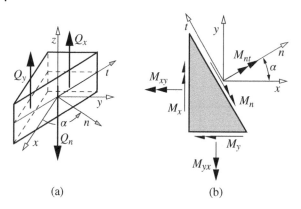

Figure 7.7 Transformation of stress resultants.

(a)   (b)

$$M_{nt} = -(M_x - M_y)\sin\alpha\cos\alpha + M_{xy}(\cos^2\alpha - \sin^2\alpha). \tag{7.36}$$

Hint: Use $M_{yx} = -M_{xy}$.

Exercise 7.14 Derive equations (7.35) and (7.36) from the definitions

$$M_n = -\int_{-d/2}^{+d/2} \sigma_n \, z \, dz \quad M_{nt} = -\int_{-d/2}^{+d/2} \tau_{nt} \, z \, dz$$

using the transformation given by eq. (K.9) and the formulas (7.32), (7.33).

Remark 7.6 The transformation of moments can be represented by a Mohr circle[4]. The maximum and minimum bending moments, denoted respectively by M_1 and M_2, called principal bending moments, occur at those values of α at which $M_{xy} = 0$. This can be seen by setting the first derivative of M_n with respect to α equal to zero. The principal moments are:

$$M_1 = \frac{M_x + M_y}{2} + R, \quad M_2 = \frac{M_x + M_y}{2} - R \tag{7.37}$$

where R is the radius of the Mohr circle:

$$R = \sqrt{\left(\frac{M_x - M_y}{2}\right)^2 + M_{xy}^2}.$$

7.2.1 The Reissner-Mindlin plate

The plate model (1, 1, 0), known as the Reissner-Mindlin plate[5], is widely used in finite element analysis. Its formulation is analogous to that of the Timoshenko beam. As in the Timoshenko beam model, the in-plane displacements are decoupled from the bending and shearing deformations. For simplicity we will be concerned with the bending and shearing deformations only, i.e., we let $u_{x|0} = u_{y|0} = 0$. We will use the notation

$$u_{x|1} = -\beta_x(x,y), \quad u_{y|1} = -\beta_y(x,y), \quad u_{z|0} = w(x,y)$$

hence the displacement vector components are of the form:

$$u_x = -\beta_x(x,y)z, \quad u_y = -\beta_y(x,y)z, \quad u_z = w(x,y) \tag{7.38}$$

---

4 Christian Otto Mohr 1835–1918.
5 Eric Reissner 1913–1996, Raymond David Mindlin 1906–1987.

and the strain terms are:

$$\epsilon_x = -\frac{\partial \beta_x}{\partial x} z \quad \epsilon_y = -\frac{\partial \beta_y}{\partial y} z \quad \epsilon_z = \frac{\partial w}{\partial z} = 0$$

$$\gamma_{xy} = -\left(\frac{\partial \beta_x}{\partial y} + \frac{\partial \beta_y}{\partial x}\right) z \quad \gamma_{yz} = -\beta_y + \frac{\partial w}{\partial y} \quad \gamma_{zx} = -\beta_x + \frac{\partial w}{\partial x}.$$

For the stress-strain law the plane stress relationships are used in the $x\,y$ plane (2.69). In the $y\,z$ and $z\,x$ planes the shear modulus is modified by the shear correction factor κ:

$$\sigma_x = \frac{E}{1-v^2}(\epsilon_x + v\epsilon_y) \quad \sigma_y = \frac{E}{1-v^2}(v\epsilon_x + \epsilon_y) \quad \sigma_z = 0$$

$$\tau_{xy} = G\gamma_{xy} \quad \tau_{yz} = \kappa G\gamma_{yz} \quad \tau_{zx} = \kappa G\gamma_{zx}.$$

Since the strain component ϵ_z and the stress component σ_z are both zero, this choice of stress-strain law may appear to be contradictory. The justification is that with this material stiffness matrix the exact solution of the Reissner-Mindlin model will approach the exact solution of the fully three-dimensional model as the thickness approaches zero:

$$\lim_{d \to 0} \frac{||\mathbf{u}_{EX}^{(3D)} - \vec{\mathbf{u}}_{EX}^{(110)}||_E}{||\vec{u}_{EX}^{(3D)}||_E} = 0. \tag{7.39}$$

The Reissner-Mindlin model would not have this property, called asymptotic consistency, if the stress-strain law of three-dimensional elasticity were used.

The strain energy is:

$$U_\kappa = \frac{1}{2}\int_V (\sigma_x \epsilon_x + \sigma_y \epsilon_y + \tau_{xy}\gamma_{xy} + \tau_{yz}\gamma_{yz} + \tau_{zx}\gamma_{zx})\,dxdydz + \frac{1}{2}\int_\Omega c_s w\,dxdy$$

where $c_s(x,y) \geq 0$ is a spring coefficient (in N/m^3 units). On substituting the expressions for stress and strain and integrating with respect to z, we have:

$$U_\kappa = \frac{1}{2}\int_\Omega D\left[\left(\frac{\partial \beta_x}{\partial x}\right)^2 + 2v\frac{\partial \beta_x}{\partial x}\frac{\partial \beta_y}{\partial y} + \left(\frac{\partial \beta_y}{\partial y}\right)^2 + \frac{1-v}{2}\left(\frac{\partial \beta_x}{\partial y} + \frac{\partial \beta_y}{\partial x}\right)^2\right.$$

$$\left. + \frac{6\kappa(1-v)}{d^2}\left(-\beta_y + \frac{\partial w}{\partial y}\right)^2 + \frac{6\kappa(1-v)}{d^2}\left(-\beta_x + \frac{\partial w}{\partial x}\right)^2\right]\,dxdy$$

$$+ \frac{1}{2}\int_\Omega c_s w\,dxdy \tag{7.40}$$

where D is the flexural rigidity. With the thickness denoted by t_z, the flexural rigidity is:

$$D \overset{\text{def}}{=} \frac{E t_z^3}{12(1-v^2)}. \tag{7.41}$$

The potential of external forces is:

$$P = \int_\Omega qw\,dxdy - \oint_\Gamma Q_n w\,ds + \oint_\Gamma M_n \beta_n\,ds + \oint_\Gamma M_{nt}\beta_t\,ds \tag{7.42}$$

where the $n\,t\,z$ coordinate system, shown in Fig. 7.7(a), is used. Noting that (by definition) $u_n = -\beta_n z$, $u_t = -\beta_t z$ and applying the rules of vector transformation we have:

$$\beta_n = \beta_x \cos \alpha + \beta_y \sin \alpha \tag{7.43}$$

$$\beta_t = -\beta_x \sin \alpha + \beta_y \cos \alpha. \tag{7.44}$$

Particular applications of the principle of minimum potential energy depend on the boundary conditions. The commonly used boundary conditions are:

(a) Fixed: $\beta_n = \beta_t = w = 0$
(b) Free: $M_n = M_{nt} = Q_n = 0$
(c) Simple support can be defined in two different ways for the Reissner-Mindlin plate model:
    (i) Soft simple support: $w = 0$, $M_n = M_{nt} = 0$
    (ii) Hard simple support: $w = 0$, $\beta_t = 0$, $M_n = 0$
(d) Symmetry: $\beta_n = 0$, $M_{nt} = 0$, $Q_n = 0$
(e) Antisymmetry: Same as hard simple support.

Shear correction for plate models

For the Reissner-Mindlin plate model either the energy or the average mid-surface deflection can be optimized with respect to the fully three-dimensional model by the choice of the shear correction factor.

$$\kappa = \begin{cases} \dfrac{5}{6(1-v)} & \text{for optimal energy} \\[2mm] \dfrac{20}{3(8-3v)} & \text{for optimal displacement.} \end{cases} \qquad (7.45)$$

For the model $(1,\ 1,\ 1)$ there is one shear correction factor:

$$\kappa = \begin{cases} \dfrac{5}{6} & \text{for } v = 0 \\[2mm] \dfrac{12-2v}{v^2}\left(-1+\sqrt{1+\dfrac{20v^2}{(12-2v)^2}}\right) & \text{for } v \neq 0. \end{cases} \qquad (7.46)$$

For all other models $(m_x,\ m_y,\ n)$ m_x, $m_y \geq 1, n \geq 2$ the shear correction factor is unity [15].

Example 7.2 Consider the domain of a plate shown in Fig. 7.8(a). The thickness is 2.5 mm. The material is an aluminum alloy, the elastic properties of which are: $E = 71.3$ GPa and $v = 0.33$. For the shear correction factor we use the value which is optimal for energy, see eq. (7.45). The plate is

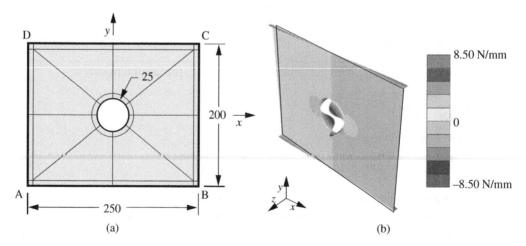

(a)   (b)

Figure 7.8 Example 7.2: (a) Definition of the domain and 28 element mesh. The dimensions are in millimeters. (b) Contours of Q_x for hard simple supports, $p = 8$, product space.

Table 7.1 Example 7.2. Estimated range of the stress resultants Q_x and Q_y (N/mm) on the region of primary interest. Mesh: 112 elements.

Boundary condition	Q_x	Q_y
Hard simple support	±8.717	±8.458
Soft simple support	±8.723	±8.465

simply supported on boundary segments BC and DA and free on segments AB and CD and on the circular boundary. It is loaded by a uniformly distributed load $q = -2.0$ kPa.

The goal is to compute the maximum and minimum values of the stress resultants Q_x and Q_y using the Reissner-Mindlin plate model with (a) soft simple supports and (b) hard simple supports. The region of primary interest is the neighborhood of the hole.

This example highlights some of the important questions that arise when using dimensionally reduced models. In the present case the questions are: how the choice of soft or hard simple support affects the quantities of interest, which shear correction factor to select, and how to design the finite element mesh to capture boundary layer effects. Other than knowing that boundary layer effects may be significant, these questions cannot be answered a priori. They have to be answered on the basis of feedback information gathered from finite element solutions.

The finite element mesh, shown in Fig. 7.8(a), was designed with the expectation that boundary layer effects may be significant. The width of the layer of elements at the boundaries was controlled by a parameter which was selected so as to minimize the potential energy at the highest polynomial degree, $p = 8$. This width was found to be 8.5 mm.

We find the quantities of interest over the elements that cover the region of primary interest, i.e. we exclude from consideration the elements adjacent to the external boundaries. In order to realize strong convergence in the quantities of interest, we subdivided each of the 28 elements shown in Fig. 7.8(a) into four elements to obtain a mesh consisting of 112 elements and performed p-extension using the product space. The results of computation are shown in Table 7.1. The minima and maxima occur along the perimeter of the hole.

If we sought the maximum of Q_x and Q_y on the entire domain, rather than on the domain of primary interest, we would have found that the maximum of Q_x diverges in the corner points of the plate as the number of degrees of freedom is increased, whereas the maximum of Q_y would have converged to the same value as shown in Table 7.1. The corner singularity is an artifact of modeling assumptions. Since the quantities of interest are not influenced by the singularity in a significant way, it can be neglected. For the hard simple support the same results as shown in Table 7.1 would have been obtained.

The contours of Q_x are plotted for the hard simple supports case in Fig. 7.8(b). The boundary layer effects are clearly visible, however the maximum value is not located in the boundary layer.

Exercise 7.15 Using the dimensions, elastic properties and boundary conditions for the plate described in Example 7.2, find the first three natural frequencies and estimate their relative error. Assume that the density is $\rho = 2780$ kg/m^3 (2.78×10^{-9} Ns^2/mm^4). For the shear correction factor use the value which is optimal for energy.

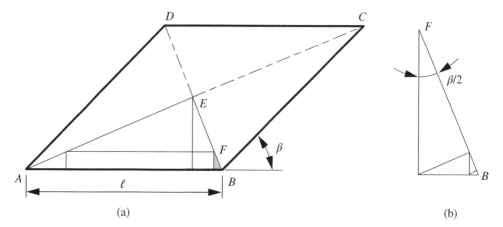

Figure 7.9 The rhombic plate problem, Exercise 7.17.

Exercise 7.16 Derive the expression (7.42) from eq. (2.97) assuming that only tractions are applied on the side surface $S = \Gamma \times (-d/2, d/2)$ of the plate. Hint: First show that

$$\int_{\partial \Omega_T} T_i u_i \, dS = \int_S (T_n u_n + T_t u_t + T_z u_z) \, dS.$$

Exercise 7.17 The mid-surface of a plate is an equilateral parallelogram (rhombus) characterized by the dimension ℓ and the angle β, as shown in Fig. 7.9(a). The plate is of uniform thickness d. The elastic properties are: $E = 2.0 \times 10^5$ MPa, $v = 0.3$. The plate is uniformly loaded, i.e., $q = q_0$ (constant) and simple support conditions are prescribed for all sides. Taking advantage of the two lines of symmetry, define the solution domain to be the triangle *ABE*. Let $\beta = \pi/6$ and $\ell/d = 100$.

This is one of the benchmark problems that have been used for illustrating the performance of various plate models and discretization schemes. The challenging aspect of the problem is the strong singularity at the obtuse corners B and D when simple support is prescribed on all boundaries.

Using the Reissner-Mindlin plate model, estimate the displacement w and the principal bending moments M_1 and M_2 in Point E, assuming (a) hard and (b) soft simple support prescribed for all sides. Report the displacement and moments in dimensionless form: $wD/(q_0\ell^4)$ and $M_i/(q_0\ell^2)$, $i = 1, 2$. Use the shear correction factor for optimal energy.

7.2.2 The Kirchhoff plate

The Kirchhoff plate model[6] is analogous to the Bernoulli-Euler beam model. The formulation is obtained by letting

$$\beta_x = \frac{\partial w}{\partial x}, \quad \beta_y = \frac{\partial w}{\partial y}$$

---

6 Gustav Robert Kirchhoff 1824–1887.

in eq. (7.40). Therefore the strain energy of the Kirchhoff plate model is:

$$U_K = \frac{1}{2}\int_\Omega D\left[\left(\frac{\partial^2 w}{\partial x^2}\right)^2 + 2v\frac{\partial^2 w}{\partial x^2}\frac{\partial^2 w}{\partial y^2} + \left(\frac{\partial^2 w}{\partial y^2}\right)^2 + 2(1-v)\left(\frac{\partial^2 w}{\partial x\partial y}\right)^2\right]\,dxdy$$

$$+ \frac{1}{2}\int_\Omega c_s w\,dxdy. \tag{7.47}$$

The potential of external forces is obtained from (7.42) with the following modification:

$$\oint_\Gamma M_{nt}\beta_t\,ds \to \oint_\Gamma M_{nt}\frac{\partial w}{\partial s}\,ds = -\oint_\Gamma \frac{\partial M_{nt}}{\partial s}w\,ds + \underbrace{\oint_\Gamma \frac{\partial(M_{nt}\,w)}{\partial s}\,ds}_{0}$$

where we used $dt \equiv ds$ and the assumption that the product $M_{nt}w$ is continuous and differentiable. Therefore we have:

$$P_K = \int_\Omega qw\,dxdy + \oint_\Gamma M_n\frac{\partial w}{\partial n}\,ds - \oint_\Gamma \left(Q_n - \frac{\partial M_{nt}}{\partial s}\right)w\,ds. \tag{7.48}$$

In the Kirchhoff model simple support has only one interpretation; that of hard simple support. By definition:

$$\Pi(w) = U_K(w) - P_K(w) \tag{7.49}$$

and

$$E(\Omega) = \{w \mid U_K(w) \le C < \infty\}.$$

Define $\tilde{E}(\Omega) \subset E(\Omega)$ to be the space of functions that satisfy the prescribed kinematic boundary conditions. The problem is to find:

$$\Pi(w_{EX}) = \min_{w\in\tilde{E}(\Omega)} \Pi(w). \tag{7.50}$$

This formulation has great theoretical and historical significance (see, for example, [104]). However, it is not well suited for computer implementation because the basis functions have to be C^1 continuous. The difficulties associated with enforcement of C^1 continuity are discussed in the following section.

Exercise 7.18 Consider the Kirchhoff plate model and assume homogeneous boundary conditions. Following the procedure of Section 7.1.2, that is, letting

$$\frac{\partial\Pi(w + \epsilon v)}{\partial\epsilon}\bigg|_{\epsilon=0} = 0$$

where $\Pi(w)$ is defined by eq. (7.49) and v is an arbitrary test function in $E^0(\Omega)$, show that the function w that minimizes Π satisfies the biharmonic equation:

$$\frac{\partial^4 w}{\partial x^4} + 2\frac{\partial^4 w}{\partial x^2\partial y^2} + \frac{\partial^4 w}{\partial y^4} = \frac{q}{D}. \tag{7.51}$$

Exercise 7.19 We have seen in Exercise 7.18 that the strong form of Kirchhoff plate model is the biharmonic equation (7.51). Assume that all sides of the equilateral triangular plate and the rhombic plate shown in Fig. 7.19 are simply supported. Considering only the symmetric part of the asymptotic expansion, characterize the smoothness of the homogeneous solution of the Kirchhoff plate model at each of the singular points. For the rhombic plate let $\beta = \pi/6$.

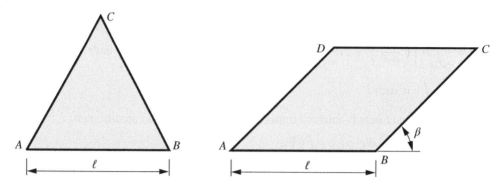

Figure 7.10 Exercise 7.19. Equilateral triangle and rhombic plate. Notation

Hints: (a) The homogeneous biharmonic equation in polar coordinates is given by eq. (G.1). (b) Letting $w = r^{\lambda+1}F(\theta)$, the symmetric eigenfunctions are of the form:

$$F_i(\theta) = a_i \cos(\lambda_i - 1)\theta + b_i \cos(\lambda_i + 1)\theta$$

see eq. (G.13). (c) The zero moment condition on the simply supported edges ($M_n = 0$) is equivalent to

$$\left(\frac{1}{r^2} \frac{\partial^2 w}{\partial \theta^2} \right)_{\theta = \pm \alpha/2} = 0$$

where $\theta = 0$ is the internal bisector of the vertex angle α as defined in Fig. 4.1.

Exercise 7.20 The exact solution of the Kirchhoff model for a simply supported and uniformly loaded equilateral triangular plate is a polynomial of degree 5 (see, for example, [104]). In the center of the plate the exact values of the displacement w and the bending moments are:

$$\frac{wD}{q_0 \ell^4} = \frac{1}{1728}, \quad \frac{M_x}{q_0 \ell^2} = \frac{1+v}{72}, \quad M_y = M_x, \quad M_{xy} = 0$$

where D is the plate constant, q_0 is the value of the constant distributed load and ℓ is the dimension shown in Fig. 7.19. Compare these results with their counterparts obtained with the Reissner-Mindlin model for (a) $\ell/d = 100$, $v = 0.3$ and (b) $\ell/d = 10$, $v = 0.3$ for hard and soft simple supports. Investigate the effects of shear correction factors for optimal energy and optimal displacement. Refer to Section 7.2.1.

Exercise 7.21 Consider a square plate, uniformly loaded, with one side fixed, the opposite side simply supported, the other two sides free. State the principle of minimum potential energy for this problem for the Kirchhoff and Reissner-Mindlin models.

Exercise 7.22 Refer to Exercise 7.21. Noting that the Reissner-Mindlin model allows distinction to be made between hard and soft simple supports whereas the Kirchhoff model does not, estimate the error of idealization of the Kirchhoff model in relation to the Reissner-Mindlin model through numerical experiments. Fix the size of the plate and vary the thickness. Compare the estimated limit values of the strain energy for soft and hard simple supports.

Enforcement of C^1 continuity

The enforcement of C^1 continuity in two dimensions requires special consideration. This is because it is not possible to enforce C^1 continuity, and not more than C^1 continuity, on polynomial basis functions. The continuity of the first and higher derivatives is enforced along the sides and at the vertices the normal derivatives must be continuous also. This means that the first derivatives in four different directions would have to be continuous. This is not possible unless the second derivatives are also continuous in the vertex points. That, however, causes problems at singular points where the second derivatives are discontinuous. To overcome this difficulty, composite elements and elements with rational basis functions have been developed.

It is best to avoid this problem by using plate and shell models that require enforcement of C^0 continuity only.

7.2.3 The transverse variation of displacements

In the formulation of plate and shell models the transverse variation of the displacement components was represented by polynomials. In this section it is shown that for homogeneous plates of small thickness the transverse variation of the displacements is indeed best represented by polynomials and for laminated plates by piecewise polynomials.

Consider the infinite strip of unit width shown in Fig. 7.11. Assume that the loading function $q = q(x)$ is periodic with period L, antisymmetric with respect to the plane $y = 0$, and satisfies the equations of equilibrium:

$$\int_{-\infty}^{\infty} q \, dx = \int_{-\infty}^{\infty} xq \, dx = 0. \tag{7.52}$$

Note that periodic loading can be generalized to nonperiodic loading using the Fourier integral method. We will be interested in the limit process $d/L \to 0$.

In plate models the transverse load q is referred to the mid-surface. We understand q to be an antisymmetric distributed load acting on the surfaces Γ^+ and Γ^-. Let $\beta = 2\pi/L$ and assume that the solution is of the form:

$$u_x = \phi(\beta, y) \sin \beta x \tag{7.53}$$

$$u_y = \psi(\beta, y) \cos \beta x \tag{7.54}$$

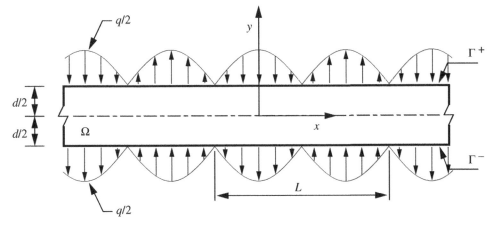

Figure 7.11 Infinite strip. Notation.

where $\phi(\beta, y)$ is antisymmetric and $\psi(\beta, y)$ is symmetric with respect to the mid-surface of the strip (the x-axis). The strain components are:

$$\epsilon_x = \frac{\partial u_x}{\partial x} = \beta\phi\cos\beta x \tag{7.55}$$

$$\epsilon_y = \frac{\partial u_y}{\partial y} = \psi'\cos\beta x \tag{7.56}$$

$$\gamma_{xy} = \frac{\partial u_x}{\partial y} + \frac{\partial u_y}{\partial x} = (\phi' - \beta\psi)\sin\beta x \tag{7.57}$$

where the primes represent differentiation with respect to y. We assume that the material is orthotropic with the material axes aligned with the x, y and z directions. Therefore the two-dimensional stress-strain relationship can be written in the form

$$\begin{Bmatrix} \sigma_x \\ \sigma_y \\ \tau_{xy} \end{Bmatrix} = \begin{bmatrix} E_1 & E_2 & 0 \\ E_2 & E_3 & 0 \\ 0 & 0 & E_6 \end{bmatrix} \begin{Bmatrix} \epsilon_x \\ \epsilon_y \\ \gamma_{xy} \end{Bmatrix} \tag{7.58}$$

where E_1, E_2, E_3 and E_6 may be functions of y. Therefore the stress components are:

$$\sigma_x = (E_1\beta\phi + E_2\psi')\cos\beta x \tag{7.59}$$

$$\sigma_y = (E_2\beta\phi + E_3\psi')\cos\beta x \tag{7.60}$$

$$\tau_{xy} = E_6(\phi' - \beta\psi)\sin\beta x. \tag{7.61}$$

The equilibrium equations with zero body force are

$$\frac{\partial\sigma_x}{\partial x} + \frac{\partial\tau_{xy}}{\partial y} = 0 \tag{7.62}$$

$$\frac{\partial\tau_{xy}}{\partial x} + \frac{\partial\sigma_y}{\partial y} = 0. \tag{7.63}$$

On substituting equations (7.59) to (7.61) into the equilibrium equations we get

$$-E_1\beta^2\phi - E_2\beta\psi' + (E_6\phi')' - (E_6\beta\psi)' = 0 \tag{7.64}$$

$$E_6\beta\phi' - E_6\beta^2\psi + (E_2\beta\phi)' + (E_3\psi')' = 0. \tag{7.65}$$

Expanding $\phi(\beta, y)$ and $\psi(\beta, y)$ into a power series with respect to β:

$$\phi(\beta, y) = \phi_0(y) + \beta\phi_1(y) + \beta^2\phi_2(y) + \cdots \tag{7.66}$$

$$\psi(\beta, y) = \psi_0(y) + \beta\psi_1(y) + \beta^2\psi_2(y) + \cdots \tag{7.67}$$

and substituting equations (7.66) and (7.66) into equations (7.64) and (7.65) we get:

$$\begin{aligned} &- E_1(\beta^2\phi_0 + \beta^3\phi_1 + \beta^4\phi_2 + \cdots) - E_2(\beta\psi_0' + \beta^2\psi_1' + \beta^3\psi_2' + \cdots) \\ &+ [E_6(\phi_0' + \beta\phi_1' + \beta^2\phi_2' + \cdots)]' - [E_6(\beta\psi_0 + \beta^2\psi_1 + \beta^3\psi_2 + \cdots)]' = 0 \end{aligned} \tag{7.68}$$

$$\begin{aligned} &E_6(\beta\phi_0' + \beta^2\phi_1' + \beta^3\phi_2' + \cdots) - E_6(\beta^2\psi_0 + \beta^3\psi_1 + \beta^4\psi_2 + \cdots) \\ &+ [E_2(\beta\phi_0 + \beta^2\phi_1 + \beta^3\phi_2 + \cdots)]' + [E_3(\psi_0' + \beta\psi_1' + \beta^2\psi_2' + \cdots)]' = 0. \end{aligned} \tag{7.69}$$

Note that these equations hold for any β.

Case A: The material properties are independent of *y*

Letting $\beta = 0$ and assuming that the material properties are independent of y, from equations (7.68), (7.69) we have

$$\phi_0'' = 0, \quad \psi_0'' = 0. \tag{7.70}$$

Therefore:

$$\phi_0(y) = a_1 y + a_2, \quad \psi_0(y) = b_1 y + b_0. \tag{7.71}$$

Since by assumption $\phi_0(y)$ is antisymmetric and $\psi_0(y)$ is symmetric:

$$\phi_0(y) = a_1 y, \quad \psi_0(y) = b_0. \tag{7.72}$$

Differentiating equations (7.68), (7.68) with respect to β, setting $\beta = 0$ and using (7.72) the following equations are obtained:

$$\phi_1'' = 0, \quad \psi_1'' = -\frac{E_2 + E_6}{E_1} a_1. \tag{7.73}$$

Solving for $\phi_1(y)$ and dropping the symmetric term; solving for $\psi_1(y)$ and dropping the antisymmetric term, we have:

$$\phi_1 = c_1 y, \quad \psi_1(y) = -\frac{E_2 + E_6}{2E_1} a_1 y^2 + d_0. \tag{7.74}$$

Equations (7.72) indicate that the solution should be in the form:

$$u_x(x, y) = u_{x|1}(x)y \tag{7.75}$$
$$u_y(x, y) = u_{y|0}(x) + u_{y|2}(x)y^2 \tag{7.76}$$

which is the assumed functional form in equations (7.1) and (7.2) modified by the additional assumptions that $u_x(x, y)$ is antisymmetric and $u_y(x, y)$ is symmetric and truncating the expressions at the quadratic term. This choice of the mode of deformation satisfies the equilibrium equations (7.68) and (7.69) up to the first power of β. By continuing this process, the equilibrium equations can be satisfied to an arbitrary power of β.

Case B: The material properties are symmetric functions of *y*

If the material properties are not independent of y then setting $\beta = 0$ in equations (7.68) and (7.69) we have:

$$(E_6 \phi_0')' = 0, \quad (E_1 \psi_0')' = 0. \tag{7.77}$$

In many practical problems the material properties are symmetric functions of y. For example, the strip may be made of laminae which are symmetrically arranged with respect to the xz plane, then, knowing that $\phi_0(y)$ is antisymmetric and $\psi_0(y)$ is symmetric, we have:

$$\phi_0(y) = a_1 F_0(y), \quad \psi_0 = b_0 \tag{7.78}$$

where

$$F_0(y) \overset{\text{def}}{=} \int_0^y \frac{1}{E_6(t)}\, dt. \tag{7.79}$$

Differentiating equations (7.68) and (7.69) with respect to β and letting $\beta = 0$ the following equations are obtained:

$$-E_2\psi_0' + (E_6\phi_1')' - (E_6\psi_0)' = 0 \tag{7.80}$$

$$E_6\phi_0' + (E_2\phi_0)' + (E_1\psi_1')' = 0. \tag{7.81}$$

Using (7.78), from (7.80) we get

$$[E_6(\phi_1' - \psi_0)]' = 0. \tag{7.82}$$

Solving for $\phi_1(y)$ and using the fact that $\phi_1(y)$ is antisymmetric:

$$\phi_1(y) = c_1 \int_0^y \frac{1}{E_6(t)} \, dt + b_0 y. \tag{7.83}$$

From equation (7.77) we have $E_6\phi_0' = a_1$ (constant) therefore equation (7.81) can be written as

$$a_1 + (E_2\phi_0)' + (E_1\psi_1')' = 0. \tag{7.84}$$

Solving for ψ_1:

$$\psi_1(y) = -a_1 \int_0^y \frac{t}{E_1(t)} \, dt + d_1 \int_0^y \frac{1}{E_1(t)} \, dt - a_1 \int_0^y \frac{E_2(t)}{E_1(t)} F_0(t) \, dt + d_0 \tag{7.85}$$

since the second term is antisymmetric, $d_1 = 0$. Defining:

$$F_1(y) = \int_0^y \frac{t}{E_1(t)} \, dt + \int_0^y \frac{E_2(t)}{E_1(t)} F_0(t) \, dt \tag{7.86}$$

we have:

$$\psi_1(y) = -a_1 F_1(y) + d_0 \tag{7.87}$$

and the solution is of the form:

$$u_x(x, y) = u_{x|1}(x)y + u_{x|2}(x)F_0(y) \tag{7.88}$$

$$u_y(x, y) = u_{y|0}(x) + u_{y|2}(x)F_1(y). \tag{7.89}$$

For additional details and examples we refer to [23].

Remark 7.7 The definition, essential properties and formulation of hierarchic models for laminated plates and shells were first addressed in [1], [23], [91]. If the goal of computation is to estimate the strength of laminated structural members then it is necessary to identify critical regions in the macromechanical model and estimate the values of predictors of failure defined at the fiber-matrix level. Predictors of failure are functionals that have been correlated with failure events through interpretation of the outcome of physical experiments. Owing to the complexity of the problem, large aspect ratios, strong boundary layer effects and the requirement of a posteriori error estimation, high order methods, coupled with proper mesh design, are expected to play an increasingly important role in this field.

Exercise 7.23 Assuming that the material properties are symmetric functions of y, construct u_x and u_y so as to satisfy the equilibrium equations (7.68) and (7.69) up to the second power of β.

7.3 Shells

The formulation of mathematical models for structural shells is a very large and rather complicated subject. Only a brief overview of some of the salient points is presented in the following.

A structural shell is characterized by a surface, called mid-surface x_i, and the thickness d. Both are given in terms of two parameters α_1, α_2:

$$x_i = x_i(\alpha_1, \alpha_2), \quad d = d(\alpha_1, \alpha_2).$$

Note that the indices of the parameters α_i take on the values $i = 1, 2$ whereas the indices of the spatial coordinates x_i range from 1 to 3. Associated with each point of the mid-surface are three basis vectors. Two of the basis vectors lie in the tangent plane at the point (α_1, α_2):

$$b_i^{(1)} = \frac{\partial x_i}{\partial \alpha_1}, \qquad b_i^{(2)} = \frac{\partial x_i}{\partial \alpha_2}.$$

Note that $b_i^{(1)}$ and $b_i^{(2)}$ are not necessarily orthogonal. However, in classical treatments of shells, the parameters α_1, α_2 are usually chosen so that the basis vectors are orthogonal. The third basis vector $b_i^{(3)}$ is the cross product of $b_i^{(1)}$ and $b_i^{(2)}$, therefore it is normal to the tangent plane. These are called curvilinear basis vectors. The normalized curvilinear basis vectors are denoted by $\mathbf{e}_\alpha, \mathbf{e}_\beta, \mathbf{e}_n$. The Cartesian unit basis vectors are denoted by $\mathbf{e}_x, \mathbf{e}_y, \mathbf{e}_z$. A vector $\mathbf{u}$ given in terms of the curvilinear basis vectors is denoted by $\mathbf{u}_{(\alpha)}$, in Cartesian coordinates by $\mathbf{u}_{(x)}$. The transformation is

$$\mathbf{u}_{(x)} = [R]\mathbf{u}_{(\alpha)} \tag{7.90}$$

where the columns of the transformation matrix $[R]$ are the unit vectors $\mathbf{e}_\alpha, \mathbf{e}_\beta, \mathbf{e}_n$.

The classical development of shell models was strongly influenced by the limitations of the methods available for solving the resulting partial differential equations. The use of curvilinear coordinates allowed the treatment of shells with simple geometric description, such as cylindrical, spherical and conical shells, by classical methods subject to the condition that the thickness of the shell is small in relation to its other dimensions. Those limitations no longer exist.

Shells should be viewed as fully three-dimensional solids that may allow a priori restrictions on the transverse variation of displacements in certain regions, usually not the regions of primary interest. In virtually all practical applications there are regions where the assumptions incorporated into shell models do not hold. Examples are the neighborhoods of nozzles, support attachments, stiffeners, cut-outs and joints where the curvature abruptly changes. From the point of view of strength analysis those are the usual regions of primary interest. The thickness of shells of engineering interest is rarely very small in relation to the other dimensions.

The hierarchic shell models are generalizations of the hierarchic plate models represented by eq. (7.29). The displacement vector components are given in the following form:

$$u_\alpha = \sum_{i=0}^{m_\alpha} u_{\alpha|i}(\alpha, \beta)\phi_i(\nu)$$

$$u_\beta = \sum_{i=0}^{m_\beta} u_{\beta|i}(\alpha, \beta)\phi_i(\nu) \tag{7.91}$$

$$u_n = \sum_{i=0}^{m_n} u_{n|i}(\alpha, \beta)\phi_i(\nu)$$

where $\phi_i(\nu)$ are called director functions. When the material is isotropic then $\phi_i(\nu)$ are polynomials; when the shell is laminated then $\phi_i(\nu)$ are piecewise polynomials (see, for example, [1]).

Eq. (7.91) represents a semi-discretization of the problem of fully three-dimensional elasticity, in the sense that $\phi_i(v)$ are fixed and thus the problem is reduced from a three-dimensional problem to a two-dimensional one. A particular shell model is characterized by the set of numbers (m_α, m_β, m_n).

In the classical treatment of shells the curvilinear basis vectors are retained throughout the analysis. In the finite element method a generalized formulation is used, most commonly the principle of virtual work. The algorithmic structure becomes simpler if the displacement components, given with reference to the curvilinear basis vectors, are transformed to the (global) Cartesian reference frame by eq. (7.90).

The generic form of the virtual work of internal stresses is given by

$$B(\mathbf{u}_{(\alpha)}, \mathbf{v}_{(\alpha)}) = \int_\omega \int_{-d/2}^{+d/2} \left([\tilde{D}][R]\mathbf{v}_{(\alpha)}\right)^T [E][\tilde{D}][R]\mathbf{u}_{(\alpha)} \, dv \, d\omega \tag{7.92}$$

where $[\tilde{D}]$ is the differential operator that transforms the displacement vector components given in terms of the curvilinear coordinates (α, β, v) to the Cartesian strain tensor components: $\{\epsilon\} = [\tilde{D}]\mathbf{u}_{(\alpha)}$. It is assumed that the material is isotropic. If the material is not isotropic, and its reference frame differs from the global Cartesian coordinate system, then $[E]$ must be transformed into the global Cartesian coordinate system.

The generic form of the virtual work of external forces is:

$$\begin{aligned}
F(\mathbf{v}_{(\alpha)}) = & \int_\omega \int_{-d/2}^{+d/2} \left([R]\mathbf{F}_{(\alpha)}\right)^T [R]\mathbf{v}_{(\alpha)} \, dv \, d\omega \\
& + \int_{\partial\omega} \int_{-d/2}^{+d/2} \left([R]\mathbf{T}_{(\alpha)}\right)^T [R]\mathbf{v}_{(\alpha)} \, dv \, ds \\
& + \int_\omega \int_{-d/2}^{+d/2} \left([\tilde{D}][R]\mathbf{v}_{(\alpha)}\right)^T [E]\{c\}\mathcal{T}_\Delta \, dv \, d\omega
\end{aligned} \tag{7.93}$$

where $\mathbf{F}_{(\alpha)}$ (resp. $\mathbf{T}_{(\alpha)}$) is the volume force (resp. surface traction vector) given in terms of the curvilinear basis, $\{c\}$ is the vector of coefficients of thermal expansion and $\mathcal{T}_\Delta$ is the temperature change.

The Naghdi shell model

The Naghdi shell model[7] [61] is analogous to the Reissner-Mindlin plate model: It is assumed that normals to the mid-surface prior to deformation remain straight lines but not necessarily normals after deformation. In other words, the kinematic assumptions account for some transverse shearing deformation. Specifically, the kinematic assumptions are the same as the kinematic assumptions for the hierarchic shell model (1,1,0):

$$\begin{aligned}
u_\alpha &= u_{\alpha|0}(\alpha, \beta) + u_{\alpha|1}(\alpha, \beta)v \\
u_\beta &= u_{\beta|0}(\alpha, \beta) + u_{\beta|1}(\alpha, \beta)v \\
u_n &= u_{n|0}(\alpha, \beta).
\end{aligned} \tag{7.94}$$

However, the definition of the material stiffness matrix is not the same as for the hierarchic shell model: Whereas in eq. (7.92) the definition of $[E]$ is that of the three-dimensional stress-strain relationship, in the Naghdi model $[E]$ is replaced by the stress-strain relationship which incorporates the assumption that plane stress conditions exist. This is necessary for asymptotic consistency.

---

7 Paul M. Naghdi 1924–1994.

The Novozhilov-Koiter shell model

The Novozhilov-Koiter shell model[8] [68] is an extension of the Euler-Bernoulli beam model and the Kirchhoff plate model to shells: It is assumed that normals to the mid-surface prior to deformation remain normals after deformation. The formulation involves replacement of the field functions u_α and u_β in Eqn. (7.94) by a linear combination of the first derivatives of u_n. Consequently the number of field functions is reduced to three and the second derivatives appear in Eqn. (7.92), which implies that the space of admissible functions has to have C^1 continuity. From the point of view of computer implementation, the advantages of using only three field functions are far outweighed by disadvantages of the requirement of C^1 continuity and restrictions on the kinematic boundary conditions. This model is mainly of theoretical and historical interest today.

7.3.1 Hierarchic thin solid models

An alternative approach to hierarchic semi-discretization is to write the Cartesian components of the displacement vector in the form

$$
\begin{aligned}
u_x &= \sum_{i=0}^{q} u_{x|i}(\alpha, \beta)\phi_i(\nu) \\
u_y &= \sum_{i=0}^{q} u_{y|i}(\alpha, \beta)\phi_i(\nu) \\
u_z &= \sum_{i=0}^{q} u_{z|i}(\alpha, \beta)\phi_i(\nu)
\end{aligned}
\tag{7.95}
$$

where q is typically a small number. The generic form of the virtual work of internal stresses is given by the expression

$$
B(\mathbf{u}_{(x)}, \mathbf{v}_{(x)}) = \int_\omega \int_{-d/2}^{+d/2} \left([D]\mathbf{v}_{(x)}\right)^T [E][D]\mathbf{u}_{(x)} \, d\nu \, d\omega.
\tag{7.96}
$$

The generic form of the virtual work of external forces is:

$$
\begin{aligned}
F(\mathbf{v}_{(\alpha)}) = &\int_\omega \int_{-d/2}^{+d/2} \mathbf{F}_{(x)} \cdot \mathbf{v}_{(x)} \, d\nu \, d\omega + \int_{\partial\Omega} \mathbf{T}_{(x)} \cdot \mathbf{v}_{(\alpha)} \, dS \\
&+ \int_\omega \int_{-d/2}^{+d/2} \left([D]\mathbf{v}_{(x)}\right)^T [E] \, \mathbf{c}\tau \, d\nu \, d\omega
\end{aligned}
\tag{7.97}
$$

where $\mathbf{F}_{(x)}$ (resp. $\mathbf{T}_{(x)}$) is the Cartesian volume force vector (resp. surface traction vector) given in terms of the curvilinear variables (α, β, ν) and $\mathbf{c}\tau$ is the thermal strain. Such models are called "thin solid" models.

In computer applications anisotropic trunk or anisotropic product spaces are used. The anisotropic trunk space on the standard hexahedral element $\Omega_{st}^{(h)}$ is defined by:

$$
\begin{aligned}
S_{tr}^{ppq}(\Omega_{st}^{(h)}) = &\operatorname{span}(\xi^k \eta^\ell \zeta^m, \ \xi^p \eta \zeta^m, \ \xi \eta^p \zeta^m, \ (\xi, \eta, \zeta) \in \Omega_{st}^{(h)}, \\
&k, \ell = 0, 1, 2, \dots, \ k + \ell \le p, \ m = 0, 1, 2 \dots, q), \quad p > 1.
\end{aligned}
\tag{7.98}
$$

---

8 Valentin Valentinovich Novozhilov 1910–1987, Warner Tjardus Koiter 1914–1997.

The parameter q is fixed, and $p > q$ is increased until convergence is realized. The anisotropic product space on $\Omega_{st}^{(h)}$ is defined by:

$$S_{pr}^{ppq}(\Omega_{st}^{(h)}) = \text{span}(\xi^k \eta^\ell \zeta^m, \ (\xi, \eta, \zeta) \in \Omega_{st}^{(h)},$$
$$k, \ell = 0, 1, 2, \ldots, p, \ m = 0, 1, 2 \ldots, q). \tag{7.99}$$

The definition of anisotropic spaces on the standard pentahedral element is analogous. The anisotropic spaces with $q = 1$ are similar but not equivalent to the Reissner-Mindlin plate model or the Naghdi shell model. The differences are in the number of field functions and the constitutive laws that have been adjusted for the Reissner-Mindlin plate and the Naghdi shell models to satisfy the requirement of asymptotic consistency.

Remark 7.8 The advantages of thin solid formulations over shell formulations are that they are easier to implement and continuity with other bodies, such as stiffeners, are easier to enforce. The disadvantages are that thin solid formulations cannot be applied to laminated shells unless each lamina is explicitly modeled and the number of field functions must be the same for each displacement component.

Example 7.3 In this example we compare the solution of the plate problem described in Example 7.2 with the solution of a thin solid model, $q = 1$. Our quantity of interest is the maximum compressive stress on the top of the plate ($z = 1.25$ mm) in the neighborhood of the circular hole. We will consider hard simple supports only. We will use the shear correction factor which is optimal for energy for the Reissner-Mindlin plate and unity for the thin solid model. For discretization we will use the 28-element mesh shown in Fig. 7.8, p-extension and product space.

Recall that the Reissner-Mindlin model satisfies the condition of asymptotic consistency but it is not in the hierarchic sequence of thin solid plates, whereas the thin solid plate with $q = 1$ is in the hierarchic sequence but does not satisfy the condition of asymptotic consistency.

The contours of the third principal stress (MPa) are shown in Fig. 7.12. The relative error in the computed stress values were verified to be well under 1%. Therefore the differences in the results are caused by the differences in the models. In terms of the maximum compressive stress, that difference is 3%. For the Reissner-Mindlin model (resp. the top surface of the thin solid model, i.e. $z = 1.25$ mm) the range of the third principal stress is $-26.61 < \sigma_1 < 0$

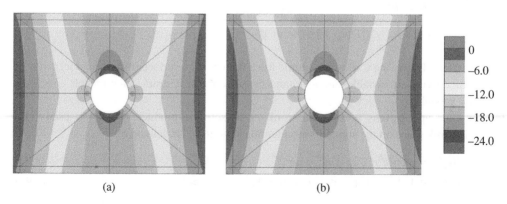

(a)   (b)

Figure 7.12 Example 7.3. Contours of the third principal stress (MPa), $p = 8$ product space. (a) Reissner-Mindlin model. (b) Top surface of the thin solid model, $q = 1$.

Table 7.2 Example 7.4. The first non-zero natural frequency (Hz) corresponding to the thin solid models characterized by $q = 1$ and $q = 3$, the fully 3-dimensional model (3D) and the Reissner-Mindlin (R-M) plate model. Product spaces.

p	$q = 1$	$q = 3$	3D	R-M
5	97.39	90.46	90.46	90.95
6	97.34	90.41	90.41	90.94
7	97.30	90.37	90.37	90.94
8	97.26	90.34	90.33	90.94
∞	97.12	90.13	90.07	90.94

(resp. $-27.40 < \sigma_3 < 4.14$). For the thin solid model characterized by $q = 3$, that range is the same as for the Reissner-Mindlin model.

The solution of the thin solid model was found to be insensitive to the width of the elements along the boundaries.

Exercise 7.24 Estimate the maximum bending moment in a plate, modeled as a thin solid, from the information that $(\sigma_3)_{\min} = -27.40$ MPa. Assume that the membrane forces are zero.

Example 7.4 Using the dimensions, elastic properties and boundary conditions for the plate described in Example 7.2, let us find the first natural frequency (Hz) predicted by the thin solid models characterized by $q = 1$ and $q = 3$ and compare the results with those computed using the fully three-dimensional model and the Reissner-Mindlin plate model. We will assume that the density is the same as in Exercise 7.15: $\rho = 2780$ kg/m^3 (2.78×10^{-9} Ns^2/mm^4) and use the 28 element mesh shown in fig. 7.12 and product space.

The results of computation for p ranging from 5 to 8 are listed in Table 7.2. The last row shows the estimated limit values obtained by extrapolation to $p = \infty$ by means of the algorithm described in Section 1.5.3. The computed natural frequencies are monotonically decreasing for each row in the first three columns. This is because the anisotropic space characterized by $q = 1$ is a subspace of the anisotropic space characterized by $q = 3$ which is a subspace of the isotropic finite element space. The Reissner-Mindlin (R-M) model is not a member of the hierarchic sequence of thin solid models because the stress-strain relationship is not consistent with the other models. However, it has the property of asymptotic consistency. The results indicate that the Reissner-Mindlin model and the thin solid model with $q = 3$ are very good approximations to the solution of the fully three-dimensional model.

Observe that the errors of approximation are negligibly small. Therefore the differences in the results are caused mainly by differences in modeling assumptions.

Exercise 7.25 The second eigenvalue of the thin solid models and the 3D model in Example 7.4 is comparable to the first eigenvalue of the Reissner-Mindlin model. Explain why.

Table 7.3 Example 7.5. Natural frequencies (Hz) for mode 20 computed using the anisotropic product space $S_{pr}^{pp3}(\Omega_{st}^{(h)})$ and 128 elements.

p	N	$t = 0.01$	$t = 0.001$	$t = 0.0001$	$t = 0.00001$
3	13,248	5019.0	1590.7	955.39	942.69
4	23,808	5002.5	1452.3	431.13	198.92
5	37,440	5002.1	1452.1	391.61	133.92
6	54,144	5002.0	1452.0	391.56	110.91
7	73,920	5002.0	1452.0	391.53	103.60
8	96,768	5002.0	1451.9	391.51	101.63
∞	∞	5002.0	1451.9	391.42	100.58

Exercise 7.26 If soft simple supports were used in the problem described in Example 7.4, how many zero eigenvalues would have been found for the thin solid models and the Reissner-Mindlin model?

Example 7.5 The following example was taken from an ADINA[9] technical brief[10] in which the natural frequencies and mode shapes of cylindrical shells are tabulated for various ratios of thickness to radius. The radius of the shell is 0.1 m, the length is 0.4 m, the ends are fixed. The material properties are $E = 2.0 \times 10^{11}$ Pa, $v = 0.3$, $\varrho = 7800$ kg/m^3. The same problem was discussed at the ASME Verification and Validation Symposium in 2012 by T. Yamada et al.[11] who presented solutions obtained with MITC4 shell elements [30] on coarse (40×20), moderate (80×40 and fine (240×120) uniform meshes.

The convergence of the natural frequencies for the 20th mode, computed using the anisotropic product space $S_{pr}^{pp3}(\Omega_{st}^{(h)})$ and 128 elements, are shown in Table 7.3. The elements were mapped using the mean optimal set T_2^5 (6×6 collocation points), see Table F.1 in Appendix F. The mode shapes for $t = 0.01$ and $t = 0.00001$ are shown in Fig. 7.13. The contours are proportional to the normal displacement.

Note that the rate of convergence of the natural frequencies is decreasing as the thickness is decreased. This is related to the decreasing regularity of the mode shapes, visible in Fig. 7.13.

Exercise 7.27 The mid-surface of a hyperboloidal shell is given by

$$\frac{x^2}{R_t^2} + \frac{y^2}{R_t^2} - \frac{z^2}{(\alpha L)^2} = 1, \quad -L \leq z \leq L, \quad \alpha^2 = \frac{R_t^2}{R_c^2 - R_t^2}$$

where R_t is the throat radius and R_c is the crown radius. Let $R_t = 1.0$ m, $L = 1.0$ m and $\alpha = 1$. Denote the thickness of the shell by d. Assume that the material is elastic with $E = 2.0 \times 10^5$ MPa, $v = 0.3$. Assume that a normal pressure $p = p_0 \cos 2\theta$ is acting on the inside surface of the shell where θ is

9 ADINA is a trademark of ADINA R&D, Inc., Watertown, Massachusetts, USA
10 http://www.adina.com/newsgH53.shtml
11 Verification of shell elements by eigenanalysis of vibration problems.

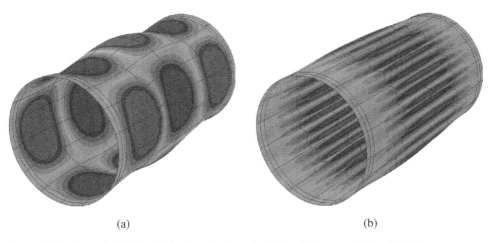

(a) (b)

Figure 7.13 Example 7.5. The 20th eigenfunctions for (a) $t = 0.01$ m and (b) $t = 0.00001$ m.

the angle measured from the positive x axis as shown in Fig. 7.14. Let $p_0 = 20.0$ kPa. The edge at $z = -L$ is fixed, i.e., all displacement components are zero, the edge at $z = L$ is free.

1. Construct a solid model of the shell with d defined as a parameter and construct a mesh similar to that shown in Fig. 7.14.
2. Using the anisotropic product spaces $S_{pr}^{ppq}(\Omega_{st}^{(h)})$ and the anisotropic trunk spaces $S_{tr}^{ppq}(\Omega_{st}^{(h)})$, investigate the rates of convergence in energy norm for $d = 0.01$ m and $d = 0.001$ m for $q = 1, 2, 3$.

This exercise demonstrates the effects of locking: The relative error is substantially larger for $d = 0.001$ m than for $d = 0.01$ m; however, the estimated asymptotic rates of convergence are close to 1.0. The strong boundary layer at the fixed edge will be clearly visible when the deformed shape is plotted.

Figure 7.14 Hyperboloidal shell.

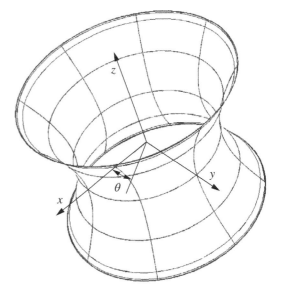

7.4 Chapter summary

In many cases it is advantageous to use dimensionally reduced models rather than a fully three-dimensional model. Whether a dimensionally reduced model should be used in a particular case depends on the goals of computation and the required accuracy. Generally speaking, dimensionally reduced models are well suited for structural analysis when the goals of computation are to determine structural stiffness, displacements, stress resultants, buckling strength natural frequencies but are not well suited for strength analysis when the region of primary interest is the vicinity of support attachments, doublers and external boundaries and the goal is to estimate strength-related QoIs.

The accuracy of the data of interest depends not only on the discretization used but also on how well the exact solution of a dimensionally reduced model approximates the exact solution of the corresponding fully three-dimensional model. The differences between the data of interest determined from the exact solution of a dimensionally reduced model and the corresponding fully three-dimensional model are model form errors. The hierarchic view of models provides a conceptual framework for the estimation and control of model form errors.

As the thickness is reduced, the exact solutions of the hierarchic beam, plate and shell models converge, respectively, to the exact solution of the Bernoulli-Euler beam model, the Kirchhoff plate model and the Novozhilov-Koiter shell model. These models, unlike the hierarchic models, require both the displacement functions and their first derivatives to be continuous. Therefore the exact solutions of the models in the hierarchic family that lie in $C^0(\Omega)$ converge to a solution that lies in $C^1(\Omega)$. Unless the polynomial degree is sufficiently high to permit close approximation of $C^1(\Omega)$ continuity, h-convergence will be slow or shear locking may occur. On the other hand, p-convergence will occur, however entry into the asymptotic range will be at $p \geq 4$.

8

Aspects of multiscale models

This chapter is concerned with finding the macromechanical properties of unidirectional fiber-matrix laminae from the mechanical properties of its constituents. We consider idealized unidirectional laminae, assuming that the fiber arrangement fits a perfect hexagonal or square pattern.

8.1   Unidirectional fiber-reinforced laminae

A representative volume element (RVE) is a sample of a heterogeneous material that has the average physical properties of the heterogeneous material. We are interested in determining the elements of the macromechanical material stiffness matrix $[E]$ for unidirectional laminae arranged in hexagonal and rectangular patterns as shown in Fig. 8.1 (a) and (b) respectively.

By definition, $\{\sigma\} = [E]\{\epsilon\}$ where $[E]$ is a 6×6 symmetric, positive-definite matrix and

$$\{\sigma\} = \{\sigma_x \ \sigma_y \ \sigma_z \ \tau_{xy} \ \tau_{yz} \ \tau_{zx}\}^T \tag{8.1}$$

$$\{\epsilon\} = \{\epsilon_x \ \epsilon_y \ \epsilon_z \ \gamma_{xy} \ \gamma_{yz} \ \gamma_{zx}\}^T \tag{8.2}$$

are the stress and strain vectors respectively. In linear elasticity the elements of $[E]$ are constants, independent of $\{\epsilon\}$.

We will be interested in the relationship between the integral average of stress $\{\overline{\sigma}\}$ and integral average of strain $\{\overline{\epsilon}\}$ with the average taken over the RVE. We will use the notation

$$\{\overline{\sigma}\} = [E_{\mathrm{RVE}}]\{\overline{\epsilon}\} \tag{8.3}$$

and determine the macroscopic material stiffness matrix $[E_{\mathrm{RVE}}]$ such that the strain energy for the heterogeneous RVE equals the strain energy of the homogenized RVE:

$$U = \frac{1}{2} \int_V \{\epsilon\}^T [E]\{\epsilon\} \ dV = \frac{1}{2}\{\overline{\epsilon}\}^T [E_{\mathrm{RVE}}]\{\overline{\epsilon}\}V \tag{8.4}$$

where V is the volume of the RVE. The elements of $[E_{\mathrm{RVE}}]$ must be substantially independent of how many RVEs are used for their determination.

We formulate an algorithm for the computation of the elements of $[E_{\mathrm{RVE}}]$ by the finite element method. The algorithm requires determination of the strain energy and the average displacements on the boundaries of the RVE, both of which converge faster than the energy norm of the solution, that is superconvergence is realized.

Finite Element Analysis: Method, Verification and Validation, Second Edition. Barna Szabó and Ivo Babuška.
© 2021 John Wiley & Sons, Inc. Published 2021 by John Wiley & Sons, Inc.
Companion Website: www.wiley.com/go/szabo/finite_element_analysis

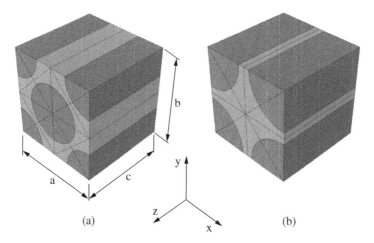

Figure 8.1 RVEs for unidirectional laminae: (a) hexagonal and (b) rectangular arrangement of fibers.

Owing to the five planes of symmetry of the RVEs shown in Fig. 8.1, the macroscopic material stiffness matrix is characterized by only six constants as indicated in eq. (8.5).

$$[E_{RVE}] = \begin{bmatrix} A & e & f & 0 & 0 & 0 \\ e & A & f & 0 & 0 & 0 \\ f & f & B & 0 & 0 & 0 \\ 0 & 0 & 0 & C & 0 & 0 \\ 0 & 0 & 0 & 0 & D & 0 \\ 0 & 0 & 0 & 0 & 0 & D \end{bmatrix}. \tag{8.5}$$

Therefore the strain energy for the RVE can be written as:

$$U = \frac{1}{2} \int_V \left(A(\bar{\epsilon}_x^2 + \bar{\epsilon}_y^2) + B\bar{\epsilon}_z^2 + 2e\,\bar{\epsilon}_x\bar{\epsilon}_y + 2f(\bar{\epsilon}_x\bar{\epsilon}_z + \bar{\epsilon}_y\bar{\epsilon}_z) \right)\, dV$$
$$+ \frac{1}{2} \int_V \left(C\bar{\gamma}_{xy}^2 + D(\bar{\gamma}_{yz}^2 + \bar{\gamma}_{zx}^2) \right)\, dV \tag{8.6}$$

where the overbars indicate integral averages over the volume of the RVE.

We determine the six constants that characterize $[E_{RVE}]$ by solving six problems in which the integral averages of the strains are either enforced by imposed displacement boundary conditions or computed by the finite element method.

Specifically, we denote the surface of the RVE to which the positive x, y and z axis is normal by X^+, Y^+ and Z^+ respectively. Similarly we denote the surface of the RVE to which the negative x, y and z axis is normal by X^-, Y^- and Z^- respectively. The displacement boundary conditions for the six problems are given in Table 8.1 where "sym" indicates symmetry, that is, the normal displacement is zero, "float" indicates "floating symmetry". The floating symmetry condition differs from the symmetry condition in that the normal displacement is a constant which is determined such that the area integral of the normal stress is zero. Boundary conditions not shown explicitly in Table 8.1 are homogeneous natural boundary conditions: For example, for $k = 1$ we have $\tau_{xy} = \tau_{xz} = 0$ on X^+. These are periodic boundary conditions defined in such a way that for each mode of deformation the number of constraints is minimal.

Table 8.1 Displacement boundary conditions.

k	X^-	X^+	Y^-	Y^+	Z^-	Z^+
1	sym	$u_x = a$	sym	float	sym	float
2	sym	float	sym	float	sym	$u_z = c$
3	sym	$u_x = a$	sym	$u_y = b$	sym	float
4	sym	$u_x = a$	sym	float	sym	$u_z = c$
5	$u_y = 0$	$u_y = a/2$	$u_x = 0$	$u_x = b/2$	sym	float
6	sym	float	$u_z = 0$	$u_z = b/2$	$u_y = 0$	$u_y = c/2$

The constants A, B, c and f are computed from the equations:

$$\left(\overline{\epsilon}_x^2 + \overline{\epsilon}_y^2\right)_k A + \left(\overline{\epsilon}_z^2\right)_k B + 2\left(\overline{\epsilon}_x\overline{\epsilon}_y\right)_k e + 2\left(\overline{\epsilon}_y\overline{\epsilon}_z + \overline{\epsilon}_z\overline{\epsilon}_x\right)_k f = 2U_k/V \tag{8.7}$$

where $k = 1, 2, 3, 4$ identifies the solution corresponding the boundary conditions shown in Table 8.1, U_k is the computed strain energy and V is the volume of the RVE.

The constants C and D are computed from

$$C = 2U_5/(\overline{\gamma}_{xy}^2 V), \quad D = 2U_6/(\overline{\gamma}_{yz}^2 V). \tag{8.8}$$

8.1.1 Determination of material constants

In the absence of thermal loading the integral average of strain is related to the integral average of stress by the transformation $\{\overline{\epsilon}\} = [C_{\mathrm{RVE}}]\{\overline{\sigma}\}$. The matrix $[C_{\mathrm{RVE}}]$ is the macroscopic material compliance matrix. Obviously $[C_{\mathrm{RVE}}] = [E_{\mathrm{RVE}}]^{-1}$.

The compliance matrix $[C_{\mathrm{RVE}}]$ has the same structure as $[E_{\mathrm{RVE}}]$ displayed in eq. (8.5). It is customary to write the material constants with reference to the longitudinal (L) and transverse (T) directions:

$$[C_{\mathrm{RVE}}] = \begin{bmatrix} \dfrac{1}{E_T} & -\dfrac{v_T}{E_T} & -\dfrac{v_{LT}}{E_L} & 0 & 0 & 0 \\[2mm] -\dfrac{v_T}{E_T} & \dfrac{1}{E_T} & -\dfrac{v_{LT}}{E_L} & 0 & 0 & 0 \\[2mm] -\dfrac{v_{TL}}{E_T} & -\dfrac{v_{TL}}{E_T} & \dfrac{1}{E_L} & 0 & 0 & 0 \\[2mm] 0 & 0 & 0 & \dfrac{1}{G_T} & 0 & 0 \\[2mm] 0 & 0 & 0 & 0 & \dfrac{1}{G_{LT}} & 0 \\[2mm] 0 & 0 & 0 & 0 & 0 & \dfrac{1}{G_{LT}} \end{bmatrix} \tag{8.9}$$

where E_L, E_T are the moduli of elasticity, v_T, v_{LT}, v_{TL} are the Poison's ratios and G_T, G_{LT} are the shear moduli. The Poison ratios v_{TL} and v_{LT} are not independent. They must satisfy the following

Table 8.2 Displacement boundary conditions for the determination of the coefficients of thermal expansion.

k	X⁻	X⁺	Y⁻	Y⁺	Z⁻	Z⁺
7	sym	float	sym	float	sym	float

condition that guarantees the symmetry of $[C_{\text{RVE}}]$:

$$\frac{v_{TL}}{E_T} = \frac{v_{LT}}{E_L}. \tag{8.10}$$

These material constants can be computed directly from (8.9).

When the elements of $[C_{\text{RVE}}]$ can be computed from five material constants then the material is said to be transversely isotropic. Specifically, for a transversely isotropic material we have:

$$G_T = \frac{E_T}{2(1 + v_T)}. \tag{8.11}$$

Although it is usually assumed that $[C_{\text{RVE}}]$ is transversely isotropic, this assumption is justified only in the special case of *regular* hexagonal arrangement of fibers.

8.1.2 The coefficients of thermal expansion

For the computation of the coefficients of thermal expansion we impose the boundary conditions shown in Table 8.2 and apply an arbitrary constant temperature change $\mathcal{T}$ to the RVE. We then compute the average displacements on X^+, Y^+ and Z^+. Denoting these displacements by $\overline{u}_x^+ . \overline{u}_y^+$ and $\overline{u}_z^+$ respectively, the coefficients of thermal expansion in the longitudinal and transverse directions are computed from

$$\alpha_X = \overline{u}_x^+(a\mathcal{T}), \quad \alpha_Y = \overline{u}_y^+(b\mathcal{T}), \quad \alpha_Z \equiv \alpha_L = \overline{u}_z^+/(c\mathcal{T}). \tag{8.12}$$

8.1.3 Examples

In the following examples we consider unidirectional fibers of diameter 7 μm (0.007 mm). The volume fraction of the fibers V_f is 60%. The fibers are transversely isotropic with the following material properties: $E_L = 2.52 \times 10^5$ MPa, $E_T = 1.65 \times 10^4$ MPa, $v_{LT} = 0.3$, $v_T = 0.2$, $G_{LT} = 4.14 \times 10^4$ MPa where the subscripts L and T refer, respectively, to the longitudinal and transverse directions. The coefficients of thermal expansion are: $\alpha_L = -1.08 \times 10^{-6}/°C$, $\alpha_T = 7.2 \times 10^{-6}/°C$. The matrix is isotropic, its modulus of elasticity is 3.79×10^3 MPa, $v = 0.3$. The coefficient of thermal expansion is: $\alpha = 5.4 \times 10^{-5}/°C$.

These material properties were taken from [110]. This will allow comparison with results obtained by other methods described in the same reference[1]. The material properties were converted to SI units.

All computed quantities of interest were verified by increasing the polynomial degree of elements from 5 to 8 and estimating their limit values.

---

1 See Exercise 4.1 on p. 85 and the solutions on p. 272.

Hexagonal pattern

Using the notation for the dimensions in Fig. 8.1 (a) we have: $a = b = c = 1.132615 \times 10^{-2}$ mm.

The algorithm for finding the macroscopic material properties was applied to the RVE shown in Fig. 8.1(a) and, in order to test for size dependence, also to a domain comprised of 27 RVEs shown in Fig. 8.2.

The results of computation are summarized in Table 8.3. We note that the number of digits displayed in the table would not be justified by the data given in the problem definition. We use four decimal places only for the purpose of comparing the results. It is seen that the results are virtually independent of the number of RVEs.

Remark 8.1 Hexagonal arrangements of fibers are usually treated as transversely isotropic materials, see for example [47], [110]. However, transverse isotropy is realized only in the special case of *regular* hexahedral arrangements. Referring to the notation in Fig. 8.1, for the regular hexahedral arrangement $b/a = \sqrt{3}$. In this example, however, $b/a = 1$.

Figure 8.2 Solution domain consisting of 27 RVEs (864 finite elements).

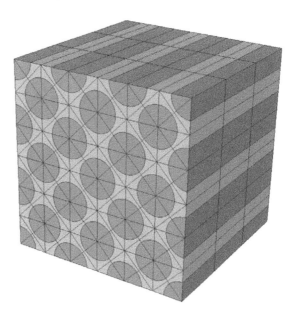

Table 8.3 Material properties of the hexagonal pattern (MPa and 1/°C).

Property	one RVE	27 RVEs	Diff. (%)
E_L	1.5272E+05	1.5272E+05	0.00
E_T	7.9143E+03	7.9143E+03	0.00
G_{LT}	5.3708E+03	5.4367E+03	1.23
G_T	3.5579E+03	3.5579E+03	0.00
ν_{LT}	0.300	0.300	0.00
ν_T	0.368	0.368	0.00
α_L	−5.3323E-07	−5.3323E-07	0.00
α_T	2.8941E-05	2.8941E-05	0.00

Using eq. (8.11) for the computation of G_T from the values of E_T and v_T in Table 8.3, we find $G_T = 2.8931 \times 10^3$ MPa which differs by -18.7% from the computed value. Therefore treating this as a transversely isotropic material will underestimate the transverse shear modulus by 18.7%. This is a systematic error that can be easily avoided.

Square pattern

Referring to the notation in Fig. 8.1(b) we have: $a = b = c = 8.0088 \times 10^{-3}$ mm. The algorithm for finding the macroscopic material properties was applied to the RVE shown in Fig. 8.1(b) and, in order to test for size dependence, also to a domain comprised of 8 RVEs shown in Fig. 8.3(a). The periodicity of the solution corresponding to transverse shear is illustrated in Fig. 8.3(b).

To determine the size of the error that would be incurred if this were treated as a transversely isotropic material, once again we use eq. (8.11) for the computation of G_T from the values of E_T and v_T in Table 8.4. We find $G_T = 3.5583 \times 10^3$ MPa which differs by 23% from the value computed by the present method but is very close to the value reported in Table 8.3.

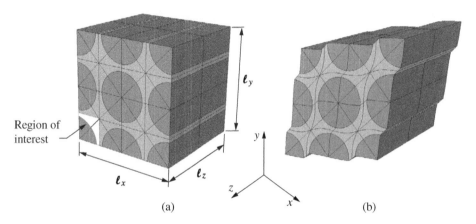

Figure 8.3 (a) Solution domain consisting of 8 RVEs (128 finite elements). (b) Deformed configuration under transverse shear.

Table 8.4 Material properties of the square pattern (MPa and 1/°C).

Property	one RVE	8 RVEs	Diff. (%)
E_L	1.5272E+05	1.5272E+05	0.00
E_T	9.0737E+03	9.0785E+03	0.05
G_{LT}	5.4639E+03	5.4698E+03	0.11
G_T	2.8932E+03	2.8932E+03	0.00
v_{LT}	0.300	0.300	0.00
v_T	2.7513E-01	2.7581E-01	0.24
α_L	−5.332E-07	−5.332E-07	0.00
α_T	2.8941E-05	2.8941E-05	0.00

Table 8.5 Comparison of material properties (psi and 1/°F).

Property	Hexagonal	Square	RM	MRM
E_L	2.220E+07	2.2201E+07	2.216E+07	2.212E+07
E_T	1.151E+06	1.3191E+06	1.02E+06	1.30E+06
G_{LT}	7.808E+05	7.9433E+05	5.0E+05	7.5E+05
G_T	5.172E+05	4.2060E+05	4.0E+05	5.2E+05
ν_{LT}	0.300	0.300	0.300	0.300
ν_T	0.368	0.275	0.27	0.26
α_L	−2.96E-07	−2.96E-07	−2.9E-07	−2.9E-07
α_T	1.61E-05	1.61E-05	1.80E-05	1.80E-05

Comparison

In Table 8.5 the computed material properties for one RVE for the hexagonal and square patterns are presented along with properties computed by the rule of mixtures (RM) and the modified rule of mixtures (MRM) as reported in [110]. The volume fraction of fibers is 60%. In order to make direct comparison possible, the data was converted to US customary units. It is seen that both RM and MRM provide a reasonable approximation for the material properties.

8.1.4 Localization

Localization is concerned with interpretation of the solution of the macroscopic problem on the scale of RVEs. The main idea is that strain computed from the solution of the macroscopic problem is virtually constant over one RVE, or a small group of RVEs, and therefore it approximates the average strain values well. Therefore the displacement components, up to rigid body displacements, can be written in terms of the average strain values:

$$\bar{u}_x = \bar{\epsilon}_x x + \bar{\gamma}_{xy} y/2 + \bar{\gamma}_{zx} z/2$$
$$\bar{u}_y = \bar{\gamma}_{xy} x/2 + \bar{\epsilon}_y y + \bar{\gamma}_{yz} z/2 \qquad (8.13)$$
$$\bar{u}_z = \bar{\gamma}_{zx} x/2 + \bar{\gamma}_{yz} y/2 + \bar{\epsilon}_z z.$$

Assuming that the coordinate system is located in the center of the RVE, or a group of RVEs, such that $-\ell_x/2 < x < \ell_x/2$, $-\ell_y/2 < y < \ell_y/2$, $-\ell_z/2 < z < \ell_z/2$, the displacement boundary conditions corresponding to equations (8.13) are as shown in Table 8.6. Because the displacement boundary conditions reduce the number of degrees of freedom and introduce local perturbations at the boundary, it is recommended that at least 8 RVEs be used for solving the local problem.

Remark 8.2 If the exact solution of the macroscopic problem is such that the elements of the strain tensor are infinity in one or more points then the maximum norm of the computed strain depends on the discretization and, since the maximum norm of the computed strain is finite, the error measured in maximum norm is infinitely large. Therefore the goal of computation cannot be the determination of maximum strain in such cases.

Table 8.6 Displacement boundary conditions for the problem of localization.

	X^-	X^+	Y^-	Y^+	Z^-	Z^+
$\bar{u}_x$	$-\bar{\epsilon}_x \ell_x/2$	$\bar{\epsilon}_x \ell_x/2$	$-\bar{\gamma}_{xy}\ell_y/2$	$\bar{\gamma}_{xy}\ell_y/2$	$-\bar{\gamma}_{zx}\ell_z/2$	$\bar{\gamma}_{zx}\ell_z/2$
$\bar{u}_y$	$-\bar{\gamma}_{xy}\ell_x/2$	$\bar{\gamma}_{xy}\ell_x/2$	$-\bar{\epsilon}_y\ell_y/2$	$\bar{\epsilon}_y\ell_y/2$	$-\bar{\gamma}_{yz}\ell_z/2$	$\bar{\gamma}_{yz}\ell_z/2$
$\bar{u}_z$	$-\bar{\gamma}_{zx}\ell_x/2$	$\bar{\gamma}_{zx}\ell_x/2$	$-\bar{\gamma}_{yz}\ell_y/2$	$\bar{\gamma}_{yz}\ell_y/2$	$-\bar{\epsilon}_z\ell_z/2$	$\bar{\epsilon}_z\ell_z/2$

8.1.5 Prediction of failure in composite materials

In fiber-reinforced composites detectable failure events occur on the scale of one or more RVEs. While the assumptions of small strain continuum mechanics do not apply to failure events, in many practically important cases the first appearance of detectable failure can be correlated with the solution of small strain continuum mechanics problems through the use of suitably defined predictors. One possible approach is to define integral averages on solution-dependent domains. This is analogous to the predictor defined in Section 6.3 in connection with fatigue failure in metals. A similar approach can be applied to composite materials. This will require coordinated experimental and analytical investigation subject to the procedures of verification, validation and uncertainty quantification (VVUQ).

Example
In this example we illustrate that the elastic strains are unbounded at fiber terminations, whereas strains averaged over the matrix phase of an RVE are well defined and converge strongly. We will consider rectangular fiber arrangement and a domain consisting of 8 RVEs shown in Fig. 8.3. The dimensions are: $\ell_x = \ell_y = \ell_z = 1.6018 \times 10^{-2}$ mm. The volume fraction is 60% and the elastic properties of the fiber and matrix are the same as those in Section 8.1.3. The region of interest is that part of the RVE which is occupied by the matrix, as indicated in Fig. 8.3(a).

The displacement boundary conditions are shown in Table 8.7. Normal displacements consistent with the average strain values $\bar{\epsilon}_x = 3.5 \times 10^{-3}, \bar{\epsilon}_y = -2.0 \times 10^{-3}, \bar{\gamma}_{xy} = 0$ are prescribed on the boundary surfaces X^+ and Y^+.

Using the finite element mesh consisting of 128 elements shown in Fig. 8.3(a), a sequence of finite element solutions was obtained corresponding to the polynomial degree p ranging from 1 to 8. For the matrix the equivalent strain is defined as:

$$\epsilon_{eq} = \sqrt{\frac{1}{2(1+v)^2}\left[(\epsilon_1 - \epsilon_2)^2 + (\epsilon_2 - \epsilon_3)^2 + (\epsilon_3 - \epsilon_1)^2\right]} \tag{8.14}$$

where ϵ_1, ϵ_2 and ϵ_3 are the principal strains and v the Poisson's ratio of the matrix. This definition of equivalent strain satisfies the condition that under uniaxial loading $\sigma_{eq} = E\epsilon_{eq}$ where σ_{eq} is the von Mises stress and E is the modulus of elasticity. Other definitions of equivalent strain can be found in the literature.

Table 8.7 Displacement boundary conditions.

X^-	X^+	Y^-	Y^+	Z^-	Z^+
sym	u_x	sym	u_y	float	free

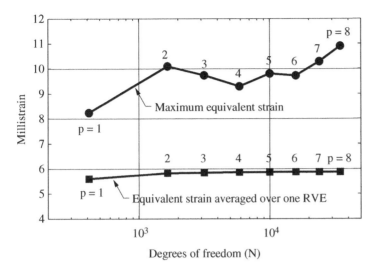

Figure 8.4 Example. Equivalent strain in the matrix computed in the region of interest shown in Fig. 8.3(a).

The maximum value and the average value of ϵ_{eq} in the matrix phase of one RVE were computed. This is the region of interest indicated in Fig. 8.3. The results of computation are shown in Fig. 8.4.

The results show that max (ϵ_{eq}) diverges. This would not be visible if p-extension would have been stopped at $p = 6$. However, it is seen that the computed data points keep increasing past $p = 6$. The divergence of max (ϵ_{eq}) was confirmed by repeating the process with a refined mesh[2].

The markedly different behaviors of the computed data between low and high N values is typical for both h- and p-extensions. For large N the sequence of values is in the asymptotic range. For low N the sequence is in the pre-asymptotic range. Extrapolation to $N \to \infty$ is justified only for high N values. This raises the obvious question: what is high and what is low? The answer to this is problem-dependent.

Because max ϵ_{eq} corresponding to the exact solution is not a finite number, it is not suitable for use as a predictor of failure. On the other hand, the first derivatives of the exact solution are square integrable, hence the integral average of ϵ_{eq} over any volume, in this example over the matrix phase of an RVE, is a finite number.

The integral average meets the other two requirements as well: small perturbations in $\mathbf{D}$ do not cause large changes in integral averages and critical values of integral averages can be inferred from observations of the outcomes of coupon experiments.

8.1.6 Uncertainties

The perfectly regular arrangement of fibers analyzed in this chapter is a highly idealized representation of reality. As reported in [11], the volume fraction in a plate fabricated under carefully controlled laboratory conditions ranged from 9.5% to 79.5%. Upon smoothing the data, the average volume fraction was 59.1% with a range of 43.8% to 69.5%. The volume fraction was lowest between the plies. The dominant uncertainties are the variations in volume fraction and longitudinal undulations (waviness) of the fibers. Properly calibrated predictors are expected to account for such variations.

---

2 These results are not presented here.

8.2 Discussion

An algorithm for finding the macroscopic thermomechanical properties of unidirectional fiber-matrix composite laminae, given the material properties and volume fractions of the constituents, was outlined in this Chapter. The quantities of interest are computed from the strain energy values of RVEs corresponding to imposed displacements. It was demonstrated that the macroscopic material properties are substantially independent of the number of RVEs.

The problem of localization was addressed. The macroscopic strain tensor in a point provides the information necessary for obtaining solutions on the microscopic scale.

The technical requirements that predictors of damage accumulation and damage propagation must meet were stated. It was noted that pointwise stresses and strains do not meet those requirements; however, many possible definitions of predictors do. One possible definition, the integral average of the equivalent strain in the matrix phase of an RVE, was discussed.

The evaluation and ranking of predictors has to be performed through a process of model development, similar to the process described in Section 6.5. Simulation governance [92] provides a framework for systematic improvement of predictors over time as new experimental information becomes available.

9

Non-linear models

The formulation of mathematical models invariably involves making restrictive assumptions such as neglecting certain geometric features, idealizing the physical properties of the material, idealizing boundary conditions, neglecting the effects of residual stresses, etc. Therefore any mathematical model should be understood to be a special case of a more comprehensive model. This is the hierarchic view of mathematical models.

In order to test whether a restrictive assumption is acceptable for a particular application, it is necessary to estimate the influence of that assumption on the quantities of interest and, if necessary, revise the model. An exploration of the influence of modeling assumptions on the quantities of interest is called virtual experimentation. Computational tools designed to support numerical simulation must have the capability to support virtual experimentation and hence solve nonlinear problems.

The formulation of nonlinear models is a very large and diverse field. Only a brief introduction is presented in this chapter, with emphasis on the algorithmic aspects and examples. For additional discussion and details we refer to other books, such as references [81], [86].

9.1 Heat conduction

Mathematical models of heat conduction often involve radiation heat transfer and the coefficients of heat conduction are typically functions of the temperature. The formulation of mathematical models that account for these phenomena is outlined in the following.

9.1.1 Radiation

When two bodies exchange heat by radiation then the flux is proportional to difference of the fourth power of their absolute temperatures:

$$q_n = \kappa f_s f_\epsilon (u^4 - u_R^4) \tag{9.1}$$

where u, u_R are the absolute temperatures of the radiating bodies, $\kappa = 5.699 \times 10^{-8}$ W/(m^2 K^4) is the Stefan-Boltzmann constant[1], $0 \leq f_s \leq 1$ is the radiation shape factor and $0 < f_\epsilon \leq 1$ is the surface emissivity which is defined as the relative emissive power of a body compared to that of an ideal blackbody. The surface emissivity is also equal to the absorption coefficient, defined as that fraction

[1] Josef Stefan 1835–1893, Ludwig Boltzmann 1844–1906.

Finite Element Analysis: Method, Verification and Validation, Second Edition. Barna Szabó and Ivo Babuška.
© 2021 John Wiley & Sons, Inc. Published 2021 by John Wiley & Sons, Inc.
Companion Website: www.wiley.com/go/szabo/finite_element_analysis

of thermal energy incident on a body which is absorbed. In general, surface emissivity is a function of the temperature[2].

Eq. (9.1) can be viewed as a convective boundary condition where the coefficient of convective heat transfer depends on the temperature of the radiating bodies. Writing:

$$q_n = \kappa f_s \, f_\epsilon (u^4 - u_R^4) = \underbrace{\kappa f_s \, f_\epsilon (u^2 + u_R^2)(u + u_R)}_{h_r(u)}(u - u_R)$$

we have the form of eq. (2.27) with h_c replaced by $h_r(u)$. Therefore radiation problems have to be solved by iteration: first the linear problem is solved, then the coefficient of convective heat transfer is updated and the solution is repeated. The stopping criterion is based on the size of temperature change. Usually very few iterations are needed.

9.1.2 Nonlinear material properties

When the conductivity or other material coefficients are functions of the temperature then solutions have to be obtained by iteration. The bilinear and linear forms are split into linear (L) and nonlinear (NL) parts:

$$B_L(u, v) + B_{NL}(u, v) = F_L(v) + F_{NL}(u, v).$$

To obtain the initial solution $u^{(1)}$ the terms $B_{NL}(u, v)$ and $F_{NL}(u, v)$ are neglected. For subsequent solutions the following problem is solved:

$$B_L(u^{(k)}, v) = F_L(v) + F_{NL}(u^{(k-1)}, v) - B_{NL}(u^{(k-1)}, v) \qquad k = 2, 3, \dots$$

The process is terminated when $u^{(k)} - u^{(k-1)}$, measured in a suitable norm, is sufficiently small. This process is not guaranteed to converge. If it fails to converge then the forcing function should be applied in small increments. In most practical problems convergence will occur when the increments are sufficiently small.

9.2 Solid mechanics

In many practical problems the assumptions that (a) the strains are small, (b) the deformation of the body is so small that the equilibrium equations written for the deformed configuration would not be significantly different from the equilibrium equations written for the undeformed configuration, (c) the stress-strain laws are linear and (d) mechanical contact can be approximated by linear boundary conditions, such as linear springs, do not hold. In this section the formulation of mathematical models that account for these phenomena is outlined.

9.2.1 Large strain and rotation

Deformation is characterized by the strain-displacement relationships. Consider an elastic body Ω in a reference state, as shown in Fig. 9.1. The material points in the reference state are identified by the position vectors X_i, called Lagrangian coordinates, and in the deformed state by the position vectors by x_i, called Eulerian coordinates. The displacement vector may be written as a function of X_i in which case the uppercase letter U_i will be used. Alternatively, the displacement vector may be written as a function of x_i, in which case the lowercase letter u_i will be used.

---

2 For example, the surface emissivity of polished stainless steel is 0.22 at 373.15 K and 0.45 at 698.15 K.

Figure 9.1 Notation.

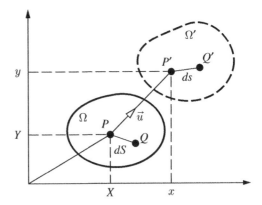

An infinitesimal "fiber" in the undeformed configuration has the length dS and in the deformed configuration it has the length ds. Let us assume first that the displacement is a function of the Eulerian coordinates x_i. In this case:

$$dX_i = dx_i - du_i = dx_i - u_{i,j}dx_j = (\delta_{ij} - u_{i,j})dx_j.$$

Therefore:

$$dS^2 = dX_k dX_k = (\delta_{ki} - u_{k,i})dx_i(\delta_{kj} - u_{k,j})dx_j$$
$$= (\delta_{ij} - u_{j,i} - u_{i,j} + u_{k,i}u_{k,j})dx_i dx_j$$

and:

$$ds^2 - dS^2 = dx_k dx_k - dX_k dX_k$$
$$= \delta_{ij}dx_i dx_j - (\delta_{ij} - u_{j,i} - u_{i,j} + u_{k,i}u_{k,j})dx_i dx_j$$
$$= (u_{i,j} + u_{j,i} - u_{k,i}u_{k,j})dx_i dx_j.$$

The Almansi strain tensor[3], denoted by e_{ij}, is defined by the relationship

$$ds^2 - dS^2 = 2e_{ij}dx_i dx_j.$$

Therefore we have the definition of the Almansi strain tensor:

$$e_{ij} = \frac{1}{2}(u_{i,j} + u_{j,i} - u_{k,i}u_{k,j}). \tag{9.2}$$

In two dimensions, using unabridged notation, this is written as

$$e_{xx} = \frac{\partial u_x}{\partial x} - \frac{1}{2}\left[\left(\frac{\partial u_x}{\partial x}\right)^2 + \left(\frac{\partial u_y}{\partial x}\right)^2\right]$$
$$e_{xy} = \frac{1}{2}\left(\frac{\partial u_x}{\partial y} + \frac{\partial u_y}{\partial x}\right) - \frac{1}{2}\left[\frac{\partial u_x}{\partial x}\frac{\partial u_x}{\partial y} + \frac{\partial u_y}{\partial x}\frac{\partial u_y}{\partial y}\right] \tag{9.3}$$
$$e_{yy} = \frac{\partial u_y}{\partial y} - \frac{1}{2}\left[\left(\frac{\partial u_x}{\partial y}\right)^2 + \left(\frac{\partial u_y}{\partial y}\right)^2\right].$$

When the displacement vector components are written in terms of the Lagrangian coordinates X_i then $x_i = X_i + U_i$, therefore

$$dx_i = dX_i + dU_i = dX_i + U_{i,j}dX_j = (\delta_{ij} + U_{i,j})dX_j$$

---

3 Emilio Almansi 1869–1948.

therefore:

$$ds^2 = dx_k dx_k = (\delta_{ki} + U_{k,i})dX_i(\delta_{kj} + U_{k,j})dX_j$$
$$= (\delta_{ij} + U_{j,i} + U_{i,j} + U_{k,i}U_{k,j})dX_i dX_j$$

and:

$$ds^2 - dS^2 = dx_k dx_k - dX_k dX_k$$
$$= (\delta_{ij} + U_{j,i} + U_{i,j} + U_{k,i}U_{k,j})dX_i dX_j - \delta_{ij}dX_i dX_j$$
$$= (U_{i,j} + U_{j,i} + U_{k,i}U_{k,j})dX_i dX_j.$$

The Green strain tensor, denoted by E_{ij}, is defined by the relationship

$$ds^2 - dS^2 = 2E_{ij}dX_i dX_j.$$

Therefore we have the definition of the Green strain tensor:

$$E_{ij} = \frac{1}{2}(U_{i,j} + U_{j,i} + U_{k,i}U_{k,j}). \tag{9.4}$$

In two dimensions, using unabridged notation, this is written as

$$E_{XX} = \frac{\partial U_X}{\partial X} + \frac{1}{2}\left[\left(\frac{\partial U_X}{\partial X}\right)^2 + \left(\frac{\partial U_Y}{\partial X}\right)^2\right]$$

$$E_{XY} = \frac{1}{2}\left(\frac{\partial U_X}{\partial Y} + \frac{\partial U_Y}{\partial X}\right) + \frac{1}{2}\left[\frac{\partial U_X}{\partial X}\frac{\partial U_X}{\partial Y} + \frac{\partial U_Y}{\partial X}\frac{\partial U_Y}{\partial Y}\right] \tag{9.5}$$

$$E_{YY} = \frac{\partial U_Y}{\partial Y} + \frac{1}{2}\left[\left(\frac{\partial U_X}{\partial Y}\right)^2 + \left(\frac{\partial U_Y}{\partial Y}\right)^2\right].$$

When the first derivatives $u_{i,j}$ (resp. $U_{i,j}$) are much smaller than unity then the product terms $u_{k,i}u_{k,j}$ (resp. $U_{k,i}U_{k,j}$) are negligible and the strain is called infinitesimal, small or linear strain. In such cases both the Almansi and Green strain tensors reduce to the definition of infinitesimal strain:

$$\epsilon_{ij} = \frac{1}{2}(u_{i,j} + u_{j,i}). \tag{9.6}$$

Example 9.1 The following example illustrates the solution of a small-strain large displacement problem. A 200 mm long (L), 1.0 mm thick (t) and 7.5 mm wide (w) elastic strip is fixed at one end and loaded at the other end by lineraly distributed normal tractions corresponding to the bending moment $M = EI/R$ where $R = L/(2\pi)$, $I = wt^3/12$, $E = 2.0 \times 10^5$ MPa is the modulus of elasticity and Poisson's ratio is zero. According to the Bernoulli-Euler beam model, this moment will bend the strip into a cylinder of radius R.

The solution, shown in Fig. 9.2, was obtained using a thin solid model, three hexahedral elements and anisotropic product space $S^{p,p,q}(\Omega_{st}^{(q)})$ with $p = 8$ and $q = 1$. The stopping criterion is given by eq. (5.28). In this example $\tau_{stop} = 0.1$ was used.

Remark 9.1 The elastic strip in the foregoing example deforms into a cylinder only when Poisson's ratio is zero. When Poisson's ratio is not zero then the deformed configuration will be doubly curved with an anticlastic curvature.

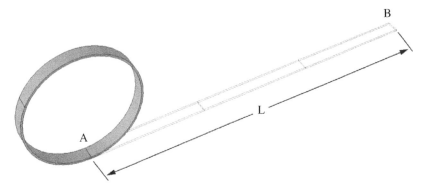

Figure 9.2 Elastic strip bent into a cylinder.

Exercise 9.1 Demonstrate in two dimensions that under arbitrary rigid body rotation all components of the Almansi strain tensor are zero. Hint: Let the reference configuration of a two-dimensional body be:

$$X = R\cos\theta, \quad Y = R\sin\theta.$$

If this body is rotated by angle α, its new position will be:

$$x = R\cos(\theta + \alpha) \quad y = R\sin(\theta + \alpha).$$

Therefore:

$$u_x = x - X = R\cos(\theta + \alpha) - R\cos\theta$$
$$u_y = y - Y = R\sin(\theta + \alpha) - R\sin\theta.$$

Use

$$\frac{\partial u_x}{\partial x} = \frac{\partial u_x}{\partial R}\frac{\partial R}{\partial x} + \frac{\partial u_x}{\partial \theta}\frac{\partial \theta}{\partial x}$$
$$= [\cos(\theta + \alpha) - \cos\theta]\cos(\theta + \alpha) + [\sin(\theta + \alpha) - \sin\theta]\sin(\theta + \alpha)$$
$$= 1 - \cos\alpha$$

etc., to complete the exercise.

Exercise 9.2 Following the procedure indicated in Exercise 9.1, demonstrate that the Green strain tensor vanishes under rigid body rotation.

Exercise 9.3 Following the procedure indicated in Exercise 9.1, compute the linear strain tensor defined by eq. (9.6).

Exercise 9.4 Consider a thin wire of length ℓ oriented along the x axis. Write expressions for the Almansi strain e_x and the Green strain E_x in terms of an imposed displacement Δ. (a) Show that $e_x \to 1/2$ as $\Delta \to \infty$. (b) Plot the Almansi strain and the Green strain in the range $-\ell < \Delta \le 10\ell$. Hint: Let

$$X = \frac{1+\xi}{2}\ell, \qquad x = \frac{1+\xi}{2}(\ell + \Delta), \qquad -1 \le \xi \le +1.$$

9.2.2 Structural stability and stress stiffening

Investigation of buckling and stress-stiffening is generally performed for structures which are beam-like and slender, or shell-like and thin. Such structures are usually stiffened and typically there are topological details, loads or boundary conditions for which the assumptions of beam, plate or shell theories do not hold. For this reason we consider the elastic stability of a fully three-dimensional body and construct a mathematical model which is not encumbered by the various restrictions that exist in beam, plate or shell models. From this model various dimensionally reduced models can be deduced as special cases.

We assume that a three-dimensional elastic body is subjected to an initial stress field σ_{ij}^0 which satisfies the equations of equilibrium of linear elasticity:

$$\sigma_{ij,j}^0 + F_i^0 = 0 \tag{9.7}$$

where F_i^0 is the body force, and the traction boundary condition:

$$\sigma_{ij}^0 n_j = T_i^0 \quad \text{on } \partial\Omega_T \cup \partial\Omega_s \tag{9.8}$$

where T_i^0 represents tractions imposed either directly on $\partial\Omega_T$ or through a displacement δ_j^0 imposed on a distributed elastic spring on $\partial\Omega_s$.

The initial stress field may be caused by body forces, surface tractions or temperature, or may be residual stress caused by manufacturing or cold working processes. The generalized form of eq. (9.7) is: For all $v_i \in E^0(\Omega)$

$$\frac{1}{2}\int_\Omega \sigma_{ij}^0(v_{i,j} + v_{j,i}) \, dV = \int_\Omega F_i^0 v_i \, dV + \int_{\partial\Omega_T \cup \partial\Omega_s} T_i^0 v_i \, dS \tag{9.9}$$

where dS represents the differential surface.

Let us assume that the reference configuration is perturbed by a small change in the body force $(\overline{F}_i)$; the temperature $(\overline{\mathcal{T}}_\Delta)$; the surface tractions $(\overline{T}_i)$ on $\partial\Omega_T$; the displacement imposed on the distributed spring $\overline{\delta}_i$ on $\partial\Omega_s$, or imposed displacement $(\overline{u}_i^*)$ on boundary segment $\partial\Omega_u$. The corresponding kinematically admissible displacement field is denoted by $\overline{u}_i$. It is assumed that the stress $\overline{\sigma}_{ij}$ caused by the perturbation is negligible in relation to the initial stress σ_{ij}^0.

When the reference configuration is not stress-free then the work done by σ_{ij}^0 due to the product terms of the Green strain tensor may not be negligible. Therefore the strain energy is:

$$U(\overline{u}_i) = \frac{1}{2}\int_\Omega C_{ijkl}\overline{\epsilon}_{ij}\overline{\epsilon}_{kl} \, dV + \frac{1}{2}\int_{\partial\Omega_s} k_{ij}\overline{u}_i\overline{u}_j \, dS$$
$$+ \frac{1}{2}\int_\Omega \sigma_{ij}^0 \overline{u}_{\alpha,i}\overline{u}_{\alpha,j} \, dV. \tag{9.10}$$

The third integral in eq. (9.10) represents the work done by σ_{ij}^0 due to the product terms of the Green strain tensor. The work done by σ_{ij}^0 due to the linear strain terms is exactly canceled by the work done by F_i^0, T_i^0 and δ_i^0 in the sense of eq. (9.9). The potential energy is then:

$$\Pi(\overline{u}_i) = U(\overline{u}_i) - \int_\Omega \overline{F}_i\overline{u}_i \, dV - \int_{\partial\Omega_T} \overline{T}_i\overline{u}_i \, dS - \int_{\partial\Omega_s} k_{ij}\overline{\delta}_j\overline{u}_i \, dS$$
$$- \int_\Omega \overline{\mathcal{T}}_\Delta C_{ijkl}\alpha_{kl}\overline{u}_{i,j} \, dV \tag{9.11}$$

where k_{ij} is a positive-semidefinite spring rate matrix, and α_{kl} represents the coefficients of thermal expansion. For isotropic materials $\alpha_{kl} = \alpha\delta_{kl}$.

We seek $\bar{u}_i \in \tilde{E}(\Omega)$ such that $\Pi(\bar{u}_i)$ is stationary; that is,

$$\delta\Pi(\bar{u}_i) = \left(\frac{\partial\Pi(\bar{u}_i + \varepsilon v_i)}{\partial\varepsilon}\right)_{\varepsilon=0} = 0 \qquad \text{for all } v_i \in E^0(\Omega). \tag{9.12}$$

From this it follows that the principle of virtual work in the presence of an initial stress field is: Find $\bar{u}_i \in \tilde{E}(\Omega)$ such that:

$$\int_\Omega C_{ijkl}\bar{u}_{k,l}v_{i,j}\,dV + \int_{\partial\Omega_s} k_{ij}\bar{u}_j v_i\,dS + \int_\Omega \sigma^0_{ij}\bar{u}_{\alpha,j}v_{\alpha,i}\,dV =$$
$$\int_\Omega \overline{F}_i v_i\,dV + \int_{\partial\Omega_T} \overline{T}_i v_i\,dS + \int_{\partial\Omega_s} k_{ij}\bar{\delta}_j v_i\,dS + \int_\Omega \mathcal{T}_\Delta C_{ijkl}\alpha_{kl}v_{i,j}\,dV \tag{9.13}$$

for all $v_i \in E^0(\Omega)$.

The effect of the initial stress σ^0_{ij} depends on its sense: If σ^0_{ij} is predominantly positive (i.e. tensile) then the stiffness increases. This is called stress stiffening. If, on the other hand, σ^0_{ij} is predominantly negative, then the stiffness decreases. Of great practical interest is the critical value of the initial stress at which the stiffness is zero. Define:

$$\sigma^0_{ij} = \lambda\sigma^\star_{ij}. \tag{9.14}$$

In stability problems $\sigma^\star_{ij}$ is the pre-buckling stress distribution. In stress stiffening problems $\sigma^\star_{ij}$ is some reference stress and λ is some fixed number.

Define the bilinear form $B_\lambda(\bar{u}_i, v_i)$ by:

$$B_\lambda(\bar{u}_i, v_i) = B(\bar{u}_i, v_i) - \lambda G(\bar{u}_i, v_i) \tag{9.15}$$

where

$$B(\bar{u}_i, v_i) = \int_\Omega C_{ijkl}\bar{u}_{i,j}v_{k,l}\,dV + \int_{\partial\Omega_s} k_{ij}\bar{u}_j v_i\,dS$$

$$G(\bar{u}_i, v_i) = -\int_\Omega \sigma^\star_{ij}\bar{u}_{\alpha,i}v_{\alpha,j}\,dV$$

$$F(v_i) = \int_\Omega \overline{F}_i v_i\,dV + \int_{\partial\Omega_T} \overline{T}_i v_i\,dS + \int_{\partial\Omega_s} k_{ij}\bar{\delta}_j v_i\,dS + \int_\Omega \mathcal{T}_\Delta C_{ijkl}\alpha_{kl}v_{i,j}\,dV.$$

The problem is then to find $\bar{u}_i \in \tilde{E}(\Omega)$ such that

$$B_\lambda(\bar{u}_i, v_i) = F(v_i) \quad \text{for all } v_i \in E^0(\Omega). \tag{9.16}$$

The set of λ for which eq. (9.16) is uniquely solvable for all F is called the resolvent set. The complement of the resolvent set is the spectrum. In addition to the point spectrum, which consists of the eigenvalues λ_i ($i = 1, 2, \ldots$), the spectrum may also include values that lie in a continuous spectrum. Fortunately, in those problems of engineering interest that require consideration of elastic stability (the analysis of thin-walled structures) the lowest values of λ lie in a point spectrum [33].

The effect of stress stiffening is illustrated by considering the free vibration of elastic structures subjected to initial stress. The mathematical model of free vibration is: Find ω and $\bar{u}_i \in E^0(\Omega)$; $\bar{u}_i \neq 0$, such that:

$$B_\lambda(\bar{u}_i, v_i) - \omega^2 \int_\Omega \rho\bar{u}_i v_i\,dV = 0 \tag{9.17}$$

where ρ is the specific density of the material and ω is the natural frequency. The natural frequency is now a function of λ. If $\sigma^\star_{ij}$ is predominantly compressive (negative) then the structural stiffness is decreased as $\lambda > 0$ is increased. If λ is in the point spectrum then there are functions $\bar{u}_i$ which satisfy eq. (9.17) for $\omega = 0$. That is, the natural frequency is zero. If $\sigma_{ij}^\star$ is tensile (positive) then the structural stiffness is increased as λ is increased. See Exercise 9.8.

Remark 9.2 Using the procedures of variational calculus, the strong form of the equations of equilibrium is found to be:

$$\overline{\sigma}_{ij,j} + \overline{F}_i + (\sigma_{kj}^0 \overline{u}_{i,k})_j = 0 \tag{9.18}$$

where $\overline{\sigma}_{ij} = C_{ijkl}(\overline{\epsilon}_{kl} - \overline{\mathcal{T}}_\Delta \alpha_{kl})$. The corresponding natural boundary conditions are:

$$(\overline{\sigma}_{ij} + \sigma_{kj}^0 \overline{u}_{i,k})n_j = \overline{T}_i \quad \text{on } \partial\Omega_T \tag{9.19}$$

and

$$(\overline{\sigma}_{ij} + \sigma_{kj}^0 \overline{u}_{i,k})n_j = k_{ij}(\overline{\delta}_j - \overline{u}_j) \quad \text{on } \partial\Omega_s. \tag{9.20}$$

Example 9.2 Consider an elastic column of uniform cross-section, area A, moment of inertia I, length ℓ and modulus of elasticity E. The centroidal axis of the column coincides with the x_1 axis. A compressive axial force P is applied, hence $\sigma_{11}^0 = -P/A$. In the classical formulation of elastic buckling the displacement field is assumed to be

$$\overline{u}_1 = -\frac{dw}{dx_1}x_2, \quad \overline{u}_2 = w(x_1), \quad \overline{u}_3 = 0 \tag{9.21}$$

where w is the transverse displacement. See, for example, [103]. Therefore the only non-zero strain component is:

$$\epsilon_{11} = \frac{d^2 w}{dx_1^2}x_2 \tag{9.22}$$

and the term $\sigma_{ij}^0 u_{\alpha,i} u_{\alpha,j}$ can be written as:

$$\sigma_{ij}^0 \overline{u}_{\alpha,i} \overline{u}_{\alpha,j} = -\frac{P}{A} \left\{ \left(\frac{d^2 w}{dx_1^2}\right)^2 x_2^2 + \left(\frac{dw}{dx_1}\right)^2 \right\}. \tag{9.23}$$

In this case the strain energy can be written in the following form:

$$U = \frac{1}{2} \int_0^\ell \left(E - \frac{P}{A}\right) I \left(\frac{d^2 w}{dx_1^2}\right)^2 dx_1 - \frac{P}{2} \int_0^\ell \left(\frac{dw}{dx_1}\right)^2 dx_1 \tag{9.24}$$

where I is the moment of inertia of the cross-section:

$$I = \int_A x_2^2 \, dx_2 dx_3. \tag{9.25}$$

The term P/A can be neglected in relation to the modulus of elasticity E in (9.24). This example serves as an illustration of how a dimensionally reduced model is derived from the three-dimensional formulation.

Example 9.3 The following example illustrates the solution of a buckling problem. An elastic strip of length $L = 200$ mm, thickness $t = 1.0$ mm, width $w = 7.5$ mm is fixed at one end and loaded at the other end by a constant normal traction T_n, corresponding to 1.0 N compressive force. The shearing tractions are zero. The modulus of elasticity is $E = 2.0 \times 10^5$ MPa and Poisson's ratio is zero. The goal is to estimate the buckling load factor, that is, the multiplier of the unit applied force corresponding to the first buckling mode.

The classical formula or the buckling load F_{cr} is

$$F_{cr} = \frac{\pi^2 EI}{4L^2} \tag{9.26}$$

where I is the moment of inertia: $I = wt^3/12$. For the data in this example $F_{cr} = 7.71$ N. Therefore the buckling load factor is 7.71.

Numerical solutions were obtained using a thin solid model, three hexahedral elements (as in Example 9.1) and anisotropic product space $S^{p,p,q}(\Omega_{st}^{(q)})$ with p ranging from 3 to 8 and $q = 1$. The computed value of the buckling load factor was 7.71. Having repeated the computations for $q = 3$, the same value was obtained.

Example 9.4   The following example illustrates the effects of stress stiffening on natural frequencies and mode shapes. A 200 mm long, 1.0 mm thick and 7.5 mm wide elastic strip is fixed at one end and loaded at the other end by normal traction T_n, the shearing tractions are zero. The modulus of elasticity is $E = 2.0 \times 10^5$ MPa Poisson's ratio is $v = 0.3$ and the mass density is $\varrho = 7.86 \times 10^{-9}$ Ns²/mm⁴ (7860 kg/m³).

Solutions were obtained for $T_n = 0$ and $T_n = 150$ MPa using a thin solid model, four hexahedral elements and anisotropic product space $S^{p,p,q}(\Omega_{st}^{(q)})$ with $p = 8$ and $q = 3$. The 20th mode shapes for $T_n = 0$ and $T_n = 150$ MPa are shown in Figures 9.3(a) and 9.3(b) respectively. The natural frequency for $T_n = 0$ converged to 7019 Hz and for $T_n = 150$ MPa it converged to 7294 Hz. Az seen in Fig. 9.3 the mode shapes are quite different: one is a torsional mode, the other is a bending mode.

Example 9.5   Consider a cylindrical shell of radius 0.10 m, length 0.40 m, wall thickness 0.001 m. The material properties are: $E = 2.0 \times 10^{11}$ Pa, $v = 0.3$. The shell is fixed at one end. On the other end (a) the radial and circumferential displacements are zero and an axial (compressive) displacement Δ is imposed; (b) the axial displacement is zero and tangential displacements are imposed consistent with rotation θ about the axis of the shell. The goals of computation are to determine the critical values of Δ and θ, the corresponding compressive force F_{crit} and twisting moment M, and to verify that the relative errors in Δ_{crit} and θ_{crit} are less than 1%.

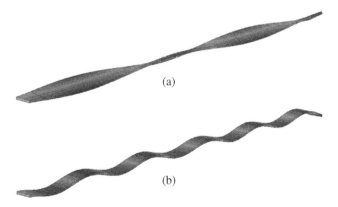

(a)

(b)

Figure 9.3   The 20th modes of vibration of the elastic strip in Example 9.4. (a) Without prestress, (b) prestress of 150 MPa in tension.

Table 9.1 Critical values of the axial displacement Δ (mm) and the rotation θ (rad) in Example 9.5.

p	N	Δ_{crit}	% Error	θ_{crit}	% Error
5	11568	2.64	11.81	1.3249×10^{-2}	2.16
6	17136	2.44	3.32	1.2983×10^{-2}	0.11
7	24240	2.38	0.62	1.2970×10^{-2}	0.01
8	32880	2.37	0.14	1.2969×10^{-2}	0.00
∞	∞	2.36	–	1.2969×10^{-2}	–

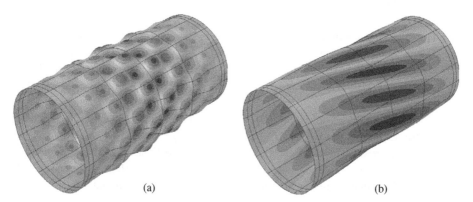

(a)   (b)

Figure 9.4 The first buckling modes of the cylindrical shell in Example 9.5. (a) Imposed axial displacement. The contour lines are proportional to the normal displacement. (b) Imposed axial rotation. The contour lines are proportional to the tangential displacement.

The computations were performed using 128 quadrilateral thin solid elements $S^{p,p,q}(\Omega_{\text{st}}^{(q)})$ with p ranging from 5 to 8 (trunk space) and q fixed at $q = 3$. The results are shown in Table 9.1 and in Fig. 9.4. The buckling mode shown in Fig. 9.4(a) happens to be an antisymmetric function.

The extrapolated value of the strain energy corresponding to the linear solution at $\Delta = 0.001$ m is $U_{\text{lin}} = 157.65$ N m. The strain energy is proportional to Δ^2. Therefore the strain energy at the critical value of Δ is

$$U_{\text{crit}} = U_{\text{lin}}(\Delta_{\text{crit}}/0.001)^2 = 157.65(2.36/1)^2 = 878 \text{ N m}$$

and the estimated value of the critical force is

$$F_{\text{crit}} = (2U/\Delta) = 2 \times 878/0.001 = 1756 \text{ kN}.$$

The twisting moment M is computed similarly. In case (b) $U_{\text{crit}} = 102.8$ N m and $M_{\text{crit}} = 15.84$ kN m.

Remark 9.3 The predicted buckling load should be understood to be an upper bound of the load carrying capacity of shells. Owing to imperfections and sensitivity to boundary conditions, experimentally obtained buckling loads tend to be substantially lower than the buckling loads predicted by models such as the model in Example 9.5. Reduction factors of $1/2$ to $1/4$ are used in engineering practice to account for such uncertainties [49].

Exercise 9.5 Consider the problem of Example 9.5 with the following boundary conditions: The shell is fixed at one end. On the other end the radial and circumferential displacements are zero and uniformly distributed axial (compressive) traction is applied the resultant of which is F. Determine the critical value of F. Explain why it is different from F_{crit} in Example 9.5.

Exercise 9.6 Neglecting the term P/A in (9.24), determine the approximate value of P at which a column fixed at both ends will buckle. Use one finite element and $p = 4$. Report the relative error. Hint: Refer to eq. (7.27) and the definition $N_5(\xi) = \psi_4(\xi)$ where $\psi_j(\xi)$ $(j = 4, 5, ...)$ is given by eq. (7.26). The exact value of the critical force is $4\pi^2 EI/\ell^2$.

Exercise 9.7 Show that the work done by σ_{ij}^0 due to the linear strain terms is exactly canceled by the work done by F_i^0, T_i^0 and δ_i^0 in the sense of eq. (9.9).

Exercise 9.8 Consider a 50 mm × 50 mm square plate, thickness: 1.0 mm. The material properties are: $E = 6.96 \times 10^4$ MPa; $v = 0.365$; $\varrho = 2.71 \times 10^{-9}$ Ns^2/mm^4. The plate is fixed on one edge, loaded by a normal traction T_n on the opposite edge and simply supported on the other edges (soft simple support). Determine the first natural frequency corresponding to $T_n = -50$ MPa, $T_n = 0$ and $T_n = 50$ MPa. (Partial answer: For $T_n = -50$ MPa the first natural frequency is 578.6 Hertz.)

Exercise 9.9 If the multi-span beam described in Exercise 7.9 had been pre-stressed by an axial force such that a constant positive initial stress σ^0 acted on the beam, would the rotation at support B be larger or smaller than if the beam had not been prestressed? Explain.

9.2.3 Plasticity

The formulation of mathematical models based on the incremental and deformational theories of plasticity are discussed in the following. In the incremental theory, as the name implies, a relationship between the increment in strain tensor and the corresponding increment in the stress tensor is defined. In the deformational theory the strain tensor (rather than the increment in the strain tensor) is related to the stress tensor. It is assumed that the strain components are sufficiently small to justify small strain representation and the plastic deformation is contained, that is, the plastic zone is surrounded by an elastic zone. Uncontained plastic flow, as in metal forming processes, is outside of the scope of the following discussion.

The formulation of plastic deformation is based on three fundamental relationships: (a) yield criterion, (b) flow rule and (c) hardening rule. We will use the von Mises yield criterion[4] and an associative flow rule known as the Prandtl-Reuss flow rule[5]. For a more detailed discussion of this subject we refer to [86].

Notation

The stress deviator tensor is defined by:

$$\tilde{\sigma}_{ij} \stackrel{\text{def}}{=} \sigma_{ij} - \frac{1}{3}\sigma_{kk}\delta_{ij}. \tag{9.27}$$

---

4 Richard von Mises 1883–1953.
5 Ludwig Prandtl 1875–1953, Endre Reuss 1900–1968.

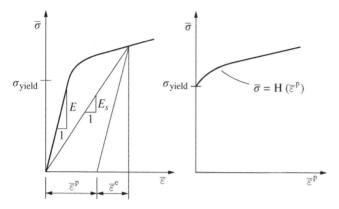

Figure 9.5 Typical uniaxial stress-strain curve.

The second invariant of the stress deviator tensor is denoted by J_2 and is defined by

$$J_2 \stackrel{\text{def}}{=} \frac{1}{2}\tilde{\sigma}_{ij}\tilde{\sigma}_{ij}$$
$$= \frac{1}{3}\left[(\sigma_{11} - \sigma_{22})^2 + (\sigma_{22} - \sigma_{33})^2 + (\sigma_{33} - \sigma_{11})^2 + 6(\sigma_{12} + \sigma_{23} + \sigma_{31})\right].$$

In a uniaxial test σ_{11} is the only non-zero stress component. Therefore $J_2 = 2\sigma_{11}^2/3$. The equivalent stress, also called von Mises stress, is $\sqrt{J_2}$ scaled such that in the special case of uniaxial stress it is equal to the uniaxial stress:

$$\bar{\sigma} \stackrel{\text{def}}{=} \sqrt{\frac{1}{2}\left[(\sigma_{11} - \sigma_{22})^2 + (\sigma_{22} - \sigma_{33})^2 + (\sigma_{33} - \sigma_{11})^2 + 6(\sigma_{12} + \sigma_{23} + \sigma_{31})\right]}. \tag{9.28}$$

We will interpret uniaxial stress-strain diagrams as a relationship between the equivalent stress and the equivalent strain. The elastic (resp. plastic) strains will be indicated by the superscript e (resp. p). The three principal strains are denoted by ϵ_1, ϵ_2, ϵ_3. The equivalent elastic strain is defined by

$$\bar{\epsilon}^e \stackrel{\text{def}}{=} \frac{\sqrt{2}}{2(1+\nu)}\sqrt{(\epsilon_1^e - \epsilon_2^e)^2 + (\epsilon_2^e - \epsilon_3^e)^2 + (\epsilon_3^e - \epsilon_1^e)^2} \tag{9.29}$$

where ν is Poisson's ratio. The definition of the equivalent plastic strain follows directly from eq. (9.29) by setting $\nu = 1/2$:

$$\bar{\epsilon}^p = \frac{\sqrt{2}}{3}\sqrt{(\epsilon_1^p - \epsilon_2^p)^2 + (\epsilon_2^p - \epsilon_3^p)^2 + (\epsilon_3^p - \epsilon_1^p)^2}. \tag{9.30}$$

A typical uniaxial stress-strain curve is shown in Fig. 9.5.

Exercise 9.10 Show that the first derivatives of J_2 with respect to σ_x, σ_y, σ_z and τ_{xy} are equal to $\tilde{\sigma}_x$, $\tilde{\sigma}_y$ $\tilde{\sigma}_z$ and $\tilde{\tau}_{xy}$ respectively.

Assumptions

The assumptions on which the formulation of the mathematical problem of plasticity is based are described in the following.

1. **Confined plastic deformation:** The strain components are much smaller than unity on the solution domain and its boundary, and the deformations are small in the sense that equilibrium equations written for the undeformed configuration are essentially the same as the equilibrium equations written for the deformed configuration.

2. **Decomposition of strain:** Assuming that the temperature remains constant, an increment in the total strain is the sum of the increment in the elastic strain and the increment in the plastic strain:

$$d\epsilon_{ij} = d\epsilon_{ij}^{e} + d\epsilon_{ij}^{p}. \tag{9.31}$$

3. **Yield criterion:** We define

$$F(\sigma_{ij}, \bar{\epsilon}^{p}) = \bar{\sigma} - H(\bar{\epsilon}^{p}). \tag{9.32}$$

When $F < 0$ then the material is elastic. Plastic deformation may occur only when $F = 0$. Any stress state for which $F > 0$ is inadmissible. This is known as the consistency condition. Therefore in plastic deformation:

$$dF = 0 : \quad \frac{\partial F}{\partial \sigma_{ij}} d\sigma_{ij} - H' d\bar{\epsilon}^{p} = 0. \tag{9.33}$$

4. **Flow rule:** The Prandtl-Reuss flow rule states that

$$d\epsilon_{ij}^{p} = \frac{\partial F}{\partial \sigma_{ij}} d\bar{\epsilon}^{p}. \tag{9.34}$$

Example 9.6 In this example we estimate the force-displacement relationship for the shear fitting described in Example 4.7 using the deformation theory of plasticity. The material is 7075-T6 aluminum alloy. The material properties are: Modulus of elasticity: $E = 1.05E7$ psi (7.24E4 MPa), Poisson's ratio: $\nu = 0.30$. The relationship between the equivalent strain $\bar{\epsilon}$ and equivalent stress $\bar{\sigma}$ is given by the Ramberg-Osgood equation:

$$\bar{\epsilon} = \frac{\bar{\sigma}}{E} + \frac{3}{7} \frac{S_{70E}}{E} \left(\frac{\bar{\sigma}}{S_{70E}} \right)^{n} \tag{9.35}$$

where S_{70E} is the stress corresponding to the point where a line that passes through the origin, and has the slope of 0.7 times the modulus of elasticity, intersects the uniaxial stress-strain curve. For the 7075-T6 aluminum alloy $S_{70E} = 58.5$ ksi (403.3 MPa), $n = 15.2$.

Incremental stress-strain relationship
An increment in stress is proportional to the elastic strain:

$$d\sigma_{ij} = C_{ijkl} d\epsilon_{kl}^{e} = C_{ijkl} (d\epsilon_{kl} - d\epsilon_{kl}^{p}).$$

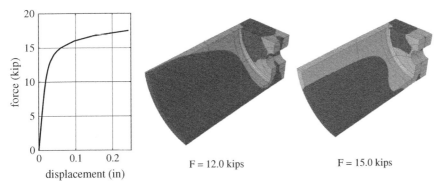

F = 12.0 kips   F = 15.0 kips

Figure 9.6 Example 9.6. The force-displacement relationship and the plastic zone for two loading conditions. The plastic zone is light grey.

Substituting eq. (9.34) we have:

$$do_{ij} = C_{ijkl}d\epsilon_{kl} - C_{ijkl}\frac{\partial F}{\partial \sigma_{kl}}d\bar{\epsilon}^p. \tag{9.36}$$

Using (9.33) we have:

$$H'd\bar{\epsilon}^p = \frac{\partial F}{\partial \sigma_{ij}}\,d\sigma_{ij} = \frac{\partial F}{\partial \sigma_{ij}}C_{ijkl}d\epsilon_{kl} - \frac{\partial F}{\partial \sigma_{pq}}C_{pqrs}\frac{\partial F}{\partial \sigma_{rs}}d\bar{\epsilon}^p \tag{9.37}$$

where the dummy indices were suitably renamed. From eq. (9.37) an expression for $d\bar{\epsilon}^p$ is obtained:

$$d\bar{\epsilon}^p = \frac{\dfrac{\partial F}{\partial \sigma_{ij}}C_{ijkl}}{H' + \dfrac{\partial F}{\partial \sigma_{pq}}C_{pqrs}\dfrac{\partial F}{\partial \sigma_{rs}}}\,d\epsilon_{kl}.$$

On substituting into eq. (9.36) we have the incremental elastic-plastic stress-strain relationship:

$$d\sigma_{ij} = \left(C_{ijkl} - \frac{C_{ijmn}\dfrac{\partial F}{\partial \sigma_{mn}}\dfrac{\partial F}{\partial \sigma_{uv}}C_{uvkl}}{H' + C_{pqrs}\dfrac{\partial F}{\partial \sigma_{pq}}\dfrac{\partial F}{\partial \sigma_{rs}}}\right)d\epsilon_{kl}. \tag{9.38}$$

The bracketed expression in (9.38) is well defined for elastic-perfectly plastic materials (i.e. materials for which $H' = 0$).

The computations involve the following: Given the current stress σ_{ij}, compute

$$d\sigma_{ij} = C_{ijkl}d\epsilon_{kl}$$

corresponding to an increment in the applied load. In each integration point compute $F(\sigma_{ij} + d\sigma_{ij})$. If $F(\sigma_{ij} + d\sigma_{ij}) \leq 0$ then nothing further needs to be done. If $F(\sigma_{ij} + d\sigma_{ij}) > 0$ then re-compute $d\sigma_{ij}$ using eq. (9.38). Repeat the process until $F(\sigma_{ij} + d\sigma_{ij}) \approx 0$. The process is started from the linear solution.

The deformation theory of plasticity
In the deformation theory of plasticity it is assumed that the plastic strain tensor is proportional to the stress deviator tensor. Referring to Fig. 9.5,

$$\bar{\epsilon}^e + \bar{\epsilon}^p = \frac{\bar{\sigma}}{E_s}$$

where E_s is the secant modulus ($0 < E_s < E$). Since the elastic part of the strain is related to the stress by Hooke's law,

$$\bar{\epsilon}^e = \frac{\bar{\sigma}}{E}$$

we have

$$\bar{\epsilon}^p = \left(\frac{1}{E_s} - \frac{1}{E}\right)\bar{\sigma}. \tag{9.39}$$

In the case of uniaxial stress $\tilde{\sigma} = 2\bar{\sigma}/3$ and hence eq. (9.39) can be written as:

$$\bar{\epsilon}^p = \frac{3}{2}\left(\frac{1}{E_s} - \frac{1}{E}\right)\tilde{\sigma}.$$

By modeling assumption this is generalized to:

$$\epsilon_{ij}^p = \frac{3}{2}\left(\frac{1}{E_s} - \frac{1}{E}\right)\tilde{\sigma}_{ij}. \tag{9.40}$$

For example, in planar problems:

$$\begin{Bmatrix} \epsilon_x^p \\ \epsilon_y^p \\ \epsilon_z^p \\ \epsilon_{xy}^p \end{Bmatrix} = \frac{3}{2}\left(\frac{1}{E_s} - \frac{1}{E}\right)\begin{Bmatrix} \tilde{\sigma}_x \\ \tilde{\sigma}_y \\ \tilde{\sigma}_z \\ \tilde{\tau}_{xy} \end{Bmatrix}. \tag{9.41}$$

In the deformation theory of plasticity the elastic-plastic compliance matrix is the matrix [C] that establishes the relationship between the total strain and stress:

$$\{\epsilon\} = [C]\{\sigma\}. \tag{9.42}$$

The elastic-plastic material stiffness matrix $[E_{ep}]$ is the inverse of the elastic-plastic compliance matrix. Using the definition of the stress deviator and the relationship between the plastic strain and deviatoric stress (9.41), we have:

$$\begin{Bmatrix} \epsilon_x^p \\ \epsilon_y^p \\ \gamma_{xy}^p \end{Bmatrix} = \frac{3}{2}\left(\frac{1}{E_s} - \frac{1}{E}\right)\begin{bmatrix} 2/3 & -1/3 & 0 \\ -1/3 & 2/3 & 0 \\ 0 & 0 & 2 \end{bmatrix}\begin{Bmatrix} \sigma_x \\ \sigma_y \\ \tau_{xy} \end{Bmatrix}.$$

Using $\{\epsilon\} = \{\epsilon^e\} + \{\epsilon^p\}$, a relationship is obtained between the total strain components and the stress tensor:

$$\begin{Bmatrix} \epsilon_x \\ \epsilon_y \\ \gamma_{xy} \end{Bmatrix} = \left(\frac{1}{E}\begin{bmatrix} 1 & -\nu & 0 \\ -\nu & 1 & 0 \\ 0 & 0 & 2(1+\nu) \end{bmatrix} + \frac{E - E_s}{E_s E}\begin{bmatrix} 1 & -1/2 & 0 \\ -1/2 & 1 & 0 \\ 0 & 0 & 3 \end{bmatrix}\right)\begin{Bmatrix} \sigma_x \\ \sigma_y \\ \tau_{xy} \end{Bmatrix}$$

where the bracketed expression is the elastic-plastic material compliance matrix. Since $E_s < E$, it is invertible.

The solution is obtained by iteration: For each integration point the equivalent stress and strain are computed. From the stress-strain relationship the secant modulus is computed and the appropriate material stiffness matrix $[E_{ep}]$ is evaluated and a new solution is obtained. The process is continued until the maximum of the difference between the equivalent stress and the uniaxial stress, given the equivalent strain, becomes less than a pre-set tolerance (which is typically 1% or less).

Example 9.7 An interesting benchmark study was performed under a research project entitled "Adaptive Finite Element Methods in Applied Mechanics" sponsored by the German Research Foundation[6] in the period 1994 to 1999. Nine academic research institutes participated in this project. Importantly, the project fostered collaboration among researchers from the applied mathematics and engineering research communities. At that time many researchers believed that high order elements cannot be used for solving material nonlinear problems. The results of this investigation served to dispel those beliefs [82].

The following challenge problem was posed: Solve a plane strain problem on a rectangular domain with a hole in the center and report certain quantities of interest. The domain and

---

6 Deutsche Forschungsgemeinschaft (DFG).

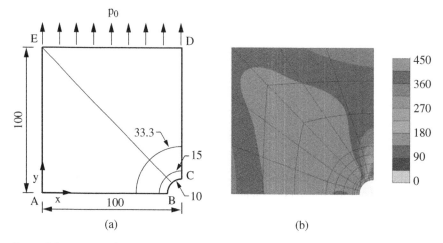

Figure 9.7 Example 9.7: (a) The domain and the initial finite element mesh. (b) The equivalent stress contours corresponding to $p_0 = 450$ computed on the 54-element mesh at $p = 8$, product space.

dimensions are shown in Fig. 9.7(a). The boundary segments AB and CD are lines of symmetry. The boundary segment DE is loaded by uniform normal traction p_0, the shearing traction is zero. On boundary segments BC and EA the normal and shearing tractions are zero. The values $p_0 = 300$ and $p_0 = 450$ were specified The investigators were instructed to assume that the material is isotropic and elastic-perfectly plastic with yield stress $\sigma_{yield} = 450$, obeying the deformation theory of plasticity and the von Mises yield criterion. The shear modulus $G = 80193.8$ and the bulk modulus[7] $K = 164206.0$ were given.

We solved the problem using a sequence of quasiuniform meshes with $p = 8$, product space, assigned to all elements. The sequence was generated starting from the six element mesh shown in Fig. 9.7(a) by uniformly dividing the standard quadrilateral element into n^2 squares. The mesh corresponding to $n = 3$ and the contours of the von Mises stress corresponding to $p_0 = 450$ are shown in Fig. 9.7(b). Starting from the linear solution, the elastic-plastic material stiffness matrix $[E_{ep}]$ was iteratively updated in each integration point until the difference between the equivalent stress and the uniaxial stress, given the equivalent strain, was less than 0.5%. The computations were performed with StressCheck.

We computed the following quantities of interest reported in Table 1 of reference [36]: (a) the integral W defined as

$$W = \frac{p_0}{2} \int_{x=0}^{100} u_y(x, 100) \, dx,$$

(b) the stress component σ_y in point B, denoted by $\sigma_y^{(B)}$ and (c) the displacement components u_y and u_x in points D and E respectively. The results of computation are shown in Table 9.2. It is seen that the computed data are virtually independent of the number of degrees of freedom N as the number of elements (denoted by M) is increased. Also, the data in Table 9.2 match the data reported in [36] to four significant digits.

---

7 The shear and bulk moduli are related E and v as $G = \dfrac{E}{2(1+v)}$ and $K = \dfrac{E}{3(1-2v)}$.

Table 9.2 Example 9.7: The results of computation for $p_0 = 300$.

M	N	W	$\sigma_y^{(B)}$	$u_y^{(D)}$	$u_x^{(E)}$
6	800	2044.983	522.27	0.140327	0.050886
54	7008	2044.986	517.52	0.140327	0.050885
150	19360	2044.984	517.44	0.140327	0.050885

Exercise 9.11 Show that in uniaxial tension or compression $\bar{\epsilon}^e = \bar{\epsilon}_1^e$, $\bar{\epsilon}^p = \bar{\epsilon}_1^p$ and $\bar{\sigma} = \sigma_1$.

Exercise 9.12 Derive eq. (9.29) by specializing the root-mean-square of the differences of principal strains to the one-dimensional case such that $\bar{\epsilon}^e = \bar{\epsilon}_1^e$.

9.2.4 Mechanical contact

Mathematical models of contact between solid bodies involve non-linear boundary conditions written in terms of a gap function $g = g(s, t) \geq 0$ where s and t are the parameters of one of the contacting surfaces: Whenever $g = 0$, the normal and shearing tractions in corresponding points must have equal value and opposite sense. When $g > 0$ then the tractions are zero. The condition $g < 0$ is not allowed.

In many practical problems the contacting surfaces are lubricated and therefore, under quasi-static conditions, the shearing tractions are negligibly small in comparison with the normal tractions.

We will be concerned with frictionless contact only. Consider two solid bodies Ω_1 and Ω_2. The surfaces are denoted by $\partial\Omega_1$ and $\partial\Omega_2$ respectively. We identify a contact zone $\partial\Omega_c$ on one of the contacting surfaces. The contact zone is a convex subset of one of the contacting surfaces. Contact is expected to occur within the contact zone. There could be several contact zones and several bodies can be in contact.

Example 9.8 Let us consider the problem of contact between two elastic bars constrained by distributed springs. The notation is shown in Fig. 9.8. We assume that the axial stiffness $(AE)_i$ and the spring coefficient c_i ($i = 1, 2$) are constants for each bar and bar 2 is fixed on its right end. The gap between the bars is $g = g_0 - U_2 + U_3$ where g_0 is the initial gap. The goal is to determine the elastic spring rate $k = F/U_1$ as a function of the applied force F.

The differential equations for the bars are:

$$-(AE)_i u_i'' + c_i u_i = 0, \quad i = 1, 2 \tag{9.43}$$

where the primes indicate differentiation with respect to x. We associate a local coordinate system with each bar, such that $x = 0$ at the left end. Therefore $U_1 = u_1(0)$, $U_2 = u_1(\ell_1)$ and $U_3 = u_2(0)$.

Figure 9.8 Example: Contact problem in 1D. Notation.

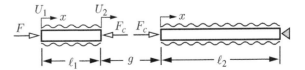

Introducing

$$\xi = x/\ell_i, \quad v_i = u_i/\ell_i, \quad \lambda_i = \sqrt{c_i\ell_i/(AE)_i}$$

equations (9.43) are written in dimensionless form:

$$-\frac{d^2 v_i}{d\xi^2} + \lambda_i^2 v_i = 0, \quad 0 \le \xi \le 1, \quad i = 1, 2. \tag{9.44}$$

The solution is:

$$v_i(x) = a_i \cosh \lambda_i \xi + b_i \sinh \lambda_i \xi. \tag{9.45}$$

Referring to the bar on the left, the boundary conditions are:

$$(AE)_1 u_1'(0) = -F, \quad (AE)_1 u_1'(\ell_1) = -F_c$$

which correspond to

$$\left(\frac{dv_1}{d\xi}\right)_{\xi=0} = -\frac{F}{(AE)_1}, \quad \left(\frac{dv_1}{d\xi}\right)_{\xi=1} = -\frac{F_c}{(AE)_1}$$

therefore the solution is

$$v_1(\xi) = \frac{F}{(AE)_1 \lambda_1} \left(\frac{\cosh \lambda_1 - F_c/F}{\sinh \lambda_1} \cosh \lambda_1 \xi - \sinh \lambda_1 \xi \right). \tag{9.46}$$

Given an initial gap $g_0 > 0$, the force needed to close the gap, denoted by F_0, can be computed from eq. (9.46) by setting $v_1(1) = g_0/\ell_1$ and $F_c = 0$. This yields:

$$F_0 = (AE)_1 \lambda_1 (g_0/\ell_1) \sinh \lambda_1 \tag{9.47}$$

where we used $\cosh^2 \lambda_1 - \sinh^2 \lambda_1 = 1$.

Therefore for $F \le F_0$ the spring rate is

$$k = \frac{F}{U_1} = \frac{F}{\ell_1 v_1(0)} = \lambda_1 \frac{(AE)_1}{\ell_1} \frac{\sinh \lambda_1}{\cosh \lambda_1}. \tag{9.48}$$

For any $F > F_0$ a contact force $F_c > 0$ will develop and the contact condition $g = 0$ must be satisfied. To the bar on the right we apply the boundary conditions

$$(AE)_2 u_2'(0) = -F_c, \quad u_2(\ell_2) = 0$$

which correspond to

$$\left(\frac{dv_2}{d\xi}\right)_{\xi=0} = -\frac{F_c}{(AE)_2}, \quad v_2(1) = 0.$$

The solution is:

$$v_2(\xi) = \frac{F_c}{(AE)_2 \lambda_2} \left(\frac{\sinh \lambda_2}{\cosh \lambda_2} \cosh \lambda_2 \xi - \sinh \lambda_2 \xi \right). \tag{9.49}$$

From eq. (9.46) we get

$$U_2 \equiv u_1(\ell_1) = \ell_1 v_1(1) = \frac{\ell_1 (F - F_c \cosh \lambda_1)}{(AE)_1 \lambda_1 \sinh \lambda_1} \tag{9.50}$$

and from eq. (9.49) we get

$$U_3 \equiv u_2(0) = \ell_2 v_2(0) = \frac{\ell_2 F_c \sinh \lambda_2}{(AE)_2 \lambda_2 \cosh \lambda_2}. \tag{9.51}$$

Using $g = g_0 - U_2 + U_3 = 0$ we can write $F_c = q(F - F_0)$ where q is a dimensionless constant that depends on the parameters that characterize the problem.

$$q = \left(\cosh \lambda_1 + \frac{\ell_2}{\ell_1} \frac{(AE)_1 \lambda_1 \sinh \lambda_1}{(AE)_2 \lambda_2 \cosh \lambda_2} \sinh \lambda_2 \right)^{-1}$$

$$g = \frac{\ell_1 F_0}{(AE)_1 \lambda_1 \sinh \lambda_1} - \frac{\ell_1 F}{(AE)_1 \ell_1 \sinh \lambda_1} + \frac{\ell_1 F_c \cosh \lambda_1}{(AE)_1 \lambda_1 \sinh \lambda_1}$$

$$+ \frac{\ell_2 F_c \sinh \lambda_2}{(AE)_2 \lambda_2 \cosh \lambda_2} = 0. \tag{9.52}$$

In this example it was possible to find F_c in two steps. In the first step the force needed to close the gap, denoted by F_0, was determined. In the second step the contact force F_c was determined for $F = \alpha F_0$ where $\alpha \geq 1$. In two and three dimensions the problem is far more complicated because it is necessary to determine the contact surfaces which depend on the contact force, the material properties and the geometric features of the contacting bodies.

Gap elements in two dimensions
Gap elements are used for the approximation of the gap function g. The gap function measures the distance between corresponding points in contacting bodies. The condition $g > 0$ indicates that corresponding points are not in contact, whereas $g < 0$ indicate that the contacting bodies intersect.

We will be concerned with frictionless mechanical contact of elastic bodies in two dimensions. The goal is to find the function T_n such that

$$gT_n = 0, \quad g \geq 0, \quad T_n^{(A)} = -T_n, \quad T_n^{(B)} = -T_n \tag{9.53}$$

where $T_n^{(A)}$ (resp. $T_n^{(B)}$) represents the normal traction acting on body A (resp. body B).

The mapping of gap elements is analogous to the mapping of quadrilateral elements. We will assume that the contacting edges correspond to sides 1 and 3. Therefore the mapping of a gap element is

$$x = \frac{1 - \eta}{2} f_x^{(A)}(\xi) + \frac{1 + \eta}{2} f_x^{(B)}(\xi) \tag{9.54}$$

$$y = \frac{1 - \eta}{2} f_y^{(A)}(\xi) + \frac{1 + \eta}{2} f_y^{(B)}(\xi) \tag{9.55}$$

where $(f_x^{(A)}(\xi), f_y^{(A)}(\xi))$ are points on the boundary of body A and, similarly, $(f_x^{(B)}(\xi), f_y^{(B)}(\xi))$ are points on the boundary of body B.

We approximate the functions f_x, f_y by Lagrange polynomials that have roots in the optimal collocation points, see Appendix F. For example:

$$f_x^{(A)}(\xi) \approx \overline{f}_x^{(A)}(\xi) = \sum_{i=1}^n X_i^{(A)} L_i(\xi), \quad X_i^{(A)} = x(\xi_i, -1) \tag{9.56}$$

$$f_y^{(A)}(\xi) \approx \overline{f}_y^{(A)}(\xi) = \sum_{i=1}^n Y_i^{(A)} L_i(\xi), \quad Y_i^{(A)} = y(\xi_i, -1) \tag{9.57}$$

where ξ_i is the ith optimal collocation point and $L_i(\xi)$ is the corresponding Lagrange interpolating polynomial.

Example 9.9 Consider a circular (planar) body B of radius R. The coordinates of the center are $x_0 = 0, y_0 = R$. Assume that body B penetrated body A, which is bounded by the line segment $y = y_0$,

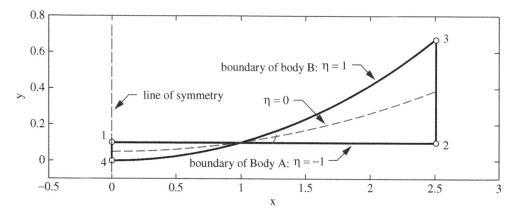

Figure 9.9 Example 9.9: Gap element with partial penetration.

$0 \le x \le x_0$. A gap element, shown in Fig. 9.9, is bounded by a circular segment at $\eta = 1$ and a line at $\eta = -1$. We denote the angle subtended by the circular arc from point (x_0, y_0) by α. Then we have:

$$X_i^{(B)} = R \sin \left(\alpha (1 + \xi_i)/2 \right) \tag{9.58}$$

$$Y_i^{(B)} = R \left(1 - \cos \left(\alpha (1 + \xi_i)/2 \right) \right). \tag{9.59}$$

The gap element characterized by the data $R = 5.0$, $\alpha = \pi/6$, $x_0 = 2.5$ and $y_0 = 0.1$ is shown in Fig. 9.9.

Outline of the algorithm

We denote the Jacobian determinant by $|J(\xi, \eta)|$ and for the ith collocation point $(i = 1, 2, \ldots, n)$ we compute

$$G_i = \mathrm{sign}(|J(\xi_i, 0)|) \sqrt{\left(X_i^{(A)} - X_i^{(B)} \right)^2 + \left(Y_i^{(A)} - Y_i^{(B)} \right)^2}. \tag{9.60}$$

The gap function is approximated by

$$g(\xi) = \sum_{i=1}^{n} G_i L_i(\xi), \qquad -1 \le \xi \le 1. \tag{9.61}$$

Explanation: When partial penetration occurs, as in Fig. 9.9, then $|J(\xi, \eta)| < 0$ over part of the gap element. Furthermore, the curve $x = x(\xi, 0), y = y(\xi, 0)$, passes through the point (or points) where sides 1 and 3 intersect and $|J(\xi, 0)| = 0$. Note that $|J(\xi, 0)|$ changes sign in the point of intersection. To show this, we write the Jacobian determinant for $\eta = 0$:

$$|J(\xi, 0)| = \frac{1}{4} \left(\frac{df_x^{(A)}}{d\xi} + \frac{df_x^{(B)}}{d\xi} \right) (-f_y^{(A)} + f_y^{(B)})$$

$$- \frac{1}{4} \left(\frac{df_y^{(A)}}{d\xi} + \frac{df_y^{(B)}}{d\xi} \right) (-f_x^{(A)} + f_x^{(B)}) \tag{9.62}$$

and observe that in the point of intersection $f_x^{(A)} = f_x^{(B)}$ and $f_y^{(A)} = f_y^{(B)}$ hence $|J(\xi, 0)| = 0$.

We assume that the contacting bodies are sufficiently constrained to prevent rigid body displacement. We apply compressive normal tractions to each of the contacting bodies over the boundary segment where $g(\xi) < 0$. Specifically,

$$T_n = -Cg(\xi), \quad T_n^{(A)} = -T_n, \quad T_n^{(B)} = -T_n, \quad \xi \in \{\xi \mid g(\xi) < 0\} \tag{9.63}$$

where C is an arbitrary positive number. We compute the absolute value of the resultant of the tractions and denote it by F. We also compute the strain energy values and denote them by $U^{(A)}$ and $U^{(B)}$. The estimated (average) spring rates $k^{(A)}$ and $k^{(B)}$ are determined from

$$k^{(A)} = \frac{F^2}{2U^{(A)}}, \quad k^{(B)} = \frac{F^2}{2U^{(B)}}. \tag{9.64}$$

The condition for closing the gap is:

$$\delta = \delta^{(A)} + \delta^{(B)} = \frac{\overline{F}}{k^{(A)}} + \frac{\overline{F}}{k^{(B)}} \tag{9.65}$$

where δ is the absolute value of the average of the gap function over the interval where $g(\xi) < 0$. From this we find:

$$\overline{F} = \frac{k^{(A)}k^{(B)}}{k^{(A)} + k^{(B)}} \, \delta. \tag{9.66}$$

We define the scaling factor $s = \overline{F}/F$ and scale T_n such that $T_n \to s \, T_n$ and update the coordinates of the points in the contact zone by the scaled displacement components. For example,

$$X_i^{(A)} \to X_i^{(A)} + s \, u_x^{(A)}(\xi_i, -1). \tag{9.67}$$

The gap function given by eq. (9.61) is re-computed with the updated coordinates and the process is repeated. In the second and subsequent steps the tractions applied in the previous step are incrementally updated. The process is continued until a stopping criterion is satisfied.

This algorithm does not require the element edges in the contact zone to be conforming, nor does it restrict the number of collocation points, except that the number of collocation points should be greater than or equal to the number of collocation points used in the mapping of elements.

Example 9.10 A classical contact problem, first solved by Hertz[8], is frictionless contact between two elastic spheres. The formulas for maximum contact pressure p_{max} and the radius of the contact area r_c can be found in standard references such as [103]. Specifically, when both spheres have the same elastic properties and $\nu = 0.3$, the maximum pressure is

$$p_{max} = 0.388 \left(PE^2 \frac{(r_1 + r_2)^2}{r_1^2 r_2^2} \right)^{1/3} \tag{9.68}$$

where P is the compressive force, E is the modulus of elasticity, r_1 and r_2 are the radii of the contacting spheres. The radius of the contact area is

$$a_c = 1.109 \left(\frac{P r_1 r_2}{E(r_1 + r_2)} \right)^{1/3}. \tag{9.69}$$

These equations are based on the assumption that $r_c \ll \min(r_1, r_2)$.

In this example we let $r_1 = 100$ mm, $r_2 = 25$ mm, $E = 7.17 \times 10^4$ MPa, $\nu = 0.3$, $P = 800$ N and solve it as an axisymmetric problem. Because the contact area, which is the region of primary interest, is very small in comparison with the radii of the spheres, the starting mesh is graded in geometric progression toward the contact area using the grading factor 0.15. It is then progressively refined by uniformly subdividing the standard elements to obtain a sequence of quasiuniform

---

8 Heinrich Rudolf Hertz 1857–1894.

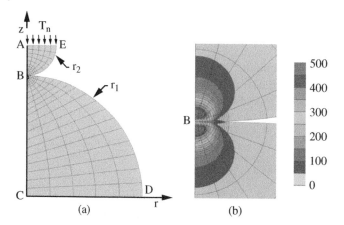

Figure 9.10 Axisymmetric model of two elastic spheres in frictionless contact. (a) Notation and 300 element mesh. (b) Contours of the von Mises stress in the neighborhood of the contact area.

mesh. One mesh in the sequence, consisting of 250 quadrilateral and 50 triangular elements, is shown in Fig. 9.10(a). The problem was solved by iteration using the stopping criterion

$$|p_{max}^{(k)} - p_{max}^{(k-1)}| \leq 0.01 |p_{max}^{(k)}|$$

where p_{max} is the maximum contact pressure and k is the iteration counter.

Referring to Fig. 9.10(a), the boundary conditions are as follows: On boundary segments AB and BC the displacement component $u_r = 0$, the shearing traction is zero. On segment CD the displacement component $u_z = 0$. The circular arcs DB and BE are traction-free and segment EA has the normal traction $T_n = T_z = -P/(r_2^2 \pi)$, the shearing traction is zero.

The deformed configuration (on 1:1 scale) and the contours of the von Mises stress in the neighborhood of the contact area are shown in Fig. 9.10(b). The maximum occurs in the point $r = 0$, $z = 99.68$ mm.

The computed values of the maximum von Mises stress $\bar{\sigma}_{max}$ (MPa), its location z (mm), the maximum contact pressure p_{max} (MPa) and the radius of the contact are r_c (mm) are listed for the quasiuniform sequence of meshes with $p = 8$ assigned to all elements in Table 9.3. It is seen that the contact pressure and the maximum von Mises stress are independent of the number of degrees of freedom to three significant digits. The radius of the contact area was determined to two significant digits only. This is because the contact pressure oscillates with a small amplitude at the boundary of the contact area.

Table 9.3 Example 9.10: Computed values of the quantities of interest with $p = 8$ assigned to all elements.

M	N	$\bar{\sigma}_{max}$	Loc. z	p_{max}	r_c
108	12985	518.79	99.683	841.9	0.717
300	35721	518.86	99.680	842.3	0.710
588	69721	519.00	99.679	842.2	0.714
Classical				843.7	0.673

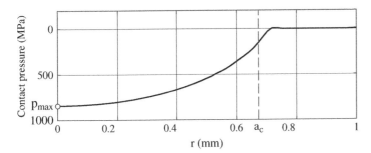

Figure 9.11 Contact pressure.

The contact pressure is plotted in Fig. 9.11. The variation in the contact pressure with respect to N is so small that the differences are not visible on this diagram. Note the small oscillation at the boundary of the contact area. The computed values of the radius of the contact zone (r_c) differ from the radius (a_c) computed from the classical solution given by eq. (9.69). The reason for this is that in the classical solution the contacting spheres are approximated by paraboloids of revolution. No such approximation was necessary in the numerical solution.

Remark 9.4 From the point of view of strength calculations the usual quantities of engineering interest in contact problems are the maximum shearing stress (for ductile materials) or the maximum tensile stress (for brittle materials). The maximum shearing stress, or maximum von Mises stress, is at a small distance from the contact area, as shown in Fig. 9.10(b). The maximum tensile stress occurs on the spherical boundary outside of the contact area.

9.3 Chapter summary

The formulation of an idea of physical reality starts with simple, usually linear, models with several restrictive assumptions. The magnitude of the influence of those assumptions on the quantities of interest have to be estimated and, when necessary, the model has to be revised. Any model should be viewed as a special case of a more comprehensive model. The objective is to identify the simplest model, given the goals of computation and the corresponding acceptable error tolerances. A practical way for doing this is to remove limitations and estimate their effects on the quantities of interest. This is feasible in practice only if the implementation allows seamless transition from linear to nonlinear models and from one nonlinear model to another.

Analysts must be mindful of the fact that nonlinear models require more information about material properties and boundary conditions than linear models. This increases the complexity of the model as well as the uncertainties associated with the additional information.

Appendix A

Definitions

Definition A.1 A function is analytic if and only if its Taylor series about any point $\mathbf{x}_0 \in \overline{\Omega}$ converges to the function in some neighborhood of $\mathbf{x}_0$.

Definition A.2 The real coordinate space of n dimensions, written as $\mathbb{R}^n$, is the generalization of the familiar one-, two- and three-dimensional Euclidean space to n dimensions.

Definition A.3 The Euclidean norm of a vector $\mathbf{a} \in \mathbb{R}^n$ is denoted by $\|\mathbf{a}\|_2$ and defined as follows:

$$\|\mathbf{a}\|_2 \overset{\text{def}}{=} \left(\sum_{i=1}^{n} a_i^2 \right)^{1/2}. \tag{A.1}$$

Definition A.4 The terms "supremum" (abbreviated sup) and "infimum" (abbreviated inf) are generalizations of maximum and minimum, respectively. They are useful for characterizing sets that do not have a maximum or minimum. For instance, the negative real numbers do not have a maximum. However, the supremum is uniquely defined: it is zero. The supremum is also called least upper bound (LUB) and the infimum is also called greatest lower bound (GLB). The terms "essential supremum" (abbreviated ess sup) and "essential infimum" (abbreviated ess inf) are used when sets of measure zero are excluded.

Definition A.5 The delta function $\delta(x)$, also called Dirac's delta function[1], has various interpretations. One such interpretation is that it is zero everywhere except in the point $x = 0$ where it is undefined. However, its integral is the unit step function. We will understand $\delta(x)$ to mean

$$\int_{-\infty}^{\infty} \delta(x - a)\, f(x)\, dx = \lim_{\epsilon \to 0} \int_{a-\epsilon}^{a+\epsilon} \delta(x - a)\, f(x)\, dx = f(a) \tag{A.2}$$

where $\epsilon > 0$.

A.1 Normed linear spaces, linear functionals and bilinear forms

In the following the definitive properties of normed linear spaces, linear functionals and bilinear forms are listed. α and β denote real numbers.

---

1 Paul Adrien Maurice Dirac 1902–1984.

Finite Element Analysis: Method, Verification and Validation, Second Edition. Barna Szabó and Ivo Babuška.
© 2021 John Wiley & Sons, Inc. Published 2021 by John Wiley & Sons, Inc.
Companion Website: www.wiley.com/go/szabo/finite_element_analysis

A.1.1 Normed linear spaces

A normed linear space X is a family of elements u, v, ... which have the following properties:

1. If $u \in X$ and $v \in X$ then $(u + v) \in X$.
2. If $u \in X$ then $\alpha u \in X$.
3. $u + v = v + u$
4. $u + (v + w) = (u + v) + w$
5. There is an unique element in X, denoted by 0, such that $u + 0 = u$ for any $u \in X$.
6. Associated with every element $u \in X$ an unique element $-u \in X$ such that $u + (-u) = 0$.
7. $\alpha(u + v) = \alpha u + \alpha v$
8. $(\alpha + \beta)\, u = \alpha u + \beta u$
9. $\alpha(\beta u) = (\alpha \beta)u$
10. $1 \cdot u = u$
11. $0 \cdot u = 0$
12. With every $u \in X$ we associate a real number $\|u\|_X$, called *the norm*. The norm has the following properties:
    (a) $\|u + v\|_X \leq \|u\|_X + \|v\|_X$. This is called the *triangle inequality*.
    (b) $\|\alpha u\|_X = |\alpha| \|u\|_X$.
    (c) $\|u\|_X \geq 0$.
    (d) $\|u\|_X \neq 0$ if $u \neq 0$.
    A *seminorm* has properties (a) to (c) of the norm but lacks property (d). See, for example, eq. (A.9).

A.1.2 Linear forms

Let X be a normed linear space and $F(v)$ a process which associates with every $v \in X$ a real number $F(v)$. $F(v)$ is called a *linear form* or *linear functional* on X if it has the following properties:

1. $F(v_1 + v_2) = F(v_1) + F(v_2)$
2. $F(\alpha v) = \alpha F(v)$
3. $|F(v)| \leq C\|v\|_X$ with C independent of v. The smallest possible value of C is called the *norm* of F.

A.1.3 Bilinear forms

Let X and Y be normed linear spaces and $B(u, v)$ a process which associates with every $u \in X$ and $v \in Y$ a real number $B(u, v)$. $B(u, v)$ is a *bilinear form on $X \times Y$* if it has the following properties:

1. $B(u_1 + u_2, v) = B(u_1, v) + B(u_2, v)$
2. $B(u, v_1 + v_2) = B(u, v_1) + B(u, v_2)$
3. $B(\alpha u, v) = \alpha B(u, v)$
4. $B(u, \alpha v) = \alpha B(u, v)$
5. $|B(u, v)| \leq C\, \|u\|_X\, \|v\|_Y$ with C independent of u and v. The smallest possible value of C is called the *norm* of B.

The space X is called *trial space* and functions $u \in X$ are called *trial functions*. The space Y is called *test space* and functions $v \in Y$ are called *test functions*. $B(u, v)$ is not necessarily symmetric.

A.2 Convergence in the space *X*

A sequence of functions $u_n \in X$ ($n = 1, 2, \ldots$) converges in the space X to the function $u \in X$ if for every $\epsilon > 0$ there is a number n_ϵ such that for any $n > n_\epsilon$ the following relationship holds:

$$\|u - u_n\|_X < \epsilon. \tag{A.3}$$

A.2.1 The space of continuous functions

Let $\Omega \in \mathbb{R}^n$ be an open bounded domain. The space of continuous functions defined on Ω is denoted by $C^0(\Omega)$. The norm associated with C^0 is

$$\|f\|_{C^0(\Omega)} = \sup_{x \in \Omega} |f(x)|. \tag{A.4}$$

The space $C^k(\Omega)$ is the set of functions defined on Ω that have the property that all derivatives, up to and including the kth derivative, lie in $C^0(\Omega)$. The set $C^k(\Omega)$ is a proper subset of $C^{k-1}(\Omega)$ for $k > 0$. The set $C^\infty(\Omega)$ is the set of analytic functions.

A.2.2 The space $L^p(\Omega)$

The space $L^p(\Omega)$ is defined as the set of functions on Ω that satisfy

$$\|f\|_{L^p(\Omega)} = \left(\int_\Omega |f|^p \, dx \right)^{1/p} < \infty \tag{A.5}$$

where the integral is the Lebesque[2] integral. The space $L^2(\Omega)$ is the set of square integrable functions. The space $L^\infty(\Omega)$ is defined by

$$\|f(\mathbf{x})\|_{L^\infty(\Omega)} = \operatorname{ess\,sup}_{\mathbf{x} \in \Omega} |f(\mathbf{x})| < \infty \tag{A.6}$$

see Definition A.4.

A.2.3 Sobolev space of order 1

Sobolev[3] spaces play a central role in the mathematical analysis of the finite element method. For basic definitions and properties of Sobolev spaces we refer to [55, 59, 84]. The Sobolev space of order 1 in two dimensions is defined as the set

$$H^1(\Omega) \overset{\text{def}}{=} \left\{ u \mid \|u\|_{L^2(\Omega)}^2 + \left\| \frac{\partial u}{\partial x} \right\|_{L^2(\Omega)}^2 + \left\| \frac{\partial u}{\partial y} \right\|_{L^2(\Omega)}^2 < \infty \right\} \tag{A.7}$$

The Sobolev norm of order one in two dimensions is defined by

$$\|u\|_{H^1(\Omega)} \overset{\text{def}}{=} \left(\|u\|_{L^2(\Omega)}^2 + \left\| \frac{\partial u}{\partial x} \right\|_{L^2(\Omega)}^2 + \left\| \frac{\partial u}{\partial y} \right\|_{L^2(\Omega)}^2 \right)^{1/2} \tag{A.8}$$

the first order seminorm is

$$|u|_{H^1(\Omega)} \overset{\text{def}}{=} \left(\left\| \frac{\partial u}{\partial x} \right\|_{L^2(\Omega)}^2 + \left\| \frac{\partial u}{\partial y} \right\|_{L^2(\Omega)}^2 \right)^{1/2} \tag{A.9}$$

---

2 Henri Léon Lebesgue 1871–1941.
3 Sergei Lvovich Sobolev 1908–1989.

and we define

$$H_0^1(\Omega) \overset{\text{def}}{=} \{u \mid u \in H^1(\Omega), \ u = 0 \text{ on } \partial\Omega\}. \tag{A.10}$$

The Sobolev norm and the energy norm are equivalent in the sense that there exist two real numbers $0 < C_1 \leq C_2$, independent of u, such that

$$C_1 \|u\|_{H^1(\Omega)} \leq \|u\|_{E(\Omega)} \leq C_2 \|u\|_{H^1(\Omega)}. \tag{A.11}$$

In the mathematical literature a priori estimates of rates of convergence are given in terms of the error measured in H^1 norm. Those estimates are valid in the energy norm also.

The standard notation for Sobolev spaces is $W^{k,s}(\Omega)$ where the index k is the number of derivatives and index s identifies the associated norm $L^s(\Omega)$. The special case $s = 2$ is a Hilbert space[4] denoted by $H^k(\Omega)$. We have given the definition for the Sobolev space $W^{1,2}(\Omega) \equiv H^1(\Omega)$.

A.2.4 Sobolev spaces of fractional index

The definition of Sobolev spaces is extended to fractional indices $k = m + \mu$ where m is an integer and $0 < \mu < 1$. Sobolev spaces of fractional index play a fundamentally important role in the theory of finite element methods. These spaces are used for characterization of the smoothness of functions. The a priori estimates of the rate of convergence (in energy norm) are given in terms of the indices of these spaces.

A detailed discussion of Sobolev spaces is well outside of the scope of this book, however an introductory discussion of special cases in one and two dimensions is necessary for the understanding of theorems pertinent to the convergence properties of the finite element method. For introductory reading on Sobolev spaces we recommend [34, 59, 84].

In one dimension the function

$$u = x^\lambda, \quad \lambda > 1/2, \quad x \in I$$

is in $H^{1+\mu}(I)$ if $x^{\lambda-1-\mu}$ is square integrable on I. For $I = (0, 1)$ we have

$$\int_0^1 x^{2\lambda-2-2\mu} \, dx < \infty$$

therefore

$$\left[\frac{x^{2\lambda-1-2\mu}}{2\lambda - 1 - 2\mu} \right]_0^1 < \infty$$

and

$$2\lambda - 1 - 2\mu > 0.$$

Selecting the largest value of μ such that $u = x^\lambda$ is square integrable we have

$$\mu = \lambda - 1/2 - \epsilon$$

where ϵ is an arbitrarily small positive number. For example, $u = x^{2/3}$ lies in $H^{7/6-\epsilon}(I)$ where $I = (0, 1)$.

In two dimensions let

$$u = r^\lambda \phi(\theta), \quad (r, \theta) \in \Omega = (0, 1) \times (-\alpha/2, \alpha/2)$$

---

4 David Hilbert 1862–1943.

where $\phi(\theta)$ is a smooth function and $0 < \alpha \leq \pi$. The function u is in $H^{1+\mu}(\Omega)$ if $r^{\lambda-1-\mu}$ is square integrable on Ω, that is,

$$\int_{-\alpha/2}^{\alpha/2} \int_0^1 r^{2\lambda-2-2\mu} \phi^2(\theta) \, rdrd\theta = \int_{-\alpha/2}^{\alpha/2} \phi^2(\theta) \, d\theta \left[\frac{r^{2\lambda-2\mu}}{2\lambda - 2\mu} \right]_0^1 < \infty$$

therefore $\lambda - \mu > 0$ and the largest μ for which this holds is $\mu = \lambda - \epsilon$ where ϵ is an arbitrarily small positive number. For example, the function $u = r^{2/3} \cos(\theta)$ lies in $H^{5/3-\epsilon}(\Omega)$.

A.3   The Schwarz inequality for integrals

Assume that $f(x) \in L^2(I)$ and $g(x) \in L^2(I)$ where $I = (a, b)$. Then:

$$\left| \int_a^b fg \, dx \right| \leq \left(\int_a^b f^2 \, dx \right)^{1/2} \left(\int_a^b g^2 dx \right)^{1/2}. \tag{A.12}$$

This is the Schwarz[5] inequality for integrals. To prove this we observe that:

$$\int_a^b (f + \lambda g)^2 dx \geq 0 \quad \text{for any } \lambda \tag{A.13}$$

and therefore:

$$\int_a^b f^2 dx + 2\lambda \int_a^b fg dx + \lambda^2 \int_a^b g^2 \, dx \geq 0 \quad \text{for any } \lambda. \tag{A.14}$$

On the left of this inequality is a quadratic expression for λ. To find the roots of this expression we need only to compute:

$$\lambda = \frac{-\int_a^b fg dx \pm \sqrt{\left(\int_a^b fg dx \right)^2 - \int_a^b g^2 dx \int_a^b f^2 dx}}{\int_a^b g^2 dx}. \tag{A.15}$$

Denoting the roots by λ_1 and λ_2, eq. (A.14) can be written as:

$$(\lambda - \lambda_1)(\lambda - \lambda_2) \geq 0$$

and we observe that the roots cannot be real and simple because then we could select any λ such that $\lambda_1 < \lambda < \lambda_2$ and we would have $(\lambda - \lambda_1)(\lambda - \lambda_2) < 0$. Therefore the radicand must be less than or equal to zero. This completes the proof.

---

5 Karl Hermann Amandus Schwarz 1843–1921.

Appendix B

Proof of *h*-convergence

A proof of h-convergence in one dimension, $p = 1$, subject to the condition that the second derivative of the exact solution is bounded, is presented here with the objective to provide an introduction, in the simplest possible setting, to the formulation of a priori error estimates.

Consider the following problem: Find $u_{EX} \in E^0(I)$ where $I = (0, \ell)$ such that:

$$\int_0^\ell (\kappa u'_{EX} v' + c u_{EX} v)\, dx = \int_0^\ell f v\, dx \quad \text{for all} \quad v \in E^0(I) \tag{B.1}$$

assuming that κ, c and f are such that u_{EX}, u'_{EX} and u''_{EX} are bounded, continuous functions with $|u''_{EX}| \le C$ on the interval I, that is, $u_{EX} \in C^2(I)$.

Subdivide the domain I into n elements of equal length. The length of each element is: $h = \ell/n$. Let u_n be the linear interpolant of u_{EX}, that is u_n is a continuous, piecewise linear function such that:

$$u_n(jh) = u_{EX}(jh), \quad j = 0, 1, \ldots, n. \tag{B.2}$$

We denote $I_k = ((k-1)h,\ kh)$ and the error of interpolation on I_k by $\bar{e}_k$:

$$\bar{e}_k(x) \overset{\text{def}}{=} u_{EX}(x) - u_n(x), \quad x \in I_k, \quad k = 1, 2, \ldots, n. \tag{B.3}$$

Because $\bar{e}_k(x)$ vanishes at the endpoints of the element, there is a point $\bar{x}_k$ where $|\bar{e}_k|$ is maximal. At this point $\bar{e}'_k = 0$, see Fig. B.1. Since u_n is linear on I_k, $u''_n = 0$ therefore:

$$\bar{e}'_k(x) = \int_{\bar{x}_k}^x \bar{e}''_k(t)\, dt = \int_{\bar{x}_k}^x u''_{EX}(t)\, dt, \quad x \in I_k \tag{B.4}$$

and since $|u''_{EX}| \le C$, we get:

$$|\bar{e}'_k(x)| \le h\, C, \quad x \in I_k. \tag{B.5}$$

Let us expand $\bar{e}_k$ into a Taylor series about the point $\bar{x}_k$ and let us assume that $\bar{x}_k$ is located such that: $kh - \bar{x}_k \le h/2$. We now have:

$$\bar{e}_k(kh) = 0 = \bar{e}_k(\bar{x}_k) + (kh - \bar{x}_k)\bar{e}'_k(\bar{x}_k) + \frac{1}{2}(kh - \bar{x}_k)^2 \bar{e}''_k(t) \tag{B.6}$$

where t is a point between $\bar{x}_k$ and $x = kh$. Because $\bar{e}'_k(\bar{x}_k) = 0$:

$$\max |e_k(\bar{x}_k)| = \frac{1}{2}|kh - \bar{x}_k|^2 |\bar{e}''_k(t)| \le \frac{h^2}{8} C. \tag{B.7}$$

If $\bar{x}_k$ is closer to $(k-1)h$ than to kh then we write the Taylor series expression for $\bar{e}_k((k-1)h)$ instead of $\bar{e}_k(kh)$ and obtain the same result. Equation (B.7) is a basic result of interpolation theory. In view

Finite Element Analysis: Method, Verification and Validation, Second Edition. Barna Szabó and Ivo Babuška.
© 2021 John Wiley & Sons, Inc. Published 2021 by John Wiley & Sons, Inc.
Companion Website: www.wiley.com/go/szabo/finite_element_analysis

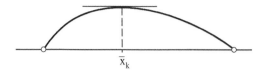

Figure B.1 The error $\bar{e}_k = u_{EX} - u_{FE}$.

of (B.4) and (B.7) the strain energy of the error of the interpolant $\bar{e}$ is:

$$U(\bar{e}) = \frac{1}{2} \int_0^{\ell} \left(\kappa(\bar{e}')^2 + c\bar{e}^2 \right) \, dx = \frac{1}{2} \sum_{k=1}^{n} \int_{(k-1)h}^{kh} \left(\kappa(\bar{e}'_k)^2 + c\bar{e}_k^2 \right) \, dx$$

$$\leq \frac{1}{2} \, nh \left(K_1 \, (Ch)^2 + K_2 \left(\frac{h^2}{8} \, C \right)^2 \right) \tag{B.8}$$

where K_1 and K_2 are constants chosen so that $\kappa(x) \leq K_1$ and $c(x) \leq K_2$ on the interval I. Observe that K_1, K_2 are independent of h. Since $nh = \ell$, there is a constant K such that:

$$U(\bar{e}) \leq \frac{1}{2} K \ell C^2 h^2. \tag{B.9}$$

Finally, because the energy norm of the error of the finite element solution $e = u_{EX} - u_{FE}$ is less than or equal to the energy norm of the error of the interpolant $\bar{e}$ (see Theorem 1.4) we have:

$$\|e\|_{E(\Omega)} \overset{\text{def}}{=} \sqrt{U(e)} \leq kCh \tag{B.10}$$

where k depends on κ, c and f but is independent of h. This is in the form of a typical a priori estimate. A priori estimates indicate how fast the error changes as the discretization is changed, given information about the smoothness of the exact solution. For example, the smoothness of the exact solution was characterized by $|u''_{EX}| \leq C$ in the foregoing discussion. In general C is not known a priori.

Appendix C

Convergence in 3D: Empirical results

The problem of elasticity was solved on the Fichera domain using the same input data as in reference [3]. The goal was to compare the empirical rates of convergence of the h- and p-versions in energy norm, with respect to the number of degrees of freedom, and to estimate the Sobolev index k of the exact solution. The computations were performed by the finite element analysis software STRIPE, developed by the Aeronautical Research Institute of Sweden.

Input data

The domain is defined by

$$\Omega \overset{\text{def}}{=} \{(X, Y, Z) \mid (X, Y, Z) \in (-50, \ 50)^3 \backslash [0, \ 50)^3\}. \tag{C.1}$$

Using the notation shown in Fig. 4.11, on the boundary surfaces that lie in the planes $X = 0, Y = 0$, $Z = 0$ and $Y = -50, Z = -50$ the tractions are zero: $T_x = T_y = T_z = 0$. On the boundary surface that lies in the $X = -50$ plane $T_x = -1, T_y = T_z = 0$. On the boundary surfaces that lie in the planes $X = 50, Y = 50$ and $Z = 50$ symmetry boundary conditions are prescribed: The normal displacements and shearing tractions are zero. The modulus of elasticity is $E = 1.0E3$ and Poisson's ratio is 0.3.

Reference solution

In order to establish a reference solution, a geometric mesh was constructed. The mesh is characterized by the sequence

$$x_i = \begin{cases} 0 & \text{for } i = 1 \\ 50q^{m+1-i} & \text{for } i = 2, 3, \dots, m+1 \end{cases} \tag{C.2}$$

which indicates the location of nodes along the x axis. In the same way nodes were located along the y and z axes. The nodes are in the points of intersection of the planes $X = \pm x_i, Y = \pm y_j, Z = \pm z_k$, $(i, j, k = 1, 2, \dots, m+1)$. The number of elements in the mesh is $M(\Delta) = 7m^3$.

The results for $m = 8$ are shown in Table C.1. It is seen that the potential energy converged to 8 digits precision. The extrapolated value of π_p was used for reference in calculating the rates of convergence for the p- and h-extensions, denoted by β_p and β_h respectively, in tables C.1 and C.2.

The estimated values of β, computed by the formula given by eq. (1.102), are shown in Fig. C.1. Uniform meshes were used for both h- and p-extensions.

The results for h-extension, $p = 2$, trunk space, are displayed in Table C.2. The parameter m represents the number of edges along the positive coordinate axes. Therefore $h = 50/m$. The values of k_h were computed using the formula (4.35).

Finite Element Analysis: Method, Verification and Validation, Second Edition. Barna Szabó and Ivo Babuška.
© 2021 John Wiley & Sons, Inc. Published 2021 by John Wiley & Sons, Inc.
Companion Website: www.wiley.com/go/szabo/finite_element_analysis

Table C.1   Fichera domain (elasticity). Estimation of the reference value of the potential energy, p-Convergence, $M(\Delta) = 3584$ ($m = 8$), geometric mesh $q = 1/7$, trunk space.

p	N	π_p	β_p	e (%)
3	84,048	$-7.1017674042E + 02$	–	1.9934
4	153,792	$-7.1037322898E + 02$	0.9853	1.0992
5	257,520	$-7.1044463079E + 02$	1.7293	0.4507
6	405,984	$-7.1045599535E + 02$	1.7004	0.2079
7	609,936	$-7.1045850634E + 02$	2.0931	0.0887
8	880,128	$-7.1045892986E + 02$	1.9361	0.0436
9	1,227,312	$-7.1045902640E + 02$	1.8878	0.0233
10	1,662,240	$-7.1045905243E + 02$	1.8294	0.0134
11	2,195,664	$-7.1045906029E + 02$	1.8294	0.0080
extrapolated:		$-7.1045906487E + 02$		

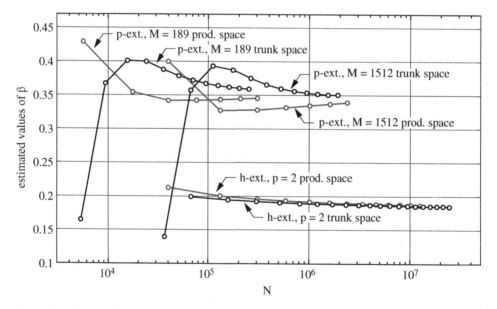

Figure C.1   Fichera domain, elasticity. Convergence paths on uniform meshes.

Discussion

The estimation of k_h is based on eq. (1.91), whereas the estimation of β_h is based on eq. (1.92). The two forms are similar and, since in three dimensions $N \propto h^{-3}$, we expect that

$$\beta_h \approx \frac{k_h - 1}{3}.$$

For example, in the entry corresponding to $m = 66$ in Table C.2, β_h estimated from k_h is 0.1838 which is close to the estimate of 0.1847 from eq. (1.92).

Table C.2   Fichera domain. h-Convergence, uniform mesh, $p = 2$ (trunk space).

m	N	π_h	k_h	β_h	$(e_r)_E$ (%)
6	21,132	$-7.08307829E + 02$	–	–	5.5027
9	67,905	$-7.09107428E + 02$	1.5731	0.1991	4.3617
12	156,960	$-7.09483558E + 02$	1.5668	0.1946	3.7055
15	301,905	$-7.09700398E + 02$	1.5633	0.1922	3.2678
18	516,348	$-7.09840746E + 02$	1.5610	0.1906	2.9501
21	813,897	$-7.09938678E + 02$	1.5593	0.1895	2.7064
24	1,208,160	$-7.10010726E + 02$	1.5580	0.1886	2.5121
27	1,712,745	$-7.10065854E + 02$	1.5570	0.1880	2.3526
30	2,341,260	$-7.10109336E + 02$	1.5561	0.1874	2.2187
33	3,107,313	$-7.10144469E + 02$	1.5554	0.1870	2.1043
36	4,024,512	$-7.10173423E + 02$	1.5548	0.1866	2.0051
39	5,106,465	$-7.10197676E + 02$	1.5543	0.1863	1.9181
42	6,366,780	$-7.10218275E + 02$	1.5538	0.1861	1.8410
45	7,819,065	$-7.10235977E + 02$	1.5534	0.1858	1.7720
48	9,476,928	$-7.10251347E + 02$	1.5530	0.1856	1.7099
51	11,353,977	$-7.10264811E + 02$	1.5527	0.1854	1.6535
54	13,463,820	$-7.10276699E + 02$	1.5524	0.1853	1.6021
57	15,820,065	$-7.10287268E + 02$	1.5521	0.1851	1.5550
60	18,436,320	$-7.10296725E + 02$	1.5519	0.1850	1.5116
63	21,326,193	$-7.10305233E + 02$	1.5517	0.1848	1.4715
66	24,503,292	$-7.10312926E + 02$	1.5514	0.1847	1.4342
reference value:		$-7.10459065E + 02$			

It is known that in two dimensions the limit value of the ratio β_p/β_h, under conditions that are normally satisfied in practice, is 2. No analogous theorem exists for three dimensions. Referring to the apparent limit values of β obtained by p- and h-extensions in Fig. C.1, we expect that the same can be proven for three dimensions as well.

A direct solver was used. It was verified that round-off errors did not influence the computations. Perturbations caused by round-off affect not fewer than the twelfth digit in the computed potential energy values.

A surprising finding of this investigation was that entry into the asymptotic range occurs at very high values of N. As seen in Table C.2, at $N = 24.5$ million, k_h and β_h are still decreasing monotonically.

Appendix D

Legendre polynomials

The *Legendre polynomials* $P_n(\xi)$ are solutions of the *Legendre differential equation* for $n = 0, 1, 2, \ldots$:

$$(1 - \xi^2)y'' - 2\xi y' + n(n+1)y = 0, \qquad -1 \leq \xi \leq 1. \tag{D.1}$$

The first eight Legendre polynomials are:

$$P_0(\xi) = 1 \tag{D.2}$$

$$P_1(\xi) = \xi \tag{D.3}$$

$$P_2(\xi) = \frac{1}{2}(3\xi^2 - 1) \tag{D.4}$$

$$P_3(\xi) = \frac{1}{2}(5\xi^3 - 3\xi) \tag{D.5}$$

$$P_4(\xi) = \frac{1}{8}(35\xi^4 - 30\xi^2 + 3) \tag{D.6}$$

$$P_5(\xi) = \frac{1}{8}(63\xi^5 - 70\xi^3 + 15\xi) \tag{D.7}$$

$$P_6(\xi) = \frac{1}{16}(231\xi^6 - 315\xi^4 + 105\xi^2 - 5) \tag{D.8}$$

$$P_7(\xi) = \frac{1}{16}(429\xi^7 - 693\xi^5 + 315\xi^3 - 35\xi). \tag{D.9}$$

Note that Legendre polynomials of even (resp. odd) degree are symmetric (resp. antisymmetric) functions:

$$P_n(-\xi) = (-1)^n P_n(\xi) \tag{D.10}$$

and $P_n(1) = 1$ for all n.

Legendre polynomials can be generated from the recursion formula:

$$(n+1)P_{n+1}(\xi) = (2n+1)\xi P_n(\xi) - nP_{n-1}(\xi), \quad n = 1, 2, \ldots \tag{D.11}$$

and Legendre polynomials satisfy the following relationship:

$$(2n+1)P_n(\xi) = P'_{n+1}(\xi) - P'_{n-1}(\xi), \quad n = 1, 2, \ldots \tag{D.12}$$

where the primes represent differentiation with respect to ξ. Legendre polynomials satisfy the following orthogonality property:

$$\int_{-1}^{+1} P_i(\xi)\, P_j(\xi)\, d\xi = \begin{cases} \dfrac{2}{2i+1} & \text{for } i = j \\ 0 & \text{for } i \neq j. \end{cases} \tag{D.13}$$

Finite Element Analysis: Method, Verification and Validation, Second Edition. Barna Szabó and Ivo Babuška.
© 2021 John Wiley & Sons, Inc. Published 2021 by John Wiley & Sons, Inc.
Companion Website: www.wiley.com/go/szabo/finite_element_analysis

All roots of Legendre polynomials lie in the interval $-1 < \xi < 1$. The n roots of $P_n(\xi)$ are the abscissas ξ_i for the n-point *Gaussian integration*:

$$\int_{-1}^{1} f(\xi) \, d\xi \approx \sum_{i=1}^{n} w_i f(\xi_i). \tag{D.14}$$

Legendre polynomials have the property $P_n(1) = 1$, $P_n(-1) = (-1)^n$.

D.1 Shape functions based on Legendre polynomials

The first five shape functions based on Legendre polynomials are displayed in Fig. 1.4. Additional shape functions are listed below.

$$N_6(\xi) = \frac{3}{8\sqrt{2}} \, \xi(\xi^2 - 1)(7\xi^2 - 3) \tag{D.15}$$

$$N_7(\xi) = \frac{1}{16} \sqrt{\frac{11}{2}} \, (\xi^2 - 1)(21\xi^4 - 14\xi^2 + 1) \tag{D.16}$$

$$N_8(\xi) = \frac{1}{16} \sqrt{\frac{13}{2}} \, \xi(\xi^2 - 1)(33\xi^4 - 30\xi^2 + 5) \tag{D.17}$$

$$N_9(\xi) = \frac{1}{128} \sqrt{\frac{15}{2}} \, (\xi^2 - 1)(429\xi^6 - 495\xi^4 + 135\xi^2 - 5) \tag{D.18}$$

$$\vdots$$

$$N_i(\xi) = \frac{1}{\sqrt{2(2i - 3)}} (P_{i-1}(\xi) - P_{i-3}(\xi)). \tag{D.19}$$

The first eight shape functions are shown in Fig. D.1.

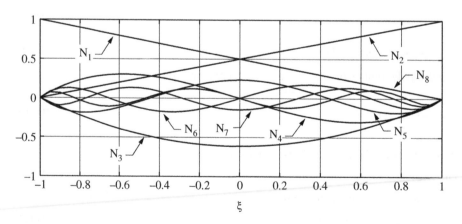

Figure D.1 The first eight shape functions based on Legendre polynomials.

Appendix E

Numerical quadrature

In the finite element method the terms of the coefficient matrices and right hand side vectors are computed by numerical quadrature. Most commonly Gaussian[1] quadrature is used, in some cases the Gauss-Lobatto quadrature is used. In one dimension the domain of integration is the standard element I_{st}. An integral expression on the standard element is approximated by the sum

$$\int_{-1}^{+1} f(\xi)\, d\xi \approx \sum_{i=1}^{n} w_i f(\xi_i) + R_n \tag{E.1}$$

where w_i are the weights; ξ_i are the abscissas and R_n is the error term. The abscissas and weights are symmetric with respect to the point $\xi = 0$.

To evaluate an integral on other than the standard domain, the mapping function defined by eq. (1.60) is used for transforming the domain of integration to the standard domain. For example:

$$\int_{x_1}^{x_2} F(x)\, dx = \int_{-1}^{+1} \underbrace{F(Q(\xi)) \frac{x_2 - x_1}{2}}_{f(\xi)}\, d\xi \quad \text{where} \quad Q(\xi) = \frac{1 - \xi}{2}x_1 + \frac{1 + \xi}{2}x_2.$$

E.1 Gaussian quadrature

In Gaussian quadrature the abscissa x_i is the ith zero of Legendre polynomial P_n. The weights are computed from:

$$w_i = \frac{2}{(1 - x_i^2)[P_n'(x_i)]^2} \tag{E.2}$$

see, for example Olver et al. [71]. The abscissas and weights for Gaussian quadrature are listed in Table E.1 up to $n = 8$. The error term is:

$$R_n = \frac{2^{2n+1}(n!)^4}{(2n + 1)[(2n)!]^3} f^{(2n)}(\zeta) - 1 < \zeta < +1$$

where $f^{(2n)}$ is $(2n)$th derivative of f. It is seen from the error term that if $f(\xi)$ is a polynomial of degree p and Gaussian quadrature is used then the integral will be exact (up to round-off errors) provided that $n \geq (p + 1)/2$. For example, to integrate a polynomial of degree 5, $n = 3$ is sufficient. For other than polynomial functions the rate of convergence depends on how well the integrand can be approximated by polynomials. It can be shown that if $f(\xi)$ is a continuous function on I_{st} then the sum in eq. (E.1) converges to the true value of the integral.

1 Johann Carl Friedrich Gauss 1777–1855.

Finite Element Analysis: Method, Verification and Validation, Second Edition. Barna Szabó and Ivo Babuška.
© 2021 John Wiley & Sons, Inc. Published 2021 by John Wiley & Sons, Inc.
Companion Website: www.wiley.com/go/szabo/finite_element_analysis

Table E.1 Abscissas and weights for Gaussian quadrature

n	$\pm\xi_i$	w_i
2	0.57735 02691 89626	1.00000 00000 00000
3	0.00000 00000 00000	0.88888 88888 88889
	0.77459 66692 41483	0.55555 55555 55556
4	0.33998 10435 84856	0.65214 51548 62546
	0.86113 63115 94053	0.34785 48451 37454
5	0.00000 00000 00000	0.56888 88888 88889
	0.53846 93101 05683	0.47862 86704 99366
	0.90617 98459 38664	0.23692 68850 56189
6	0.23861 91860 83197	0.46791 39345 72691
	0.66120 93864 66265	0.36076 15730 48139
	0.93246 95142 03152	0.17132 44923 79170
7	0.00000 00000 00000	0.41795 91836 73469
	0.40584 51513 77397	0.38183 00505 05119
	0.74153 11855 99394	0.27970 53914 89277
	0.94910 79123 42759	0.12948 49661 68870
8	0.18343 46424 95650	0.36268 37833 78362
	0.52553 24099 16329	0.31370 66458 77887
	0.79666 64774 13627	0.22238 10344 53374
	0.96028 98564 97536	0.10122 85362 90376

The integration procedure can be extended directly to the standard quadrilateral element and the standard hexahedral element. For the standard quadrilateral element:

$$\int_{-1}^{+1} \int_{-1}^{+1} f(\xi,\eta) \, d\xi d\eta = \sum_{i=1}^{n_\xi} \sum_{j=1}^{n_\eta} w_i w_j f(\xi_i, \eta_j) \tag{E.3}$$

where n_ξ (resp. n_η) is the number of quadrature points along the ξ (resp. η axis). Usually but not necessarily $n_\xi = n_\eta$ is used. Analogously, for the standard hexahedral element:

$$\int_{-1}^{+1} \int_{-1}^{+1} \int_{-1}^{+1} f(\xi,\eta,\zeta) \, d\xi d\eta d\zeta = \sum_{i=1}^{n_\xi} \sum_{j=1}^{n_\eta} \sum_{k=1}^{n_\zeta} w_i w_j w_k f(\xi_i, \eta_j, \zeta_k). \tag{E.4}$$

E.2 Gauss-Lobatto quadrature

In the Gauss-Lobatto quadrature the abscissas are as follows: $x_1 = -1$, $x_n = 1$ and for $i = 2, 3, \ldots, n-1$ the $(i-1)$th zero of $P'_{n-1}(x)$ where $P_{n-1}(x)$ is the $(n-1)$th Legendre polynomial. The weights are:

$$w_i = \begin{cases} \dfrac{2}{n(n-1)} & \text{for } i = 1 \text{ and } i = n \\[2ex] \dfrac{2}{n(n-1)[P_{n-1}(x_i)]^2} & \text{for } i = 2, 3, \ldots, (n-1). \end{cases} \tag{E.5}$$

Table E.2 Abscissas and weights for Gauss-Lobatto quadrature

n	$\pm\xi_i$	w_i
2	1.00000 00000 00000	1.00000 00000 00000
3	0.00000 00000 00000	1.33333 33333 33333
	1.00000 00000 00000	0.33333 33333 33333
4	0.44721 35954 99958	0.83333 33333 33333
	1.00000 00000 00000	0.16666 66666 66667
5	0.00000 00000 00000	0.71111 11111 11111
	0.65465 36707 07977	0.54444 44444 44444
	1.00000 00000 00000	0.10000 00000 00000
6	0.28523 15164 80645	0.55485 83770 35486
	0.76505 53239 29465	0.37847 49562 97847
	1.00000 00000 00000	0.06666 66666 66667
7	0.00000 00000 00000	0.48761 90476 19048
	0.46884 87934 70714	0.43174 53812 09863
	0.83022 38962 78567	0.27682 60473 61566
	1.00000 00000 00000	0.04761 90476 19048
8	0.20929 92179 02479	0.41245 87946 58704
	0.59170 01814 33142	0.34112 26924 83504
	0.87174 01485 09607	0.21070 42271 43506
	1.00000 00000 00000	0.03571 42857 14286

The abscissas and weights for Gauss-Lobatto quadrature are listed in Table E.2 up to $n = 8$.

The error term is:

$$R_n = \frac{-n(n-1)^3 2^{2n-1}[(n-2)!]^4}{(2n-1)[(2n-2)!]^3} f^{(2n-2)}(\zeta), \quad -1 < \zeta < +1$$

from which it follows that if $f(\xi)$ is a polynomial of degree p and Gauss-Lobatto quadrature is used then the integral will be exact (up to round-off errors) provided that $n \geq (p+3)/2$.

Appendix F

Polynomial mapping functions

The use of isoparametric mapping of elements for polynomial degrees 1 and 2 was introduced in the 1960s. The algorithm for two-dimensional elements was outlined in Section 3.4.1. In this section we address extension of the idea of isoparametric mapping to polynomial functions of degree greater than 2 and its combination with the blending function method.

Specifically, we are interested in approximating arbitrary continuous functions such as $x_2(\eta)$ and $y_2(\eta)$ in equations (3.46) and (3.47) by polynomial interpolating functions on the interval $-1 < \eta < 1$.

In the interest of simplicity in notation we will seek a polynomial interpolation function of degree n to $f(\xi) \in C^0(I_{st})$ that is nearly optimal in maximum norm. We denote the polynomial interpolation functions of degree p by $w_p(\xi)$ and the set of interpolation points, called the nodal set, by

$$T = \{\xi_1 = -1, \ \xi_2, \ \xi_3, \ \dots \ \xi_p, \ \xi_{p+1} = 1\}$$

and the interpolating polynomial is

$$w_p = \sum_{k=1}^{p+1} f(\xi_k) N_k(\xi) \equiv \mathcal{L}_T f$$

where $N_k(\xi)$ are the Lagrange shape functions of degree p defined by eq. (1.50) and $\mathcal{L}_T$ is the linear projection operator that maps continuous functions defined on I_{st} onto polynomials of degree p defined on I_{st}. We define the Lebesque constant $\lambda(T)$

$$\lambda(T) = \max_{\xi \in I_{st}} \sum_{i=1}^{p+1} |N_i(\xi)| \tag{F.1}$$

and note that $|\mathcal{L}_T f| \leq \lambda(T) \, \|f\|_{max}$.

Let $w_p^{\star}$ be the best polynomial approximation of f in maximum norm. Then

$$|f - \mathcal{L}_T f| = |f - w_p^{\star} + w_p^{\star} - \mathcal{L}_T f| \leq |f - w_p^{\star}| + |w_p^{\star} - \mathcal{L}_T f| \tag{F.2}$$

where we used the triangle inequality. Noting that $w_p^{\star}$ can be written as $w_p^{\star} = \mathcal{L}_T w_p^{\star}$, we have

$$w_p^{\star} - \mathcal{L}_T f = \mathcal{L}_T (w_p^{\star} - f)$$

and eq. (F.2) can be written as

$$\| f - \mathcal{L}_T f \|_{max} \leq \| f - w_p^{\star} \|_{max} + \| \mathcal{L}_T (f - w_p^{\star}) \|_{max} \tag{F.3}$$

from which it follows that the Lebesque constant bounds the interpolation error:

$$\| f - \mathcal{L}_T f \|_{max} \leq (1 + \lambda(T)) \, \| f - w_p^{\star} \|_{max} \tag{F.4}$$

Finite Element Analysis: Method, Verification and Validation, Second Edition. Barna Szabó and Ivo Babuška.
© 2021 John Wiley & Sons, Inc. Published 2021 by John Wiley & Sons, Inc.
Companion Website: www.wiley.com/go/szabo/finite_element_analysis

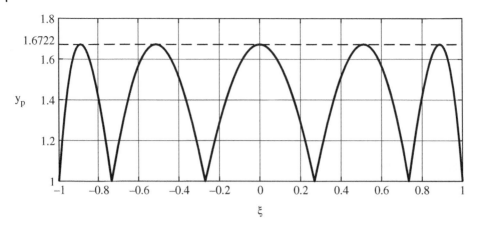

Figure F.1 The function y_p on the optimal nodal set for $p = 5$.

that is, the interpolation polynomial is at most a factor $(1 + \lambda(T))$ worse than the best possible polynomial approximation of degree p. The set of interpolation points of degree p that minimizes the Lebesque constant will be denoted by T_1^p. This set is also called the optimal nodal set or the set of optimal collocation points.

Example F.1 The optimal nodal set for $p = 5$ is:
$$T_1^5 = \{-1 - 0.734127 - 0.268907\ 0.268907\ 0.734127\ 1\}. \tag{F.5}$$

The function
$$y_p = \sum_{i=1}^{p+1} |N_i(\xi)|$$

is plotted for the optimal nodal set in Fig. F.1. The maximum of y_p is the Lebesque constant: $(y_p)_{\max} = \lambda(T) = 1.6722$.

Extension of the optimal interpolation set T_1^p to two dimensions is technically difficult. For this reason another optimal set, known as the mean optimal set, is used. The mean optimal set in one dimension, denoted by T_2^p, minimizes
$$\eta^2 = \int_{-1}^{1} \sum_{i=1}^{p+1} |N_i(\xi)|^2\ d\xi. \tag{F.6}$$

The optimal sets T_1^p and T_2^p are listed in Table F.1.

F.1 Interpolation on surfaces

In three-dimensional problems the boundary surfaces are covered by triangles and/or rectangles that are mapped from the standard quadrilateral and triangular elements. In this section the optimal interpolation points are described for the standard quadrilateral and triangular elements.

Table F.1 Coordinates of one-dimensional optimal sets T_1^p and T_2^p from reference [31]. Only the positive interior coordinates are listed.

p	T_1^p	T_2^p
3	0.4177913013559897	0.4306648
4	0.6209113046899123	0.6363260
5	0.2689070447719729	0.2765187
	0.7341266671891752	0.7485748
6	0.4461215299911067	0.4568660
	0.8034402382691066	0.8161267
7	0.1992877299056662	0.2040623
	0.5674306027472533	0.5790145
	0.8488719610366557	0.8598070
8	0.3477879716116667	0.3551496
	0.6535334790799030	0.6649023
	0.8802308527184540	0.8896327

F.1.1 Interpolation on the standard quadrilateral element

On the standard quadrilateral element the coordinates of the interpolation points are the tensor product of the one-dimensional interpolation set T_2^p.

F.1.2 Interpolation on the standard triangle

Certain restrictions apply to the definition of optimal sets for the standard triangle. These restrictions are made necessary by the requirement that mapped surfaces have to be continuous. Therefore the interpolation points on the edges of the standard triangle must have the same locations as on the standard quadrilateral element. Therefore on each edge of the standard triangle the interpolation points are the same as the one-dimensional T_2^p interpolation set. Only the interior points are determined by minimizing the function

$$\eta_\Delta^2 = \int_{\Omega_{st}^{(t)}} \sum_{i=1}^{n_p} |N_i(\xi, \eta)|^2 \, d\xi d\eta \tag{F.7}$$

where $n_p = (p+1)(p+2)/2$. Furthermore, since the standard triangle has three-fold symmetry, the interpolation points also have three-fold symmetry. The triangular coordinates of the six interior interpolation points of the optimal nodal set for $p = 5$ are listed in Table F.2. Note that the coordinates in the second and third lines are even permutations of the coordinates in the first line, indicating three-fold symmetry. Lines 4 to 6 are also even permutations of the triangular coordinates.

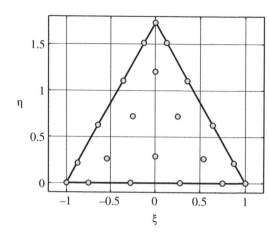

Figure F.2 Optimal interpolation points for the standard triangle, $p = 5$.

Table F.2 Triangular coordinates of the interior interpolation points for $p = 5$.

L_1	L_2	L_3
0.152754	0.152754	0.694493
0.694493	0.152754	0.152754
0.152754	0.694493	0.152754
0.416888	0.416888	0.166225
0.166225	0.416888	0.416888
0.416888	0.166225	0.416888

Appendix G

Corner singularities in two-dimensional elasticity

The following discussion of corner singularities in two-dimensional elasticity is analogous to the discussion of corner singularities in Section 4.2.2. The corner problem in two-dimensional elasticity is complicated by the fact that the eigenvalues may be complex numbers.

G.1 The Airy stress function

It is assumed in the following that the material is isotropic and elastic, hence the material properties are characterized by the two material constants E and ν, and the volume forces are zero. We examine the solution in the neighborhood of corner points when the intersecting edges are stress free. The treatment of other cases is analogous.

The stress fields in planar elasticity can be derived from the Airy stress function[1] denoted by $U(r, \theta)$. The Airy stress function satisfies the biharmonic equation:

$$\left(\frac{\partial^2}{\partial r^2} + \frac{1}{r}\frac{\partial}{\partial r} + \frac{1}{r^2}\frac{\partial^2}{\partial \theta^2}\right)\left(\frac{\partial^2}{\partial r^2} + \frac{1}{r}\frac{\partial}{\partial r} + \frac{1}{r^2}\frac{\partial^2}{\partial \theta^2}\right) U = 0. \tag{G.1}$$

The components of the stress tensor in polar coordinates are related to U by the following formulas (see, for example, [105]):

$$\sigma_r = \frac{1}{r}\frac{\partial U}{\partial r} + \frac{1}{r^2}\frac{\partial^2 U}{\partial \theta^2}, \quad \sigma_\theta = \frac{\partial^2 U}{\partial r^2}, \quad \tau_{r\theta} = -\frac{\partial}{\partial r}\left(\frac{1}{r}\frac{\partial U}{\partial \theta}\right). \tag{G.2}$$

The Cartesian components of the stress tensor are:

$$\sigma_x = \frac{\partial^2 U}{\partial y^2}, \quad \sigma_y = \frac{\partial^2 U}{\partial x^2}, \quad \tau_{xy} = -\frac{\partial^2 U}{\partial x \partial y}. \tag{G.3}$$

The stress function can be written in complex variable form[2] :

$$U = \Re(\bar{z}\varphi(z) + \chi(z)) \tag{G.4}$$

where $\Re(\cdot)$ means the real part of the expression in the bracket, $\varphi(z)$ and $\chi(z)$, called complex potentials, are analytic functions of the complex variable z. The overbar indicates the complex conjugate. The stresses components in polar coordinates are related to $\varphi(z)$ and $\chi(z)$ as follows:

$$\sigma_r + \sigma_\theta = 2\left(\varphi'(z) + \overline{\varphi'(z)}\right) = 4\Re(\varphi'(z)) \tag{G.5}$$

$$\sigma_\theta - \sigma_r + 2i\tau_{r\theta} = 2z\bar{z}^{-1}\left(\bar{z}\varphi''(z) + \chi''(z)\right). \tag{G.6}$$

1 Sir George Biddell Airy 1801–1892.
2 This formula is attributed to Edouard Jean-Baptiste Goursat 1858–1936.

Finite Element Analysis: Method, Verification and Validation, Second Edition. Barna Szabó and Ivo Babuška.
© 2021 John Wiley & Sons, Inc. Published 2021 by John Wiley & Sons, Inc.
Companion Website: www.wiley.com/go/szabo/finite_element_analysis

The stresses components in Cartesian coordinates are related to $\varphi(z)$ and $\chi(z)$ as follows:

$$\sigma_x + \sigma_y = 2\left(\varphi'(z) + \overline{\varphi'(z)}\right) = 4\Re(\varphi'(z)) \tag{G.7}$$

$$\sigma_y - \sigma_x + 2i\tau_{xy} = 2\left(\bar{z}\varphi''(z) + \chi''(z)\right). \tag{G.8}$$

The components of the displacement vector in polar coordinates (up to rigid body displacement and rotation) are related to $\varphi(z)$ and $\chi(z)$ as follows:

$$2G(u_r + iu_\theta) = z^{-1/2}\bar{z}^{1/2}\left(\kappa\varphi(z) - z\,\overline{\varphi'(z)} - \overline{\chi'(z)}\right) \tag{G.9}$$

and in Cartesian coordinates:

$$2G(u_x + iu_y) = \kappa\varphi(z) - z\,\overline{\varphi'(z)} - \overline{\chi'(z)} \tag{G.10}$$

where κ is defined as

$$\kappa \stackrel{\text{def}}{=} \begin{cases} \dfrac{3-\nu}{1+\nu} & \text{for plane stress} \\ 3 - 4\nu & \text{for plane strain} \end{cases} \tag{G.11}$$

This is known as the Kolosov-Muskhelishvili method[3] . Details and derivations are available in books on elasticity, such as [60, 105].

We will be interested in solutions corresponding to

$$\varphi(z) = (a_1 - ia_2)z^\lambda, \quad \chi(z) = (a_3 - ia_4)z^{\lambda+1}, \quad \lambda \geq 0, \; \lambda \neq 1 \tag{G.12}$$

where a_i $(i = 1, 2, 3, 4)$ and λ are real numbers. The corresponding stress function is:

$$U = r^{\lambda+1}(a_1\cos(\lambda-1)\theta + a_2\sin(\lambda-1)\theta + a_3\cos(\lambda+1)\theta + a_4\sin(\lambda+1)\theta). \tag{G.13}$$

In the case of $\lambda = 1$,

$$\varphi(z) = a_1 z - ia_2 z\,\log z, \quad \chi(z) = (a_3 - ia_4)z^2 \tag{G.14}$$

the corresponding stress function is:

$$U = r^2(a_1 + a_2\theta + a_3\cos 2\theta + a_4\sin 2\theta). \tag{G.15}$$

G.2 Stress-free edges

We refer to Fig. 4.1 and assume that the boundary segments Γ_{AB} and Γ_{BC} are stress free, that is, $\sigma_\theta = \tau_{r\theta} = 0$ at $\theta = \pm\alpha/2$. Using equations (G.13) and (G.2) we find:

$$\sigma_\theta = r^{\lambda-1}\lambda(\lambda+1)[a_1\cos(\lambda-1)\theta + a_2\sin(\lambda-1)\theta$$
$$+ a_3\cos(\lambda+1)\theta + a_4\sin(\lambda+1)\theta]$$

$$\tau_{r\theta} = r^{\lambda-1}\lambda(\lambda-1)[a_1\sin(\lambda-1)\theta - a_2\cos(\lambda-1)\theta)]$$
$$+ r^{\lambda-1}\lambda(\lambda+1)[a_3\sin(\lambda+1)\theta - a_4\cos(\lambda+1)\theta].$$

On setting $\sigma_\theta = \tau_{r\theta} = 0$ at $\theta = \pm\alpha/2$, following straightforward algebraic manipulation, we have:

$$\begin{bmatrix} \cos(\lambda-1)\dfrac{\alpha}{2} & \cos(\lambda+1)\dfrac{\alpha}{2} \\ -\Lambda \, \sin(\lambda-1)\dfrac{\alpha}{2} & \sin(\lambda+1)\dfrac{\alpha}{2} \end{bmatrix} \begin{Bmatrix} a_1 \\ a_3 \end{Bmatrix} = 0 \tag{G.16}$$

---

3 Gury Vasilievich Kolosov 1867–1936; Nikolai Ivanovich Muskhelishvili 1891–1976.

and

$$
\begin{bmatrix}
\sin(\lambda-1)\dfrac{\alpha}{2} & \sin(\lambda+1)\dfrac{\alpha}{2} \\[3mm]
-\Lambda\ \cos(\lambda-1)\dfrac{\alpha}{2} & \cos(\lambda+1)\dfrac{\alpha}{2}
\end{bmatrix}
\begin{Bmatrix} a_2 \\[2mm] a_4 \end{Bmatrix} = 0
\tag{G.17}
$$

where:

$$
\Lambda \overset{\text{def}}{=} \frac{1-\lambda}{1+\lambda}.
$$

Note that a_1 and a_3 (resp. a_2 and a_4) are coefficients of symmetric (resp. antisymmetric) functions in eq. (G.13). Therefore, analogously to eq. (4.7), $U(r,\theta)$ can be written in terms of sums of symmetric and antisymmetric functions. The symmetric functions associated with the eigenvalues of eq. (G.16) are called Mode I eigenfunctions and the antisymmetric functions associated with the eigenvalues of eq. (G.17) are called Mode II eigenfunctions.

Nontrivial solutions exist if either the determinant of eq. (G.16) or the determinant of eq. (G.17) vanishes. This will occur if either

$$
\cos(\lambda-1)\frac{\alpha}{2}\sin(\lambda+1)\frac{\alpha}{2} + \Lambda\sin(\lambda-1)\frac{\alpha}{2}\cos(\lambda+1)\frac{\alpha}{2} = 0
$$

which can be simplified to

$$
\sin\lambda\,\alpha + \lambda\,\sin\alpha = 0, \quad \lambda \neq 0, \pm 1
\tag{G.18}
$$

or

$$
\sin(\lambda-1)\frac{\alpha}{2}\cos(\lambda+1)\frac{\alpha}{2} + \Lambda\cos(\lambda-1)\frac{\alpha}{2}\sin(\lambda+1)\frac{\alpha}{2} = 0
$$

which can be simplified to:

$$
\sin\lambda\,\alpha - \lambda\,\sin\alpha = 0, \quad \lambda \neq 0, \pm 1.
\tag{G.19}
$$

We denote

$$
Q(\lambda\alpha) \overset{\text{def}}{=} \frac{\sin\lambda\alpha}{\lambda\alpha} \quad \text{and} \quad q(\alpha) \overset{\text{def}}{=} \frac{\sin\alpha}{\alpha}
\tag{G.20}
$$

and discuss the eigenvalues corresponding to the symmetric and antisymmetric eigenfunctions separately in the following. For a detailed treatment of this subject we refer to [53].

Example G.1 Let us find the first eigenvalue for $\alpha = 3\pi/2\ (270°)$ from eq. (G.18). In this case we need to find $t \overset{\text{def}}{=} \lambda\alpha$ such that

$$
f(t) \overset{\text{def}}{=} \frac{\sin t}{t} - \frac{2}{3\pi} = 0.
$$

Using a root finding routine, we get $t = 2.56581916$ and $\lambda = 0.54448374$.

G.2.1 Symmetric eigenfunctions

Equation (G.18) can be written as

$$
Q(\lambda\alpha) + q(\alpha) = 0.
\tag{G.21}
$$

The function $Q(\lambda\alpha)$ is plotted on the interval $0 < \lambda\alpha < 4\pi$ in Fig. G.1. The problem is to find the roots of eq. (G.21) for a given α.

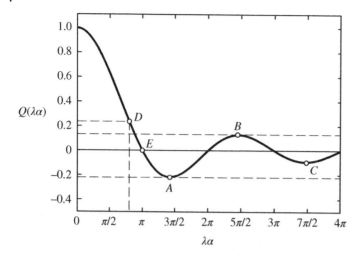

Figure G.1 The function $Q(\lambda\alpha)$ on the interval $0 < \lambda\alpha < 4\pi$.

Observe the following:

1. In the interval $0 < \alpha < \alpha_A$ where $\alpha_A = 2.553591$ (146.31°) the line $-q(\alpha)$ has no points in common with $Q(\lambda\alpha)$, therefore there are no real roots.
2. At $\alpha = \alpha_A$ the line $-q(\alpha_A)$ is tangent to $Q(\lambda\alpha)$ in point A. Therefore there is a double root at this angle. There are double roots at $\alpha = \alpha_B^{(1)} = 3.625739$ (207.74°), $\alpha = \alpha_B^{(2)} = 5.499379$ (315.09°) and also at $\alpha = \alpha_C = 2.875839$ (164.77°).
3. In the interval $\alpha_A < \alpha < \alpha_B$ there are at least two real and simple roots.
4. At $\alpha = \pi$ and $\alpha = 2\pi$ there are infinitely many real roots. Furthermore, at $\alpha = \pi$ all roots are integers.
5. Point D corresponds to $\alpha = 3\pi/2$ (270°) where $\lambda\alpha = 2.565819$, hence $\lambda = 0.544484$. See Example G.1. There are no other symmetric real roots at this angle.
6. Point E corresponds to $\alpha = 2\pi$ where $\lambda\alpha = \pi$, hence $\lambda = 1/2$. Point E also corresponds to $\alpha = \pi$ which is a special case, discussed next.

In formulating eq. (G.21) we excluded $\lambda = 1$ from consideration. When $\lambda = 1$ then U is given by eq. (G.15). Considering the symmetric terms only and using eq. (G.2), we have

$$\sigma_\theta = 2(a_1 + a_3 \cos 2\theta), \quad \tau_{r\theta} = 2a_3 \sin 2\theta.$$

Letting $\sigma_\theta(\pm\alpha/2) = \tau_{r\theta}(\pm(\alpha/2) = 0$, we find that non-trivial solution exists only if

$$\det \begin{bmatrix} 1 & \cos\alpha \\ 0 & \sin\alpha \end{bmatrix} = 0.$$

Therefore $\alpha = n\pi$, $(n = 1, 2, \dots)$. Point E in Fig. G.1 represents $\alpha = \pi$.

Remark G.1 To find the complex roots of eq. (G.21) we write $\lambda = \xi + i\eta$. Therefore eq. (G.21) becomes

$$\frac{\sin(\xi\alpha + i\eta\alpha)}{\xi\alpha + i\eta\alpha} = q(\alpha) \tag{G.22}$$

which is equivalent to the following system of two equations:

$$\sin \xi \alpha \cosh \eta \alpha = \xi q(\alpha) \tag{G.23}$$

$$\cos \xi \alpha \sinh \eta \alpha = \eta q(\alpha). \tag{G.24}$$

For details we refer to [108].

G.2.2 Antisymmetric eigenfunctions

Equation (G.19) can be written as

$$Q(\lambda \alpha) - q(\alpha) = 0. \tag{G.25}$$

Note that $\lambda = 1$ trivially satisfies eq. (G.25) for all α and recall that we have excluded $\lambda = 1$ from consideration when we formulated eq. (G.25). There are no real roots in the interval $0 < \alpha < \alpha_B$ where $\alpha_B = 2.777068 \ (159.11°)$. There are at least two real roots in the interval $\alpha_B < \alpha < \alpha_C$ where $\alpha_C^{(1)} = 3.463416 \ (198.44°)$; $\alpha_C^{(2)} = 5.732235 \ (328.43°)$. As in the case of Mode I, there are infinitely many real roots at $\alpha = \pi$ and $\alpha = 2\pi$ and at $\alpha = \pi$ all roots are integers. There is only one real root in the interval $\alpha_C < \alpha < \alpha_A$ where $\alpha_A = 4.493409 \ (257.45°)$.

The angle α_A is a special angle, which corresponds to $\lambda = 1$. To show this, we consider the antisymmetric terms in eq. (G.15). Using eq. (G.2), we have

$$\sigma_\theta = 2(a_2 \theta + a_4 \sin 2\theta), \quad \tau_{r\theta} = -(a_2 + 2a_4 \cos 2\theta).$$

Letting $\sigma_\theta(\pm \alpha/2) = \tau_{r\theta}(\pm \alpha/2) = 0$, we find that non-trivial solution exists when

$$\det \begin{bmatrix} \alpha/2 & \sin \alpha \\ 1 & 2 \cos \alpha \end{bmatrix} = 0.$$

Therefore $\alpha = \tan \alpha$. In the interval $0 < \alpha < 2\pi$ there is one root; $\alpha = \alpha_A = 4.493409 \ (257.45°)$. This corresponds to point A in Fig. G.1.

G.2.3 The L-shaped domain

The L-shaped domain problem ($\alpha = 3\pi/2$) described here is a widely used benchmark problem in finite element analysis. It is representative of a geometric detail that frequently occurs in finite element analysis, typically because fillets are omitted for convenience. The resulting singularity may influence the accuracy of the quantities of interest.

It is assumed that the edges that intersect in the singular point are stress-free. In Example G.1 it was found that the lowest positive eigenvalue of eq. (G.18) is $\lambda_1 = 0.544483737$. The Airy stress function corresponding to λ_1 can be written as:

$$U = a_1 r^{\lambda_1 + 1}(\cos(\lambda_1 - 1)\theta + Q_1 \cos(\lambda_1 + 1)\theta) \tag{G.26}$$

where a_1 is an arbitrary real number and $Q_1 = 0.543075579$. Equation (G.26) is equivalent to

$$U = a_1 \Re(\bar{z}z^{\lambda_1} + Q_1 z^{\lambda_1 + 1}). \tag{G.27}$$

Using eq. (G.27) and eq. (G.3) it is possible to show that the stress components corresponding to λ_1 are:

$$\begin{aligned}
\sigma_x &= a_1 \lambda_1 \, r^{\lambda_1 - 1} \left[\left(2 - Q_1(\lambda_1 + 1)\right) \cos(\lambda_1 - 1) \, \theta - (\lambda_1 - 1) \cos(\lambda_1 - 3) \, \theta \right] \\
\sigma_y &= a_1 \lambda_1 \, r^{\lambda_1 - 1} \left[\left(2 + Q_1(\lambda_1 + 1)\right) \cos(\lambda_1 - 1) \, \theta + (\lambda_1 - 1) \cos(\lambda_1 - 3) \, \theta \right] \\
\tau_{xy} &= a_1 \lambda_1 \, r^{\lambda_1 - 1} \left[(\lambda_1 - 1) \sin(\lambda_1 - 3) \, \theta + Q_1(\lambda_1 + 1) \sin(\lambda_1 - 1) \, \theta \right].
\end{aligned} \tag{G.28}$$

Using eq. (G.10) and eq. (G.27) it is possible to show that the displacement components corresponding to λ_1, up to rigid body displacement and rotation terms, are:

$$u_x = \frac{a_1}{2G} \, r^{\lambda_1} \left[\left(\kappa - Q_1(\lambda_1 + 1) \right) \cos \lambda_1 \theta - \lambda_1 \cos(\lambda_1 - 2)\,\theta \right]$$
$$u_y = \frac{a_1}{2G} \, r^{\lambda_1} \left[\left(\kappa + Q_1(\lambda_1 + 1) \right) \sin \lambda_1 \theta + \lambda_1 \sin(\lambda_1 - 2)\,\theta \right]$$

(G.29)

where κ is defined by eq. (G.11).

Complex eigenvalues

In planar elasticity, in contrast to the Laplace equation, λ can be complex and it can be either a simple or a multiple root. If λ is complex then its conjugate is also a root. In the case of multiple roots special treatment is necessary which is not discussed here. We refer to [72] for details.

Consider the Airy stress function in the form: $U = r^{\lambda+1} F(\theta)$. If λ is complex we write $\lambda = \xi + i\eta$ and $F = f + ig$. Therefore

$$U = r^{(\xi+1+i\eta)}(f + ig) = r^{(\xi+1)} e^{(\ln r^{i\eta})}(f + ig).$$

Writing

$$e^{(\ln r^{i\eta})} = e^{(i\eta \ln r)} = \cos(\eta \ln r) + i \sin(\eta \ln r)$$

we have:

$$U = r^{\xi+1}[(f \cos(\eta \ln r) - \eta \sin(\eta \ln r)) + i(f \sin(\eta \ln r) - \eta \cos(\eta \ln r))].$$

(G.30)

Both the real and imaginary parts of U are solutions of the biharmonic equation (G.1). Since $\ln r \to -\infty$ as $r \to 0$, the sinusoidal terms oscillate with a wavelength approaching zero.

Remark G.2 Note that the number of square integrable derivatives, and hence the rate of convergence, is independent of the imaginary part.

Table G.1 Lowest positive values of $\mathfrak{R}(\lambda)$ for three types of homogeneous boundary conditions prescribed on intersecting straight boundary segments.

α	free-free		fixed-free	fixed-fixed	
	$\mathfrak{R}(\lambda_1^{(s)})$	$\mathfrak{R}(\lambda_1^{(a)})$	$\mathfrak{R}(\lambda_1)$	$\mathfrak{R}(\lambda_1^{(s)})$	$\mathfrak{R}(\lambda_1^{(a)})$
45°	5.39053	9.56271	1.30434	5.57328	2.60831
90°	2.73959	4.80825	0.75835	2.82579	1.49046
135°	1.88537	3.24281	0.69339	1.57323	1.16088
180°	1.00000	2.00000	0.50000	1.00000	1.00000
225°	0.67358	1.30209	0.40594	0.73554	0.87723
270°	0.54448	0.90853	0.34032	0.60404	0.74446
315°	0.50501	0.65970	0.28784	0.53793	0.60945
360°	0.50000	0.50000	0.25000	0.50000	0.50000

G.2.4   Corner points

The lowest values of $\Re(\lambda)$ associated with the points of intersection of straight boundary segments in two-dimensional elasticity are listed in Table G.1 for three types of homogeneous boundary conditions prescribed on the intersecting boundary segments. The angle intersection α has the same sense as in Fig. 4.1.

1. Free-free conditions: On both edges $\sigma_\theta = \tau_{r\theta} = 0$. In this case the eigenvalues are independent of Poisson's ratio v.
2. Fixed-free condition (plane stress, $v = 0.3$): on one edge $u_r = u_\theta = 0$, on the other edge $\sigma_\theta = \tau_{r\theta} = 0$
3. Fixed-fixed condition (plane stress, $v = 0.3$): $u_r = u_\theta = 0$ on both edges.

   In the free-free and fixed-fixed cases the eigenfunctions are either symmetric or antisymmetric. These are indicated in Table G.1 with the superscripts s and a, respectively. At $\alpha = 180°$ the eigenvalues are integers.

Appendix H

Computation of stress intensity factors

The goal of computation in linear elastic fracture mechanics is to estimate stress intensity factors. The contour integral method and the energy release rate method, outlined in the following, are widely used for this purpose.

H.1 Singularities at crack tips

At a crack tip $\alpha = 2\pi$, see equations (G.16) and (G.17). The Airy stress function corresponding to the symmetric part of the asymptotic expansion is:

$$U_s = \sum_{i=1}^{\infty} a_i \Re(\bar{z} z^{\lambda_i} + Q_i z^{\lambda_i+1}), \qquad \lambda_i = i/2 \tag{H.1}$$

where $Q_i = (2 - i)/(2 + i)$ when i is odd and $Q_i = -1$ when i is even. The first two terms of this asymptotic expansion play a central role in linear elastic fracture mechanics. The stress components are obtained from eq. (G.3). For example:

$$\sigma_x = \frac{\partial^2 U_s}{\partial y^2} = \frac{\partial}{\partial y}\left(\frac{\partial U_s}{\partial z}\frac{\partial z}{\partial y} + \frac{\partial U_s}{\partial \bar{z}}\frac{\partial \bar{z}}{\partial y}\right) = i\frac{\partial}{\partial y}\left(\frac{\partial U_s}{\partial z} - \frac{\partial U_s}{\partial \bar{z}}\right)$$

$$= \frac{\partial}{\partial z}\left(-\frac{\partial U_s}{\partial z} + \frac{\partial U_s}{\partial \bar{z}}\right) + \frac{\partial}{\partial \bar{z}}\left(\frac{\partial U_s}{\partial z} - \frac{\partial U_s}{\partial \bar{z}}\right) \tag{H.2}$$

therefore:

$$\sigma_x = a_1 \lambda_1 \Re(2z^{\lambda_1-1} - (\lambda_1 - 1)\bar{z} z^{\lambda_1-2} - Q_1(\lambda_1 + 1)z^{\lambda_1-1})$$

$$= a_1 \lambda_1\, r^{\lambda_1-1}\left[(2 - Q_1(\lambda_1 + 1))\cos(\lambda_1 - 1)\,\theta - (\lambda_1 - 1)\cos(\lambda_1 - 3)\,\theta\right]$$

$$= \frac{a_1}{\sqrt{r}}\left(\frac{3}{4}\cos(\theta/2) + \frac{1}{4}\cos(5\theta/2)\right). \tag{H.3}$$

Letting $a_1 = K_I/\sqrt{2\pi}$ and using trigonometric identities the functional form used in linear elastic fracture mechanics can be obtained:

$$\sigma_x = \frac{K_I}{\sqrt{2\pi r}}\cos\frac{\theta}{2}\left(1 - \sin\frac{\theta}{2}\sin\frac{3\theta}{2}\right) \tag{H.4}$$

see eq. (4.42). It is left to the reader to show that eq. (H.3) and eq. (H.4) are equivalent.

The stress components corresponding to the second term of eq. (H.1) are: $\sigma_x = 4a_2$, $\sigma_y = \tau_{xy} = 0$. The stress $\sigma_x = 4a_2$ (constant) is called the T-stress and denoted by T. This is the second term in eq. (4.42). The other stress components are obtained analogously.

Finite Element Analysis: Method, Verification and Validation, Second Edition. Barna Szabó and Ivo Babuška.
© 2021 John Wiley & Sons, Inc. Published 2021 by John Wiley & Sons, Inc.
Companion Website: www.wiley.com/go/szabo/finite_element_analysis

The Airy stress function corresponding to the antisymmetric part of the asymptotic expansion is:

$$U_a = \sum_{i=1}^{\infty} b_i \mathfrak{I}(\bar{z}z^{\lambda_i} + Q_i z^{\lambda_i+1}), \qquad \lambda_i = i/2 \tag{H.5}$$

where $\mathfrak{I}(\cdot)$ represents the imaginary part of $(\cdot)$ and $Q_i = -1$ when i is odd, $Q_i = (2-i)/(2+i)$ when i is even. This can be verified by comparing eq. (H.5) with the antisymmetric terms in eq. (G.13). The stress components corresponding to the second term of eq. (H.5) are: $\sigma_x = \sigma_y = \tau_{xy} = 0$.

Using equations (H.1), (H.5) and (G.2), (G.3) the stress distribution in the neighborhood of a crack tip can be determined up to the coefficients a_i, b_i ($i = 1, 2, \ldots, \infty$). Procedures for the determination of these coefficients from finite element solutions are described in Sections H.2 and H.3.

H.2 The contour integral method

The contour integral method was described in connection with the Laplace equation in Section 4.2.4. In this section the contour integral method for the computation of the Mode I stress intensity factor in planar elasticity is outlined.

The Airy stress function corresponding to the symmetric part of the asymptotic expansion is given by eq. (H.1). The extraction function for the coefficient of the first term, denoted by **w**, corresponds to the first negative eigenvalue $\lambda_1 = -1/2$. Therefore $\varphi(z) = z^{-1/2}$ and $\chi(z) = 3z^{1/2}$ in eq. (G.4) and the Airy stress function is

$$U = C\mathfrak{R}(\bar{z}z^{-1/2} + 3z^{1/2}) \tag{H.6}$$

where C is an arbitrary real number in units of MPa m$^{1/2}$. The stress components $\sigma_x^{(\mathbf{w})}$, $\sigma_y^{(\mathbf{w})}$, $\tau_{xy}^{(\mathbf{w})}$ are determined from equations (G.7) and (G.8):

$$\sigma_x^{(\mathbf{w})} = -\frac{C}{4}r^{-3/2}(\cos(3\theta/2) + 3\cos(7\theta/2)) \tag{H.7}$$

$$\sigma_y^{(\mathbf{w})} = -\frac{C}{4}r^{-3/2}(7\cos(3\theta/2) - 3\cos(7\theta/2)) \tag{H.8}$$

$$\tau_{xy}^{(\mathbf{w})} = \frac{3C}{4}r^{-3/2}(\sin(3\theta/2) - \sin(7\theta/2)) \tag{H.9}$$

and the traction vector components are computed from

$$T_x^{(\mathbf{w})} = \sigma_x^{(\mathbf{w})} \cos\theta + \tau_{xy}^{(\mathbf{w})} \sin\theta \tag{H.10}$$

$$T_y^{(\mathbf{w})} = \tau_{xy}^{(\mathbf{w})} \cos\theta + \sigma_y^{(\mathbf{w})} \sin\theta. \tag{H.11}$$

The components of **w** are determined from eq. (G.10):

$$w_x = C\frac{r^{-1/2}}{2G}\left[\left(\kappa - \frac{3}{2}\right)\cos(\theta/2) + \frac{1}{2}\cos(5\theta/2)\right] \tag{H.12}$$

$$w_y = -C\frac{r^{-1/2}}{2G}\left[\left(\kappa + \frac{3}{2}\right)\sin(\theta/2) - \frac{1}{2}\sin(5\theta/2)\right] \tag{H.13}$$

where G is the shear modulus. The path-independent integral, evaluated on a circle of radius r, is:

$$I_\Gamma(\mathbf{u}, \mathbf{w}) \overset{\text{def}}{=} \int_{-\pi}^{\pi} (T_x^{(\mathbf{u})}w_x + T_y^{(\mathbf{u})}w_y)r \, d\theta - \int_{-\pi}^{\pi} (T_x^{(\mathbf{w})}u_x + T_y^{(\mathbf{w})}u_y)r \, d\theta. \tag{H.14}$$

Table H.1 Values of the function $F(\kappa)$.

ν	plane stress	plane strain
0	-4π	-4π
0.1	-11.4240	-11.3097
0.2	-10.4720	-10.0531
0.3	-9.66644	-8.79646
0.4	-8.97598	-7.53982
0.5	-8.37758	-6.28319

This integral is analogous to eq. (4.19). The stress intensity factor K_I is defined by convention to be $a_1\sqrt{2\pi}$ where a_1 is the coefficient of the first term in eq. (H.1). Using the orthogonality property of eigenfunctions we get

$$I_\Gamma(\mathbf{u}, \mathbf{w}) = \frac{a_1 C}{G} F(\kappa) \tag{H.15}$$

where $F(\kappa)$ is defined by

$$F(\kappa) \overset{\text{def}}{=} GI_\Gamma(\mathbf{u}_1/a_1, \mathbf{w}/C). \tag{H.16}$$

The function $\mathbf{u}_1$ is the displacement field corresponding to the first term in eq. (H.1). It is determined using eq. (G.10):

$$u_x^{(1)} = \frac{a_1 r^{1/2}}{2G}\left[\left(\kappa - \frac{1}{2}\right)\cos(\theta/2) - \frac{1}{2}\cos(3\theta/2)\right] \tag{H.17}$$

$$u_y^{(1)} = \frac{a_1 r^{1/2}}{2G}\left[\left(\kappa + \frac{1}{2}\right)\sin(\theta/2) - \frac{1}{2}\sin(3\theta/2)\right]. \tag{H.18}$$

The corresponding stress field is given by equations (4.42) to (4.44) with a_1 substituted for $K_I/\sqrt{2\pi}$. Note that $F(\kappa)$ is dimensionless and can be computed explicitly. Values of $F(\kappa)$ are listed in Table H.1.

From eq. (H.15), using $K_I = a_1\sqrt{2\pi}$, we get

$$K_I = \frac{\sqrt{2\pi}G}{CF(\kappa)}I_\Gamma(\mathbf{u}, \mathbf{w}) = \frac{\sqrt{2\pi}G}{F(\kappa)}I_\Gamma(\mathbf{u}, \mathbf{w}/C). \tag{H.19}$$

Note that since $I_\Gamma(\mathbf{u}, \mathbf{w}/C)$ has the dimension $\text{m}^{1/2}$, the dimension of K_I is MPa $\text{m}^{1/2}$. Note further that since C is arbitrary, it can be chosen to be 1 MPa $\text{m}^{3/2}$. To obtain an approximation for K_I, we substitute the finite element solution $\mathbf{u}_{FE}$ for $\mathbf{u}$. The derivation of the extraction function for K_{II} is analogous.

H.3 The energy release rate

The relationship between the stress intensity factors and the energy release rate $\mathcal{G}$ is derived in the following for a plane elastic body of thickness t_z. We assume that the body force and thermal load are zero. By definition,

$$\mathcal{G} \overset{\text{def}}{=} -\frac{\partial\Pi}{\partial a} \tag{H.20}$$

where Π is the potential energy defined by eq. (2.102) and a is the crack length.

H.3.1 Symmetric (Mode I) loading

Consider the state of stress at the crack tip. In the coordinate system centered on the crack tip, as shown in Fig. H.1, neglecting terms of higher order,

$$\sigma_y = \frac{K_I}{\sqrt{2\pi x}} \qquad 0 < x. \tag{H.21}$$

The displacement of the crack face is

$$u_y = \frac{(1+v)(\kappa+1)}{E} \frac{K_I}{\sqrt{2\pi}} \sqrt{-x} \qquad x \le 0 \tag{H.22}$$

where κ is defined by eq. (G.11).

Assume now that the crack length increases by a small amount Δa, as shown in Fig. H.1. In this case

$$u_y = \frac{(1+v)(\kappa+1)}{E} \frac{K_I}{\sqrt{2\pi}} \sqrt{\Delta a - x} \qquad 0 \le x \le \Delta a. \tag{H.23}$$

The work required to return the crack to its original length, that is to close the length increment Δa, is:

$$\Delta W = 2 \int_0^{\Delta a} \frac{1}{2} \sigma_y u_y \, t_z dx = \frac{(1+v)(\kappa+1)}{E} \frac{K_I^2 t_z}{2\pi} \underbrace{\int_0^{\Delta a} \sqrt{\frac{\Delta a - x}{x}} \, dx}_{\Delta a \pi / 2}. \tag{H.24}$$

Hence:

$$\Delta W = \frac{(1+v)(\kappa+1)}{4E} K_I^2 t_z \Delta a. \tag{H.25}$$

In order to restore the crack to its initial length, energy equal to ΔW had to be imparted to the elastic body. This is the energy expended in crack growth, called Griffith's surface energy[1].

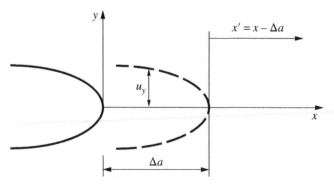

Figure H.1 Notation.

---

[1] Alan Arnold Griffith 1893–1963.

The potential energy had to decrease by the same amount when the crack increment occurred. Hence:

$$\mathcal{G} = -\lim_{\Delta a \to 0} \frac{\Delta \Pi}{\Delta a} = -\frac{\partial \Pi}{\partial a} = \frac{(1 + v)(\kappa + 1)}{4E} K_I^2 t_z. \tag{H.26}$$

H.3.2 Antisymmetric (Mode II) loading

When the loading is purely antisymmetric then the relationship between the energy release rate and the stress intensity factor is analogous to the symmetric case, however instead of equations (H.21) and (H.22) we have:

$$\tau_{xy} = \frac{K_{II}}{\sqrt{2\pi x}} \qquad 0 < x \tag{H.27}$$

and

$$u_x = \frac{(1 + v)(\kappa + 1)}{E} \frac{K_{II} t_z}{\sqrt{2\pi}} \sqrt{-x} \qquad x \leq 0. \tag{H.28}$$

The derivation of the relationship between $\mathcal{G}$ and K_{II}, under the assumption of perfectly antisymmetric loading, is left to the reader.

H.3.3 Combined (Mode I and Mode II) loading

In view of the fact that the solutions corresponding to Mode I and Mode II loadings are *energy orthogonal*, we have:

$$\Pi(\mathbf{u}_I + \mathbf{u}_{II}) = \Pi(\mathbf{u}_I) + \Pi(\mathbf{u}_{II}) \tag{H.29}$$

where $\mathbf{u}_I$ and $\mathbf{u}_{II}$ are the Mode I and Mode II solutions respectively. Therefore in the case of combined loading we have:

$$\mathcal{G} = \frac{(1 + v)(\kappa + 1)}{4E} (K_I^2 + K_{II}^2) t_z. \tag{H.30}$$

Consequently the stress intensity factors are related to $\mathcal{G}$ as follows:

$$(K_I^2 + K_{II}^2) = \begin{cases} \dfrac{E\mathcal{G}}{t_z} & \text{for plane stress} \\[2ex] \dfrac{E\mathcal{G}}{(1 - v^2)t_z} & \text{for plane strain.} \end{cases} \tag{H.31}$$

H.3.4 Computation by the stiffness derivative method

In the following the vector of coefficients of the basis functions computed by the finite element method is denoted by x which is a function of the crack length a. Let us assume that we have computed $x = x(a)$ for a problem of linear elastic fracture mechanics. The potential energy is

$$\Pi(a) = \frac{1}{2} x^T K x - x^T r$$

where $K = K(a)$ is the stiffness matrix and $r = r(a)$ is the load vector. Following crack extension, the potential energy is:

$$\Pi(a + \Delta a) = \frac{1}{2}(x^T + \Delta x^T)(K + \Delta K)(x + \Delta x) - (x^T + \Delta x^T)(r + \Delta r)$$

$$= \Pi(a) + \Delta x^T \underbrace{(Kx - r)}_{\text{this is zero}} + \frac{1}{2}\Delta x^T K \Delta x + \frac{1}{2} x^T \Delta K x +$$

$$x^T \Delta K \Delta x + \frac{1}{2}\Delta x^T \Delta K \Delta x - x^T \Delta r - \Delta x^T \Delta r.$$

Therefore:

$$\mathcal{G} = -\lim_{\Delta a \to 0} \frac{\Pi(a + \Delta a) - \Pi(a)}{\Delta a} = -\frac{1}{2} x^T \frac{\partial K}{\partial a} x + x^T \frac{\partial r}{\partial a}. \tag{H.32}$$

In finite element computations $\partial K/\partial a$ and $\partial r/\partial a$ are approximated by finite differences:

$$\frac{\partial K}{\partial a} \approx \frac{K(a + \Delta a) - K(a - \Delta a)}{2\Delta a}$$

$$\frac{\partial r}{\partial a} \approx \frac{r(a + \Delta a) - r(a - \Delta a)}{2\Delta a}.$$

This involves re-computation of the stiffness matrices for those elements only which have a vertex on the crack tip. In most cases $\partial r/\partial a$ is either zero or negligibly small.

Appendix I

Fundamentals of data analysis

"Without data, you're just another person with an opinion." W. E. Deming[1].

Parametric data analysis, also called statistical inference, is a process by which the statistical attributes of a large population are estimated from observed data which are viewed as samples taken from the large population.

The fundamental concepts and algorithmic procedures of parametric data analysis needed for understanding validation and ranking of mathematical models are introduced. It is assumed that the reader is already familiar with the basic terminology and concepts of statistics, such as those that can be found in the first few chapters in introductory texts: random variables, statistical independence, correlation, mean, variance, commonly used probability density functions and conditional probability.

I.1 Statistical foundations

Frequentist probability or frequentism is one interpretation of probability; it defines an event's probability as the limit of its relative frequency in a large number of trials. This interpretation supports the statistical needs of experimental scientists and pollsters; probabilities can be found in principle by a repeatable objective process and are thus devoid of subjectivity.

Bayesian probability is another interpretation of probability: instead of frequency or propensity of some phenomenon, probability is interpreted as reasonable expectation representing a state of knowledge, or as quantification of a personal belief. Validation and ranking of mathematical models involves arguments based on Bayesian probability. In the following discussion probability should be understood in this sense.

Data analysis is largely based on three basic theorems of statistics, known as the product rule, Bayes' theorem and marginalization. These theorems are stated in this section and their applications in data analysis are illustrated in Section I.3. For detailed introductory treatment of data analysis we recommend references [87] and [52].

The product rule

The product rule states that the probability of X and Y both being true is equal to the probability of X being true, given Y, times the probability of Y being true.

$$\Pr(X, Y) = \Pr(X \mid Y) \Pr(Y) = \Pr(Y \mid X) \Pr(X). \tag{I.1}$$

1 William Edwards Deming 1900–1993.

Finite Element Analysis: Method, Verification and Validation, Second Edition. Barna Szabó and Ivo Babuška.
© 2021 John Wiley & Sons, Inc. Published 2021 by John Wiley & Sons, Inc.
Companion Website: www.wiley.com/go/szabo/finite_element_analysis

The second equality follows from the fact that the ordering of X and Y is immaterial.

Bayes' theorem

Bayes' theorem follows directly from the product rule:

$$\Pr(X \mid Y) = \frac{\Pr(Y \mid X)\Pr(X)}{\Pr(Y)}. \tag{I.2}$$

The importance of this theorem becomes obvious if we replace X with M, representing the statistical model, and Y with D, representing the available data:

$$\Pr(M \mid D) = \frac{\Pr(D \mid M)\Pr(M)}{\Pr(D)}. \tag{I.3}$$

Thus Bayes' theorem relates the probability that a statistical model is the true representation of the (unknown) statistical properties of the population from which data D was sampled, to the probability that we would have observed data D if the model were true.

The term $\Pr(M \mid D)$ is called posterior probability. It represents our degree of belief in the validity of model M in the light of the data D. The term $\Pr(D \mid M)$ is the likelihood function. The term $\Pr(M)$ is called prior probability which represents our degree of belief in the validity of model M before the current data is analyzed and the term $\Pr(D)$ is called marginal likelihood or evidence. It is a scaling factor needed to ensure that the integral of the right hand side over the entire range of parameters is unity.

Marginalization

Marginalization is used in data analysis to eliminate parameters that enter into the analysis but are not of primary interest. Those parameters are called nuisance parameters. The probability that X is true equals the probability that X and Y is true integrated over the entire range of possible values for Y:

$$\Pr(X) = \int_{-\infty}^{\infty} \Pr(X, Y)\, dY. \tag{I.4}$$

In this instance Y is the nuisance parameter.

I.2 Test data

The examples presented in Chapter 6 are based on fatigue test records for 24S-T3 aluminum alloy[2] specimens extracted from references [40–44].

Surface finish is known to be of major importance in determining fatigue strength. In order to avoid introducing scratches and residual stresses, the surfaces of the specimens were electropolished.

The average static properties of the material are listed in Table I.1. The heading "Grain (L)" (resp. "Cross (T)") indicates that the specimen was cut such that the long dimension was oriented in the direction of the grain (resp. across the grain).

The fatigue tests were conducted at 1100 cycles per minute (18.3 Hz). The estimated precision of setting and maintaining loads was approximately $\pm 3\,\%$ for tension-tension tests and $\pm 5\,\%$ for tension-compression tests.

---

2 The current designation of this alloy is 2024-T3.

Table I.1 Average static properties of 24S-T3 (2024-T3) aluminum from reference [40].

Property	Grain (L)	Cross (T)
Modulus of elasticity in compression	10,650 ksi	10,450 ksi
Elongation in 2 inches, percent	18.2 %	18.3 %
Yield strength (0.2 % offset) in tension	54.0 ksi	50.0 ksi
Yield strength (0.2 % offset) in compression	44.5 ksi	50.5 ksi
Ultimate strength in tension	73.0 ksi	71.0 ksi

The test records contain the following information: (a) Specimen dimensions in inches[3], (b) specimen label, (c) the maximum stress in the test section or notch root S_{max} in psi units[4], (d) the cycle ratio R, (e) the number of cycles at the end of the test N and (f) notes indicating remarkable observations relating to the test, such as whether failure occurred outside of the test section or the test was stopped prior to failure (run-out). Two of the notched test specimens are shown in Fig. I.1.

The test data are summarized in Table I.2. For the notch-free specimens the range of σ_{max} was (24, 58) ksi, the range of σ_{min} was (−50, 29) ksi.

In the column labeled $(K_t)_{act}$ the actual stress concentration factors, defined as the ratio of maximum stress to the average stress in the test section, are shown. In the column labeled A, the highly stressed area is shown. This area, multiplied by the thickness, approximates the highly stressed volume defined in eq. (6.31) for $\gamma = 0.85$. Errors in the computed values of A have been verified to be not greater than 1%.

In the last column the specimen count is listed with and without run-outs. The number of specimens without counting run-outs is in brackets. This count does not include specimens that were tested at stress levels higher than 58.0 ksi and specimens that were disregarded by the investigators because of flaws, buckling or failure in the grips.

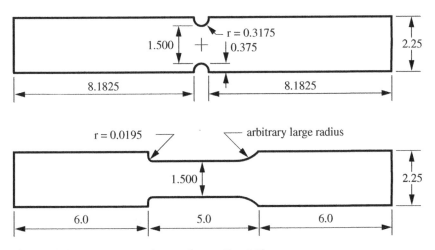

Figure I.1 Notched test specimens of type #3 and #6.

---

3 1 inch = 25.40 mm.
4 pounds per square inch; 1 psi = 6.895 kPa.

Table I.2 Summary of test records extracted from NACA & NASA technical reports for 24S-T3 (2024-T3) aluminum specimens.

k	Specimen	Ref.	r_k (in)	$(K_t)_{act}$	A (in^2)	Count
1	Notch free	[40]	12.0	1.00	test section	53 (44)
2	Open hole	[41]	1.5000	2.11	2.898 E−02	39 (34)
3	Edge notch	[41]	0.3175	2.17	4.325 E−03	42 (38)
4	Fillet	[41]	0.1736	2.19	9.561 E−04	32 (27)
5	Edge notch	[41]	0.0570	4.43	1.827 E−04	34 (29)
6	Fillet	[41]	0.0195	4.83	1.618 E−05	36 (32)
7	Edge notch	[42]	0.0313	5.83	5.888 E−05	46 (40)
8	Edge notch	[43]	0.7600	1.62	2.875 E−02	31 (27)
9	Edge notch	[44]	0.0035	4.48	5.908 E−07	17 (13)
10	Edge notch	[44]	0.0710	4.41	3.324 E−04	19 (15)

I.3 Statistical models

In this section we formulate three statistical models that could reasonably be expected to represent the population from which the data collected from constant cycle fatigue experiments of metal specimens was sampled and estimate their parameters. Evaluation of the relative merit of these models is discussed in Section I.4.

The available information comprises a detailed description of the testing apparatus, the geometric configuration and surface finish of the test specimens, and a set of records consisting of the number of cycles at which failure occurred or the test was stopped, the maximum and minimum load values and notes concerning unusual events, such as failure outside of the test section, buckling and so on. A representative set of test data, called S-N data, taken from [40], is shown in Fig. I.2. There are 53 data points of which 9 are runouts, that is, the test was stopped before failure occurred. The statistical term for runouts is "right-censored data".

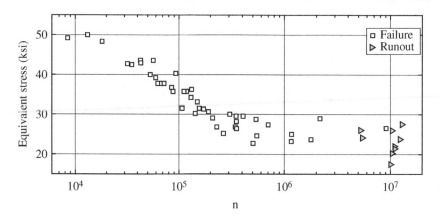

Figure I.2 S-N Data for 24S-T3 (2024-T3) aluminum alloy.

In these tests the equivalent stress σ_{eq} was controlled. Various definitions of equivalent stress are possible, see Sections 6.1.1, 6.3. The number of cycles to failure N is assumed to be an independent and identically distributed random variable. In this context "independent" should be understood to mean that the occurrence of one outcome does not affect the probability of occurrence of other outcomes. This assumption is justified noting that each outcome corresponds to a different test specimen. The term "identically distributed" refers to an assumption that there exists a probability distribution and each recorded number of cycles n_i is a particular realization of that probability distribution. A statistical model is a precise statement of the assumed statistical distribution of N, given σ_{eq}. In other words, a statistical model is a statement of an idea of what is expected to happen, whereas the test results tell us what in fact did happen.

For example, we may assume that $W = \log_{10} N$ is normally distributed with mean $\mu = \mu(\sigma_{eq})$ and standard deviation $s = s(\sigma_{eq})$. Conventionally, upper case letters denote random variables, the corresponding lower case letters denote their realizations. Therefore the probability density function of W can be written as

$$\Pr(w \mid \mu, s) = \frac{1}{\sqrt{2\pi}\, s} \exp\left(-\frac{(w-\mu)^2}{2s^2}\right). \tag{I.5}$$

which has the property

$$\frac{1}{\sqrt{2\pi}\, s} \int_{-\infty}^{\infty} \exp\left(-\frac{(w-\mu)^2}{2s^2}\right) dw = 1. \tag{I.6}$$

Introducing the change of variables $w = \log_{10} n = \ln n / \ln(10)$ we have $dw = dn/(n\ln(10))$ and hence

$$\Pr(\log_{10} n \mid \mu, s) = \frac{1}{\sqrt{2\pi}\, s\, n \ln(10)} \exp\left(-\frac{(\log_{10} n - \mu)^2}{2s^2}\right) \tag{I.7}$$

which also has the essential property of probability density functions that their integral over the domain of definition is unity:

$$\frac{1}{\sqrt{2\pi}\, s\, n\, \ln(10)} \int_{0}^{\infty} \exp\left(-\frac{(\log_{10} n - \mu)^2}{2s^2}\right) dn = 1.$$

The cumulative distribution function of the standard normal distribution is:

$$\Phi\left(\frac{\log_{10} n - \mu}{s}\right) = \frac{1}{2}\left(1 + \mathrm{erf}\left(\frac{\log_{10} n - \mu}{s}\right)\right) \tag{I.8}$$

where erf is the error function.

Various plausible assumptions can be made concerning the functional forms of $\mu(\sigma_{eq})$ and $s(\sigma_{eq})$. We will consider two functional forms for $\mu(\sigma_{eq})$: The bilinear form

$$\mu_1(\sigma_{eq}) = \begin{cases} \log_{10} N_0 - m_1(\sigma_{eq} - S_0) & \text{for } \sigma_{eq} \geq S_0 \\ \log_{10} N_0 + m_2(S_0 - \sigma_{eq}) & \text{for } \sigma_{eq} < S_0 \end{cases} \tag{I.9}$$

and the logarithmic form

$$\mu_2(\sigma_{eq}) = \begin{cases} A_1 - A_2 \log_{10}(\sigma_{eq} - A_3) & \text{for } \sigma_{eq} - A_3 > 0 \\ \infty & \text{for } \sigma_{eq} - A_3 \leq 0 \end{cases} \tag{I.10}$$

where σ_{eq}, S_0 and A_3 are in ksi units[5]. Since the argument of the log function has to be dimensionless, the argument should be understood to have been scaled by 1 ksi. The subscript on μ will be dropped when there is no possibility of confusion.

---

5 kilo-pounds per square inch. 1 ksi = 6.895 MPa.

This functional form implies that there is a material property, called fatigue limit, denoted by A_3 in eq. (I.10). Fatigue failure can occur only for $\sigma_{eq} > A_3$. For $\sigma_{eq} \le A_3$ the fatigue life is infinity. We will refer to this as the fatigue limit model. In reference [73] it was proposed that A_3 should be treated as a random variable. We will consider that possibility as well and will refer to that model as the random fatigue limit (RFL) model.

We assume, for the sake of simplicity, that the standard deviation s is constant. Therefore the bilinear model is characterized by the five parameters $\theta_1 = \{S_0 \ N_0 \ m_1 \ m_2 \ s\}^T$ and the assumption that $\log_{10}N$ is normally distributed.

The fatigue limit model is characterized by the four parameters $\theta_2 = \{A_1 \ A_2 \ A_3 \ s\}^T$ and the random fatigue limit model is characterized by the five parameters $\theta_3 = \{A_1 \ A_2 \ s \ \mu_f \ s_f\}^T$ where μ_f and s_f are the mean and standard deviation of $\log_{10} A_3$ respectively.

Statistical models are calibrated against the available data by maximizing the likelihood function. By definition, the likelihood function is the joint probability density of a random sample. For the bilinear and fatigue limit models being considered here the likelihood function is

$$L = \prod_{i=1}^{n} \left(\frac{\exp(-(\log_{10} n_i - \mu(\sigma_{eq}^{(i)}))^2/2s^2)}{\sqrt{2\pi}sn_i \ln(10)} \right)^{1-\delta_i} \left(1 - \Phi\left(\frac{\log_{10} n_i - \mu(\sigma_{eq}^{(i)})}{s} \right) \right)^{\delta_i} \quad (I.11)$$

where $\delta_i = 0$ if the test specimen failed at n_i cycles, $\delta_i = 1$ if a runout was recorded, that is, the test was stopped at n_i cycles before failure occurred. The runout data carry the information that the specimen survived n_i cycles. Maximizing L with respect to the parameters of the statistical model yields the combination of model parameters that maximize the probability of observations actually obtained. In other words, calibration is based on the idea that, assuming the functional form of a statistical model to be correct, the set of model parameters that would have produced the calibration data with the greatest probability is the best estimate for the model parameters. We will denote the parameters that maximize the likelihood function by $\hat{\theta}_k$.

Since logarithm is a monotonically increasing function, the parameters that maximize L also maximize $LL \equiv \ln L$. Therefore LL, called the log likelihood function, can be used in place of the likelihood function when computing the statistical parameters. Solving for the statistical parameters involves solving a multivariate nonlinear problem which is usually easier to do when the function being maximized is the log-likelihood function rather than the likelihood function. The log likelihood function corresponding to eq. (I.11) is

$$LL = -\sum_{i=1}^{n}(1-\delta_i)\left(\frac{(\log_{10} n_i - \mu(\sigma_{eq}^{(i)}))^2}{2s^2} + \ln\left(\sqrt{2\pi}sn_i \ln(10) \right) \right)$$
$$+ \sum_{i=1}^{n}\delta_i \ln\left(1 - \Phi\left(\frac{\log_{10} n_i - \mu(\sigma_{eq}^{(i)})}{s} \right) \right). \quad (I.12)$$

The bilinear model

The estimated parameters for the bilinear model and the maximum value of LL are shown in Table I.3. The parameter S_0 is in ksi units, the parameters m_1, m_2 are in ksi^{-1} units, the standard deviation s is dimensionless. The function $\mu(\sigma_{eq})$ characterized by the parameters in Table I.3, known as the S-N curve, is shown in Fig. I.3.

Remark I.1 The parameters were computed by means of the Matlab function *mle* (maximum likelihood estimate) using several seed values to ascertain, with a high degree of confidence, that the parameters correspond to the global maximum of the likelihood function.

Table I.3 Estimated parameters $\hat{\theta}_1$ for the bilinear model.

Model	S_0	$\log_{10} N_0$	m_1	m_2	s	$LL(\hat{\theta}_1)$
Bilinear	31.305	5.26352	0.0596	0.1779	0.4583	−601.931

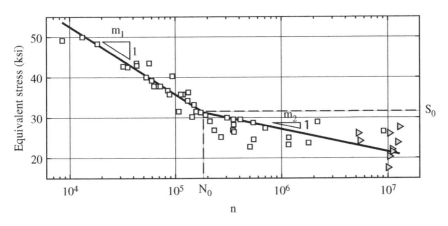

Figure I.3 Bilinear S-N curve for 24S-T3 aluminum alloy.

Table I.4 Estimated parameters $\hat{\theta}_2$ for the fatigue limit model.

Model	A_1	A_2	A_3	s	$LL(\hat{\theta}_2)$
Fatigue limit	9.019	3.198	17.856	0.4662	−602.112

The fatigue limit model

The estimated parameters for the fatigue limit model are shown in Table I.4. The parameter A_3 is in ksi units, the other parameters are dimensionless. The S-N curve is shown in Fig. I.4.

Remark I.2 Fatigue limit and endurance limit are expressions used to describe a property of materials: the amplitude of cyclic stress at full reversal ($R = -1$) that can be applied to the material without causing fatigue failure at any number of cycles. Fatigue strength is the value of σ_{eq} at a fixed number of cycles (usually $n = 5 \times 10^8$ cycles) read from the S-N curve: $\sigma_{eq} = \mu^{-1}(\log_{10} n)$.

Ferrous alloys and titanium alloys appear to have fatigue limits in the megacycle range. However, as noted in [26], for some low carbon steels the difference between the fatigue strength at 10^6 and 10^9 cycles is less than 50 MPa, whereas for other steels this difference ranges from 50 to 200 MPa. Non-ferrous metals, such as aluminium and copper alloys, do not have fatigue limits. Nevertheless, fatigue limit models are being used for non-ferrous metals, see (for example) the MMPDS Handbook [58].

Since fatigue limit is a material property, it is a random number that has to be inferred from a statistical model. The estimated mean value of the fatigue limit or fatigue strength tends to be sensitive to the choice of the statistical model. See Example 6.1.

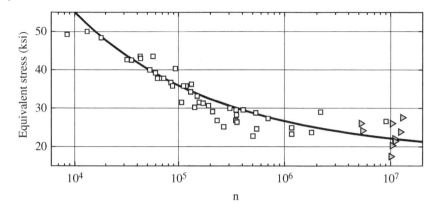

Figure I.4 MMPDS Fatigue limit model, S-N curve for 24S-T3 aluminum alloy.

The random fatigue limit model

We introduce the notation $v = \log_{10} A_3$ and assume that v is normally distributed with mean μ_f and standard deviation s_f. We invoke two fundamental theorems in statistics; the product rule and marginalization. The product rule takes the form:

$$\Pr(w, v \mid \sigma_{eq}) = \Pr(w \mid \sigma_{eq}, v)\Pr(v \mid \sigma_{eq}) \tag{I.13}$$

where $w = \log_{10} n$ as defined previously. The equation of marginalization is

$$\Pr(w \mid \sigma_{eq}) = \int_{-\infty}^{\log_{10}\sigma_{eq}} \Pr(w, v \mid \sigma_{eq})\, dv \tag{I.14}$$

where the upper limit of the integral is determined from the condition that $\sigma_{eq} - A_3 > 0$. Therefore:

$$\Pr(w \mid \sigma_{eq}) = \int_{-\infty}^{\log_{10}\sigma_{eq}} \Pr(w \mid \sigma_{eq}, v)\Pr(v \mid \sigma_{eq})\, dv. \tag{I.15}$$

Since by hypothesis w is normally distributed with mean μ and standard deviation s, the first term of the integrand in eq. (I.15) is the probability density function:

$$\Pr(w \mid \sigma_{eq}, v) = \phi(w, \sigma_{eq}, v) = \frac{\exp(-(w - \mu(\sigma_{eq}, 10^v))^2/(2s^2))}{s\sqrt{2\pi}}. \tag{I.16}$$

The second term of the integrand follows from the assumption that v is normally distributed. Therefore the probability density of v is:

$$\Pr(v \mid \sigma_{eq}) = f(v) = \frac{\exp(-(v - \mu_f)^2/(2s_f^2))}{s_f\sqrt{2\pi}}, \qquad v < \log_{10}(\sigma_{eq}) \tag{I.17}$$

The marginal probability density function of w, given σ_{eq}, is:

$$\Pr(w \mid \sigma_{eq}) = \phi_M(w, \sigma_{eq}) = \int_{-\infty}^{\log_{10}\sigma_{eq}} \phi(w, 10^v)f(v)\, dv \tag{I.18}$$

and the marginal cumulative distribution function (CDF) of w, given σ_{eq}, is:

$$\Phi_M(w, \sigma_{eq}) = \frac{1}{2}\int_{-\infty}^{\log_{10}\sigma_{eq}} \left(1 + \mathrm{erf}\left(\frac{w - \mu(\sigma_{eq}, 10^v)}{s\sqrt{2}}\right)\right)f(v)\, dv. \tag{I.19}$$

Table I.5 Estimated parameters $\hat{\theta}_3$ for the random fatigue limit model.

Model	A_1	A_2	s	μ_f	s_f	$LL(\hat{\theta}_3)$
RFL	7.191	1.991	0.1255	1.3438	0.0488	−576.734

Given a set of independent observations $(w_i, \sigma_{eq}^{(i)})$, $(i = 1, 2, \ldots, n)$ the likelihood function is

$$L(\theta) = \prod_{i=1}^{n} [\phi_M(w_i, \sigma_{eq}^{(i)})]^{1-\delta_i} [1 - \Phi_M(w_i, \sigma_{eq}^{(i)})]^{\delta_i} \tag{I.20}$$

where $\delta_i = 0$ if the test resulted in failure, $\delta_i = 1$ if the test was stopped before failure occurred. The maximum likelihood estimate of θ, denoted by $\hat{\theta}_3$, maximizes $L(\theta)$, or equivalently, the corresponding log likelihood function. The estimated parameters for the random fatigue limit model are shown in Table I.5. The S-N curve is shown in Fig. I.5.

Another way to visualize the relationship between the random fatigue limit model and the S-N data is to display the empirical cumulative distribution function of the S-N data and compare that distribution with the median corresponding to the random fatigue limit model.

Given σ_{eq}, the p-quantile corresponding to the statistical model can be computed from the inverse of the cumulative distribution function. Specifically, we find $w_i = \log_{10} n_i$ such that

$$Q(p) \equiv \Phi_M(w_i, \sigma_{eq}) - p = 0, \quad 0 < p < 1 \tag{I.21}$$

where Φ_M is the marginal cumulative distribution function defined by eq. (I.19). Letting $p = 0.5$ we get the predicted median. The results are shown in Fig. I.6. There were nine runouts recorded which are indicated by open circles.

A runout is predicted when the function $Q(p) = 0$ does not have a real root at $p = 0.5$ or the predicted number of cycles is greater than 100 million. See Remark I.3.

A recorded runout indicates the number of cycles at which a test was stopped and the specimen did not fail. Therefore a predicted runout is not necessarily a prediction of a recorded runout. Nine runouts were recorded for the S-N data, four runouts are predicted by the random fatigue limit model.

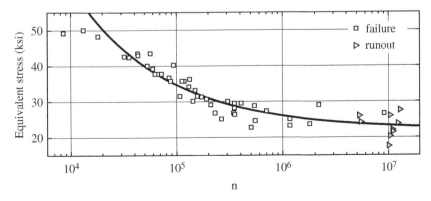

Figure I.5 Random fatigue limit model: S-N data for 24S-T3 aluminum alloy.

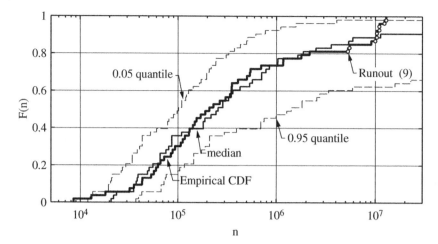

Figure I.6 Empirical CDF of the S-N data compared with the median predicted by the random fatigue limit model.

Remark I.3 The probability that failure will occur at $n < 10^w$ cycles, given σ_{eq}, is $\Phi_M(w, \sigma_{eq})$, see eq. (I.19). Letting $w \to \infty$ we find:

$$\Pr(\text{failure} \mid n < \infty, \sigma_{eq}) = \int_{-\infty}^{\log_{10}\sigma_{eq}} f(v) \, dv \tag{I.22}$$

therefore

$$\Pr(\text{no failure} \mid n < \infty, \sigma_{eq}) = 1 - \int_{-\infty}^{\log_{10}\sigma_{eq}} f(v) \, dv. \tag{I.23}$$

A feature of the random fatigue limit model is that there is a non-zero probability that failure will not occur at any stress level. At high stress levels this probability is negligibly small. When σ_{eq} is equal to the mean fatigue limit then this probability is 50%.

For example, the estimated mean of the random fatigue limit of 24S-T3 aluminum is $10^{\mu_f} = 22.07$ ksi. The marginal cumulative distribution functions are shown in Fig. I.7 for various values of σ_{eq}.

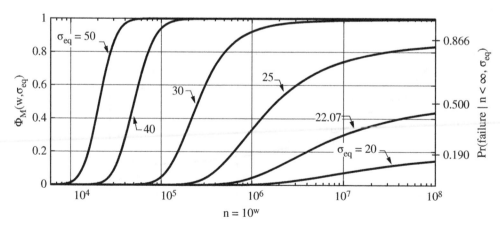

Figure I.7 Marginal cumulative distribution functions of the random fatigue limit model for 24S-T3 aluminum. The stress values are in ksi units.

The limit values for $\sigma_{eq} = 25$, 22.07, 20 ksi with respect to $n \to \infty$ are shown on the right. It is seen that in the neighborhood of $\sigma_{eq} = 10^{\mu_j}$ a small change in σ_{eq} results in a large change in the predicted value of n. Therefore the random fatigue limit model is not a reliable predictor of n when σ_{eq} is close to 10^{μ_j}.

Remark I.4 The integrals in equations (I.18) and (I.19) have to be evaluated numerically.

Remark I.5 The assumption that $f(v)$ has the functional form of eq. (I.17) is one of many possible assumptions. In reference [17] $f(v)$ was assumed to have normal and smallest extreme value distributions.

I.4 Ranking

Of necessity, several assumptions have to be made in the formulation of statistical models. These assumptions are based on prior experience, insight and intuition. It is possible to objectively rank models based on their predictive performance, measured by the likelihood function, with reference to the available data. Since the available data tend to increase over time and new statistical models may be proposed, it is necessary to establish a process for systematic revision and updating models over time. In industrial and research organizations this falls under the administration of simulation governance and management [92]. Ranking, based on Bayes' theorem, is discussed in this section.

The Bayes factor

Ranking statistical models can be based on assessment of the relative merit of model M_i with respect to model M_j as quantified by the Bayes factor BF_{ij} which is defined by

$$BF_{ij} \overset{\text{def}}{=} \frac{\Pr(D \mid M_i)}{\Pr(D \mid M_j)}. \tag{I.24}$$

The term $\Pr(D)$ cancels and, unless we have some reason to assign to the ratio of prior probabilities $\Pr(M_i)/\Pr(M_j)$ a number other than unity, we have

$$\frac{\Pr(M_i \mid D)}{\Pr(M_j \mid D)} = \frac{\Pr(D \mid M_i)}{\Pr(D \mid M_j)}$$

$$= \frac{L(D \mid \theta_i)}{L(D \mid \theta_j)} = \exp(LL(D \mid \theta_i) - LL(D \mid \theta_j)). \tag{I.25}$$

Let us assign the index 1, 2, 3 to the bilinear, fatigue limit and random fatigue limit models respectively. From the log likelihood values in tables I3, I4 and I5 we find:

$$BF_{31} \approx 8.8 \times 10^{10}, \qquad BF_{32} \approx 1.1 \times 10^{11}$$

that is, the random fatigue limit model is very strongly preferred over the other two models.

I.5 Confidence intervals

Confidence intervals can be estimated by means of the profile likelihood function. Let θ_k be a parameter for which we are interested in computing the confidence interval and θ_k the vector of all

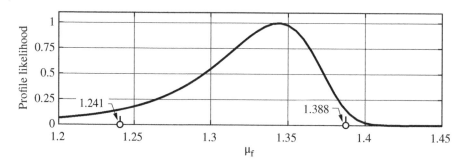

Figure I.8 Profile likelihood and estimated confidence interval for μ_f.

other parameters associated with a statistical model. In other words, θ_k comprises all parameters of the statistical model except θ_k. The profile likelihood function for θ_k is defined by

$$R(\theta_k) = \max_{\theta_k} \frac{L(\theta_k, \theta_k)}{L(\hat{\theta})} \tag{I.26}$$

where $\hat{\theta}$ is the set of parameters that maximizes the likelihood function. Wilks' theorem[6] states that, as the sample size approaches infinity, the statistic $-2\ln R(\theta_k)$ will approach the chi-squared distribution with one degree of freedom provided that the statistical model is correct. Therefore the confidence interval can be estimated from the formula

$$-2\ln(R(\theta_k)) \leq \chi^2_{1;1-\alpha} \tag{I.27}$$

where $\chi^2_{1;1-\alpha}$ is the $100(1 - \alpha)$ percentile of the chi-square distribution with one degree of freedom. For example, referring to the parameters of the random fatigue limit model in Table I.5, the estimated 95% confidence interval for μ_f is (1.241, 1.388). The profile likelihood function and the confidence interval for μ_f are shown in Fig. I.8.

---

6 Samuel Stanley Wilks 1906–1964.

Appendix J

Estimation of fastener forces in structural connections

A method for estimating forces in individual fasteners in a fastener group is described in the following. Discussion of this method, used in traditional structural engineering practice, can be found in textbooks on the mechanics of materials, see for example [79]. It is an example of a highly simplified mathematical model applied to a rather complicated problem.

The simplified model is based on the assumption that a fastened structural component behaves as a perfectly rigid body. The fasteners behave as linearly elastic springs. Therefore the displacement and rotation of the fastened component is determined by the displacements of the linear springs.

Consider the solution domain of a rigid planar body of constant thickness t. The domain $\overline{\Omega}$ is bounded by the contour $\overline{\Gamma}$. Inside $\overline{\Gamma}$ are n circular holes of radius ϱ_k, denoted by $\Omega_o^{(k)}$, $k = 1, 2, \dots, n$. The solution domain and notation are shown in Fig. J.1. Formally,

$$\Omega \overset{\text{def}}{=} \overline{\Omega} - \Omega_o^{(1)} \cup \Omega_o^{(2)} \cup \cdots \cup \Omega_o^{(n)}.$$

The boundary of $\Omega_o^{(k)}$ is denoted by $\Gamma_\varrho^{(k)}$. For simplicity we will use $\varrho_k = \varrho$ in the following.

Tractions T_x, T_y are prescribed on $\overline{\Gamma}$. The forces F_x, F_y and the moment M are defined such that the equations of equilibrium are satisfied:

$$F_x + \int_{\overline{\Gamma}} T_x \, t \, ds = 0 \tag{J.1}$$

$$F_y + \int_{\overline{\Gamma}} T_y \, t \, ds = 0 \tag{J.2}$$

$$M + \int_{\overline{\Gamma}} \left(-T_x y + T_y x\right) \, t \, ds = 0 \tag{J.3}$$

where F_x, F_y and M are the resultants of forces imposed by the fasteners on the rigid planar body.

The goal is to estimate the forces $f_x^{(k)}, f_y^{(k)}$ in a fastened structural connection. We will denote the spring rate of the kth fastener by κ_k and the displacement vector components in the center of the hole by $u_x^{(k)}$ and $u_y^{(k)}$. Therefore $f_x^{(k)} = \kappa_k u_x^{(k)}$ and $f_y^{(k)} = \kappa_k u_y^{(k)}$. The origin of the coordinate system is in the center of rotation which is defined such that the following equations are satisfied:

$$\sum_{k=1}^{n} \kappa_k x_k = 0, \qquad \sum_{k=1}^{n} \kappa_k y_k = 0. \tag{J.4}$$

Denoting the displacement vector components of the center of rotation by $u_x^{(c)}$ and $u_y^{(c)}$ and the rotation by $\theta^{(c)}$, we have:

$$u_x^{(k)} = u_x^{(c)} - \theta^{(c)} r_k \sin \alpha_k = u_x^{(c)} - \theta^{(c)} y_k \tag{J.5}$$

$$u_y^{(k)} = u_y^{(c)} + \theta^{(c)} r_k \cos \alpha_k = u_y^{(c)} + \theta^{(c)} x_k. \tag{J.6}$$

Finite Element Analysis: Method, Verification and Validation, Second Edition. Barna Szabó and Ivo Babuška.
© 2021 John Wiley & Sons, Inc. Published 2021 by John Wiley & Sons, Inc.
Companion Website: www.wiley.com/go/szabo/finite_element_analysis

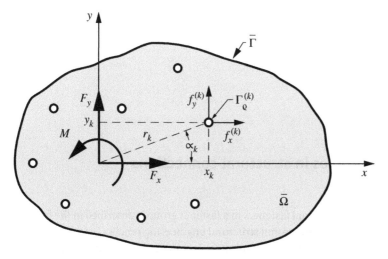

Figure J.1   Notation.

Therefore:

$$f_x^{(k)} = \kappa_k(u_x^{(c)} - \theta^{(c)}y_k) \tag{J.7}$$

$$f_y^{(k)} = \kappa_k(u_y^{(c)} + \theta^{(c)}x_k). \tag{J.8}$$

The resultants can be written in terms of the fastener forces:

$$F_x = \sum_{k=1}^{n} f_x^{(k)} = u_x^{(c)} \sum_{k=1}^{n} \kappa_k \tag{J.9}$$

$$F_y = \sum_{k=1}^{n} f_y^{(k)} = u_y^{(c)} \sum_{k=1}^{n} \kappa_k \tag{J.10}$$

$$M = \sum_{k=1}^{n} (f_y^{(k)} x_k - f_x^{(k)} y_k) = \theta^{(c)} \sum_{k=1}^{n} \kappa_k(x_k^2 + y_k^2) \tag{J.11}$$

where we used eq. (J.4). Writing the displacements of the center of rotation in terms of the stress resultants:

$$u_x^{(c)} = \frac{F_x}{\sum_{k=1}^{n} \kappa_k}, \quad u_y^{(c)} = \frac{F_y}{\sum_{k=1}^{n} \kappa_k}, \quad \theta^{(c)} = \frac{M}{\sum_{k=1}^{n} \kappa_k(x_k^2 + y_k^2)}$$

equations (J.7) and (J.8) can be written for the *j*th fastener as:

$$f_x^{(j)} = \frac{\kappa_j}{\sum_{k=1}^{n} \kappa_k} F_x - \frac{\kappa_j y_j}{\sum_{k=1}^{n} \kappa_k(x_k^2 + y_k^2)} M \tag{J.12}$$

$$f_y^{(j)} = \frac{\kappa_j}{\sum_{k=1}^{n} \kappa_k} F_y + \frac{\kappa_j x_j}{\sum_{k=1}^{n} \kappa_k(x_k^2 + y_k^2)} M. \tag{J.13}$$

In the special case when κ_j has the same value for all j we have:

$$f_x^{(j)} = \frac{F_x}{n} - \frac{y_j}{\sum\limits_{k=1}^{n}(x_k^2 + y_k^2)}M \tag{J.14}$$

$$f_y^{(j)} = \frac{F_y}{n} + \frac{x_j}{\sum\limits_{k=1}^{n}(x_k^2 + y_k^2)}M. \tag{J.15}$$

Remark J.1 Note that no discretization is necessary, therefore the errors in predictions based on this model are due entirely to model form errors.

Appendix K

Useful algorithms in solid mechanics

Some of the frequently used algorithms in finite element analysis applied to problems in solid mechanics are summarized. The basic properties of the stress tensor and traction vector and their transformation are reviewed. The equations of static equilibrium of forces and moments are stated.

K.1 The traction vector

Let us assume that the state of stress at a point is known and let us determine the components of the traction vector T_x, T_y, T_z acting on the inclined face of the infinitesimal tetrahedral element shown in Fig. K.1. By definition, the traction vector, also known as the stress vector, is

$$\mathbf{T} = \lim_{\Delta A \to 0} \frac{\Delta \mathbf{F}}{\Delta A}$$

where $\Delta \mathbf{F}$ is the differential force, acting on the inclined face of the tetrahedron, the area of which is ΔA. This force is in equilibrium with the resultants of the stresses acting on the other three faces of the tetrahedron.

We first show that the unit normal to the inclined plane, denoted by $\mathbf{n}$, has the components

$$n_x = \frac{\Delta A_x}{\Delta A}, \quad n_y = \frac{\Delta A_y}{\Delta A}, \quad n_z = \frac{\Delta A_z}{\Delta A} \tag{K.1}$$

where ΔA_x (resp. ΔA_y, ΔA_z) is the area of that face of the tetrahedron to which the x axis (resp. y, z axis) is normal, see Fig. K.1(a). Consider the cross product:

$$\mathbf{c} = \mathbf{a} \times \mathbf{b} = (\Delta y \mathbf{e}_y - \Delta x \mathbf{e}_x) \times (\Delta z \mathbf{e}_z - \Delta y \mathbf{e}_y)$$
$$= \Delta y \Delta z \mathbf{e}_x + \Delta x \Delta z \mathbf{e}_y + \Delta x \Delta y \mathbf{e}_z = 2\Delta A_x \mathbf{e}_x + 2\Delta A_y \mathbf{e}_y + 2\Delta A_z \mathbf{e}_z$$

where $\mathbf{e}_x$, $\mathbf{e}_y$, $\mathbf{e}_z$ are the unit basis vectors. The vector $\mathbf{c}$ is normal to the inclined plane and its absolute value is $2\Delta A$. Therefore $\mathbf{n} = \mathbf{c}/|\mathbf{c}| = \mathbf{c}/2\Delta A$ and the components of the unit normal are as given by eq. (K.1).

The equations of equilibrium are:

$$\sum \Delta F_x = 0: \quad T_x \Delta A - \sigma_x \Delta A_x - \tau_{yx} \Delta A_y - \tau_{zx} \Delta A_z = 0$$
$$\sum \Delta F_y = 0: \quad T_y \Delta A - \tau_{xy} \Delta A_x - \sigma_y \Delta A_y - \tau_{zy} \Delta A_z = 0$$
$$\sum \Delta F_z = 0: \quad T_z \Delta A - \tau_{xz} \Delta A_x - \tau_{yz} \Delta A_y - \sigma_z \Delta A_z = 0.$$

Finite Element Analysis: Method, Verification and Validation, Second Edition. Barna Szabó and Ivo Babuška.
© 2021 John Wiley & Sons, Inc. Published 2021 by John Wiley & Sons, Inc.
Companion Website: www.wiley.com/go/szabo/finite_element_analysis

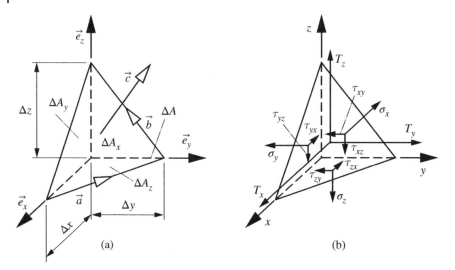

Figure K.1 Notation.

On dividing by ΔA, and making use of the fact that the stress tensor is symmetric, we have:

$$\begin{Bmatrix} T_x \\ T_y \\ T_z \end{Bmatrix} = \begin{bmatrix} \sigma_x & \tau_{xy} & \tau_{xz} \\ \tau_{yx} & \sigma_y & \tau_{yz} \\ \tau_{zx} & \tau_{zy} & \sigma_z \end{bmatrix} \begin{Bmatrix} n_x \\ n_y \\ n_z \end{Bmatrix}. \tag{K.2}$$

In index notation this can be written as:

$$T_i = \sigma_{ij} n_j. \tag{K.3}$$

K.2 Transformation of vectors

Consider a Cartesian coordinate system x_i' rotated relative to the x_i system as shown in Fig. K.2. Let α_{ij} be the angle between the axis x_i' and the axis x_j and let

$$g_{ij} = \cos \alpha_{ij}. \tag{K.4}$$

In other words, the ith row of g_{ij} is the unit vector in the direction of axis x_i' in the unprimed system. Therefore if a_i is an arbitrary vector in the unprimed system then the same vector in the primed system is:

$$a_i' = g_{ij} a_j. \tag{K.5}$$

Conversely, the jth column of g_{ij} is the unit vector in the direction of axis x_j in the primed system. Therefore if a_r' is an arbitrary vector in the primed system then the same vector in the unprimed system is:

$$a_r = g_{kr} a_k'. \tag{K.6}$$

Given the definition of g_{ij} and the orthogonality of the coordinate systems, g_{ij} multiplied by its transpose must be the identity matrix:

$$g_{ri} g_{rj} = g_{is} g_{js} = \delta_{ij}. \tag{K.7}$$

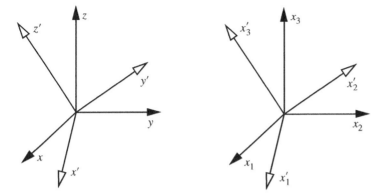

Figure K.2 Coordinate transformation. Notation.

In other words g_{ij} is an orthogonal matrix. This can be proven formally as follows:

$$a_i = g_{ri}a'_r$$
$$a'_s = g_{sj}a_j$$

and

$$a'_r = \delta_{rs}a'_s = \delta_{rs}g_{sj}a_j.$$

Therefore,

$$a_i \equiv \delta_{ij}a_j = g_{ri}\underbrace{\delta_{rs}g_{sj}}\, a_j$$
$$\underbrace{\phantom{g_{ri}\delta_{rs}g_{sj}}}_{g_{ri}g_{rj}}$$

and, for an arbitrary vector a_j we have:

$$(\delta_{ij} - g_{ri}g_{rj})a_j = 0.$$

Consequently the bracketed term must be zero. This completes the proof.

K.3 Transformation of stresses

Referring to the definition of traction vectors given in Section K.1, we have:

$$T'_i = \sigma'_{ik}n'_k = \sigma'_{ik}g_{ks}n_s$$

and applying the transformation rule (K.5):

$$T'_i = g_{ir}T_r = g_{ir}\sigma_{rs}n_s.$$

Therefore we have:

$$(g_{ks}\sigma'_{ik} - g_{ir}\sigma_{rs})n_s = 0.$$

Since this equation holds for arbitrary n_s, the bracketed term must vanish. Consequently:

$$g_{ks}\sigma'_{ik} = g_{ir}\sigma_{rs}$$

multiplying by g_{js},

$$\underbrace{g_{js}g_{ks}}_{\delta_{jk}}\, \sigma'_{ik} = g_{ir}g_{js}\sigma_{rs}$$

and using eq. (K.7), we have the transformation rule for stresses:

$$\sigma'_{ij} = g_{ir}g_{js}\sigma_{rs}. \tag{K.8}$$

Remark K.1 Denoting the stress tensor by $[\sigma]$ and g_{ij} by $[g]$, eq. (K.8) is the symmetric matrix triple product

$$[\sigma'] = [g][\sigma][g]^T. \tag{K.9}$$

K.4 Principal stresses

It is possible to find a plane such that the traction vector, acting on that plane, is normal to the plane (i.e., the shearing components are zero):

$$\sigma_{ij}n_j = Tn_i \equiv T\delta_{ij}n_j$$

where T is the magnitude of the traction vector. Therefore:

$$(\sigma_{ij} - T\delta_{ij})n_j = 0. \tag{K.10}$$

This is a eigenvalue problem. Since σ_{ij} is symmetric, all eigenvalues are real. The eigenvalues are called principal stresses and the eigenvectors define the directions of the principal stresses. Since the eigenvectors are mutually orthogonal, in every point there is an orthogonal coordinate system in which the stress state is characterized by normal stresses only. This coordinate system is uniquely defined only when all eigenvalues are simple.

It follows from eq. (K.10) that the principal stresses are the roots of the following characteristic equation:

$$T^3 - I_1 T^2 - I_2 T - I_3 = 0 \tag{K.11}$$

where

$$I_1 = \sigma_{kk}, \quad I_2 = \frac{1}{2}(\sigma_{ij}\sigma_{ij} - \sigma_{ii}\sigma_{jj}), \quad I_3 = \det(\sigma_{ij}). \tag{K.12}$$

The principal stresses are denoted by a single index: $\sigma_1, \sigma_2, \sigma_3$ and ordered such that $\sigma_1 \geq \sigma_2 \geq \sigma_3$. The principal stresses characterize the state of stress in a point.

The principal stresses do not depend on the coordinate system in which the stress components are given. Therefore the coefficients I_1, I_2 and I_3 are invariant with respect to rotation of the coordinate system. These coefficients are called the first, second and third stress invariant respectively.

K.5 The von Mises stress

The stress deviator tensor is defined by

$$\hat{\sigma}_{ij} = \sigma_{ij} - \frac{1}{3}\delta_{ij}\sigma_{kk}. \tag{K.13}$$

It has the property that its first invariant is zero. The second invariant of the stress deviator tensor, denoted by J_2, is particularly important in materials science and engineering because, according to a widely used phenomenological model proposed by von Mises[1], ductile materials begin to yield when J_2 reaches a critical value. This is known as the von Mises yield criterion or the distortion energy criterion. Since $\hat{\sigma}_{kk} = 0$, the second invariant is

$$J_2 = \frac{1}{2}\hat{\sigma}_{ij}\hat{\sigma}_{ij}. \tag{K.14}$$

Writing J_2 in terms of the principal stresses we have:

$$\begin{aligned} J_2 &= \frac{1}{2}\left((\sigma_1 - \frac{1}{3}\sigma_{kk})^2 + (\sigma_2 - \frac{1}{3}\sigma_{kk})^2 + (\sigma_3 - \frac{1}{3}\sigma_{kk})^2\right) \\ &= \frac{1}{6}\left((\sigma_1 - \sigma_2)^2 + (\sigma_2 - \sigma_3)^2 + (\sigma_3 - \sigma_1)^2\right). \end{aligned} \tag{K.15}$$

According to the von Mises yield criterion at the onset of yielding in ductile materials J_2 has a critical value which is a material property found though uniaxial tensile tests of specimens, such as the specimen shown in Fig. 6.1. At the onset of yielding $\sigma_{11} = \sigma_{\text{yld}}, \sigma_{22} = \sigma_{33} = 0$ in the test section. Therefore the critical value of J_2 is

$$J_2^{\text{crit}} = \frac{1}{3}\sigma_{\text{yld}}^2.$$

The von Mises stress, denoted by $\overline{\sigma}$, is defined by

$$\overline{\sigma} = \sqrt{\frac{(\sigma_1 - \sigma_2)^2 + (\sigma_2 - \sigma_3)^2 + (\sigma_3 - \sigma_1)^2}{2}}. \tag{K.16}$$

At the onset of yielding $\overline{\sigma} = \sigma_{\text{yld}}$. The von Mises stress is one possible generalization of the uniaxial yield stress to arbitrary stress conditions.

Example K.1 Let us predict the onset of yielding in pure shear, that is, $\sigma_{12} = \sigma_{21} = \tau$, all other components of the stress tensor are zero. We will use eq. (K.16) to find the value of τ at which the material begins to yield. The principal stresses are: $\sigma_1 = \tau, \sigma_2 = 0, \sigma_3 = -\tau$. Therefore

$$\overline{\sigma} = \sqrt{3}\tau_{yld} = \sigma_{\text{yld}}$$

hence $\tau_{yld} = \sigma_{\text{yld}}/\sqrt{3}$.

K.6 Statically equivalent forces and moments

Suppose that a force F_i and moment M_i act on a rigid body. The line of action of F_i passes through point P. Force F_i and moment M_i are statically equivalent to force F_i' whose line of action passes through point P' and moment M_i' if the following condition is satisfied:

$$F_i' = F_i \tag{K.17}$$

$$M_i' = M_i - e_{ijk}r_j F_k \tag{K.18}$$

where r_j is the position vector of point P' with reference to point P as indicated in Fig. K.3. The moment vector components are indicated by double arrows.

---

1 Richard Edler von Mises 1883–1953.

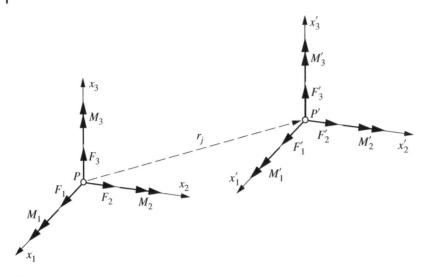

Figure K.3 Statically equivalent forces and moments. Notation.

Statically equivalent force-moment pairs are used in assigning traction boundary conditions such that the conditions of equilibrium are satisfied.

Example K.2 Consider a segment of a coil spring shown in Fig. 5.13. The centerline is given by

$$x_1 = x = r_c \cos\theta, \quad x_2 = y = r_c \sin\theta, \quad x_3 = z = \frac{d}{2\pi}\theta, \quad 0 < \theta < \pi/3. \tag{K.19}$$

The spring is compressed by two equal and opposite forces F acting along the z-axis. We are interested in finding the components of the statically equivalent forces on cross-sections A and B in the local coordinate systems indicated in Fig. K.4 where the coordinate axes x^A, y^A, z^A and x^B, y^B, z^B are coincident respectively with the tangent, normal and binormal of the centerline at cross-sections A and B.

The solution involves two steps: First, the statically equivalent forces and moments acting on cross-sections A and B are determined in local coordinate systems the axes of which are parallel to the x, y, z system. For this we use equations (K.17), (K.18). Second, the the force components are transformed to the local coordinate systems using eq. (K.5).

Step 1: The forces and moments acting on cross-sections A and B, in coordinate systems the origins of which are in the points of intersection of the centerline with the cross-sections and the axes are parallel with the global coordinate axes, in conventional notation, are:

$$\mathbf{F}^A = F\{0\ 0\ 1\}^T$$
$$\mathbf{M}^A = -\{r_c\ 0\ 0\}^T \times \{0\ 0\ F\}^T = r_c F\{0\ 1\ 0\}^T$$
$$\mathbf{F}^B = F\{0\ 0\ -1\}^T$$
$$\mathbf{M}^B = -\{r_c \cos(\pi/3)\ \ r_c \sin(\pi/3)\ \ d/6\}^T \times \{0\ 0\ -F\}^T$$
$$\quad = r_c F\{\sin(\pi/3)\ \ -\cos(\pi/3)\ \ 0\}^T.$$

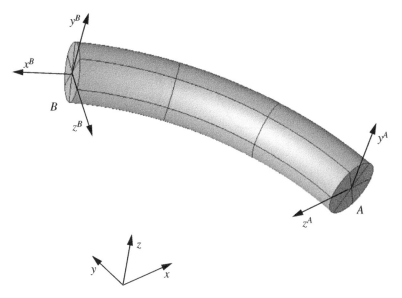

Figure K.4 Example K.2: Segment of a coil spring.

Step 2: To transform the forces and moments to the local coordinate systems we need the transformation matrices g_{ij}^A and g_{ij}^B, defined by eq. (K.4). We denote the tangent, normal and binormal unit vectors by t_i, n_i and b_i respectively. By definition, b_i is the cross product of t_i and n_i, that is, $b_i = e_{ijk}t_jn_k$. Using eq. (K.19) and defining $\beta = d/(2\pi r_c)$, these unit vectors are:

$$t_i = \frac{1}{\sqrt{1+\beta^2}}\{-\sin\theta \quad \cos\theta \quad \beta\} \tag{K.20}$$

$$n_i = \{-\cos\theta \quad -\sin\theta \quad 0\} \tag{K.21}$$

$$b_i = \frac{1}{\sqrt{1+\beta^2}}\{\beta\sin\theta \quad -\beta\cos\theta \quad 1\}. \tag{K.22}$$

Therefore the transformation matrix g_{ij} is

$$g_{ij} = \frac{1}{\sqrt{1+\beta^2}}\begin{bmatrix} -\sin\theta & \cos\theta & \beta \\ -\sqrt{1+\beta^2}\cos\theta & -\sqrt{1+\beta^2}\sin\theta & 0 \\ \beta\sin\theta & -\beta\cos\theta & 1 \end{bmatrix}. \tag{K.23}$$

In this example cross-section A is located at $\theta = 0$. Therefore the local force and moment vectors denoted by $\mathbf{F}_L^A$ and $\mathbf{M}_L^A$ are:

$$\mathbf{F}_L^A = \frac{1}{\sqrt{1+\beta^2}}\begin{bmatrix} 0 & 1 & \beta \\ -\sqrt{1+\beta^2} & 0 & 0 \\ 0 & -\beta & 1 \end{bmatrix}\begin{Bmatrix} 0 \\ 0 \\ F \end{Bmatrix} = \frac{F}{\sqrt{1+\beta^2}}\begin{Bmatrix} \beta \\ 0 \\ 1 \end{Bmatrix}$$

$$\mathbf{M}_L^A = \frac{1}{\sqrt{1+\beta^2}}\begin{bmatrix} 0 & 1 & \beta \\ -\sqrt{1+\beta^2} & 0 & 0 \\ 0 & -\beta & 1 \end{bmatrix}\begin{Bmatrix} 0 \\ r_cF \\ 0 \end{Bmatrix} = \frac{r_cF}{\sqrt{1+\beta^2}}\begin{Bmatrix} 1 \\ 0 \\ -\beta \end{Bmatrix}.$$

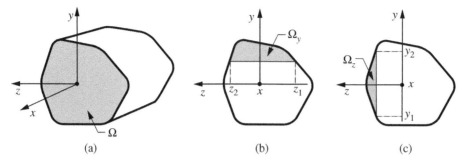

Figure K.5 Notation.

Similarly, for cross-section B located at $\theta = \pi/3$, we have:

$$
\mathbf{F}_L^B = \frac{1}{\sqrt{1+\beta^2}}
\begin{bmatrix}
-\sqrt{3}/2 & 1/2 & \beta \\
-\sqrt{1+\beta^2}/2 & -\sqrt{3(1+\beta^2)}/2 & 0 \\
\sqrt{3}\beta/2 & -\beta/2 & 1
\end{bmatrix}
\begin{Bmatrix}
0 \\ 0 \\ -F
\end{Bmatrix}
= \frac{-F}{\sqrt{1+\beta^2}}
\begin{Bmatrix}
\beta \\ 0 \\ 1
\end{Bmatrix}
$$

$$
\mathbf{M}_L^B = \frac{1}{\sqrt{1+\beta^2}}
\begin{bmatrix}
-\sqrt{3}/2 & 1/2 & \beta \\
-\sqrt{1+\beta^2}/2 & -\sqrt{3(1+\beta^2)}/2 & 0 \\
\sqrt{3}\beta/2 & -\beta/2 & 1
\end{bmatrix}
\begin{Bmatrix}
r_c F\sqrt{3}/2 \\ -r_c F/2 \\ 0
\end{Bmatrix}
$$

$$
= \frac{-r_c F}{\sqrt{1+\beta^2}}
\begin{Bmatrix}
1 \\ 0 \\ -\beta
\end{Bmatrix}.
$$

Remark K.2 Note that in the foregoing example $\mathbf{F}_L^B = -\mathbf{F}_L^A$ and $\mathbf{M}_L^B = -\mathbf{M}_L^A$. This is because in this special case the transformation of the force and moment vectors to the local system is independent of θ.

K.6.1 Technical formulas for stress

The term "rod" will be used as a generic name for any slender body such as arch, bar, beam, column, spring and shaft in the following. In the formulation of mathematical models for rods it is often necessary to apply traction boundary conditions that have known resultants. The question is: Given the stress resultants, what surface tractions should be prescribed? The answer to this question is not unique; however, in the technical theory of rods there are commonly used relationships between stress distributions corresponding to moments and forces acting on a cross-section. These distributions are based on assumptions concerning the mode of deformation of straight rods supplemented by equilibrium considerations. A brief summary is presented in the following. For detailed discussion we refer to introductory texts on the strength of materials.

The origin of the local coordinate system is in the centroid of the cross-section of a rod, as indicated in Fig. K.5. The locus of the centroids is the axis of the rod. In the large majority of problems of practical interest the axis of the rod is a continuous and differentiable function. We assume this to be the case in the following. For curved rods the tangent, normal and binormal unit vectors associated with the axis of the rod define the local coordinate system. The notation is shown in Fig. K.5.

Normal traction

The stress σ_x is assumed to be a linear function over a cross-section. Therefore it can be written as

$$\sigma_x = a_0 + a_1 y + a_2 z. \tag{K.24}$$

By definition, the axial force F_x and the moments acting about the y and z axes, denoted respectively by M_y and M_z are:

$$F_x = \int_\Omega \sigma_x \, dydz, \quad M_y = \int_\Omega \sigma_x z \, dydz, \quad M_z = -\int_\Omega \sigma_x y \, dydz. \tag{K.25}$$

On substituting eq. (K.24) into eq. (K.25) we get the following three equations:

$$\begin{bmatrix} A & 0 & 0 \\ 0 & I_z & I_{yz} \\ 0 & I_{yz} & I_y \end{bmatrix} \begin{Bmatrix} a_0 \\ a_1 \\ a_2 \end{Bmatrix} = \begin{Bmatrix} F_x \\ -M_z \\ M_y \end{Bmatrix} \tag{K.26}$$

where A is the area of the cross-section and

$$I_y = \int_\Omega z^2 \, dydz, \quad I_z = \int_\Omega y^2 \, dydz, \quad I_{yz} = \int_\Omega yz \, dydz \tag{K.27}$$

are the moments of inertia of the cross-section. The zero entries in the coefficient matrix occur because the y and z axes are centroidal axes and hence the first moments of the area are zero. On solving eq. (K.26) we get the normal traction denoted by T_x:

$$T_x = \frac{F_x}{A} - \frac{M_z I_y + I_{yz} M_y}{I_y I_z - I_{yz}^2} y + \frac{M_y I_z + I_{yz} M_z}{I_y I_z - I_{yz}^2} z. \tag{K.28}$$

Remark K.3 For straight rods the y and z axes are usually chosen such that $I_{yz} = 0$. When a cross-section has one or more axis of symmetry then $I_{yz} = 0$.

Shearing tractions

The formulas for shearing tractions corresponding to the shearing forces F_y and F_z are developed from considerations of equilibrium of straight rods. Details of the formulation can be found in any introductory text on the strength of materials. The formulas are:

$$T_y^{(F)} = \frac{F_y Q_z(y)}{I_z t_z}, \qquad T_z^{(F)} = \frac{F_z Q_y(z)}{I_y t_y} \tag{K.29}$$

where $Q_z(y)$ and $Q_y(z)$ are the first moments of the shaded areas identified are Ω_y and Ω_z in figures K.5(b) and K.5(c) about the z and y axis respectively, $t_z = z_2 - z_1$ and $t_y = y_2 - y_1$.

Example K.3 For a circular cross-section of radius r_w

$$Q_z = 2 \int_y^{r_w} y \sqrt{r_w^2 - y^2} \, dy = \frac{2}{3}(r_w^2 - y^2)^{3/2}$$

and $t_z = 2\sqrt{r_w^2 - y^2}$, $I_z = r_w^4/4$. Therefore

$$T_y^{(F)} = F_y \frac{4}{3} \frac{r_w^2 - y^2}{r_w^4 \pi} \quad \text{and similarly} \quad T_z^{(F)} = F_z \frac{4}{3} \frac{r_w^2 - z^2}{r_w^4 \pi}. \tag{K.30}$$

Finding the shearing tractions corresponding to the twisting moment M_x for general cross-sections involves the solution of Poisson's equation, usually by numerical means. However, in the important special case of straight rods with circular cross-section a closed form solution exists, the formulation of which is also available in introductory texts. In polar coordinates $(r\ \theta)$, the origin being coincident with the center of the circular cross-section, $\tau_r = 0$ and

$$\tau_\theta = \frac{M_x r}{J} \tag{K.31}$$

where τ_θ is normal to the radius, $J = r_w^4 \pi / 2$ is the polar moment of inertia and r_c is the radius of the circular cross-section. The shearing tractions corresponding to M_x are

$$T_y^{(M)} = -\frac{2 M_x z}{r_w^4 \pi}, \qquad T_z^{(M)} = \frac{2 M_x y}{r_w^4 \pi}. \tag{K.32}$$

Remark K.4 The formulas (K.28), (K.30) and (K.32) are based on assumptions that are exactly satisfied for straight rods of constant cross-section under special loading conditions only. These formulas are useful in the general case because the corresponding tractions are in equilibrium with the applied forces and moments and therefore the resultants of the difference between the exact tractions and the tractions given by these formulas is zero. Therefore the error in the imposed tractions causes local perturbations in the solution which decay in accordance with Saint-Venants's principle[2].

Example K.4 For the problem described in Example K.2 the tractions acting on section A in the local coordinate system indicated in Fig. K.4 are:

$$T_x^{(A)} = \frac{F_x^A}{A} - \frac{M_z^A y}{I} + \frac{M_y^A z}{I}$$

$$T_y^{(A)} = \frac{F_y^A Q_z}{I t_z} - \frac{M_x^A z}{J}$$

$$T_z^{(A)} = \frac{F_z^A Q_y}{I t_y} + \frac{M_x^A y}{J}$$

where

$$F_x^A = \frac{F\beta}{\sqrt{1 + \beta^2}}, \quad F_y^A = 0, \quad F_z^A = \frac{F}{\sqrt{1 + \beta^2}}$$

$$M_x^A = \frac{r_c F}{\sqrt{1 + \beta^2}}, \quad M_y^A = 0, \quad M_z^A = -\frac{r_c F \beta}{\sqrt{1 + \beta^2}}$$

$$A = r_w^2 \pi, \quad I = \frac{r_w^4 \pi}{4}, \quad J = \frac{r_w^4 \pi}{2}, \quad \frac{Q_z}{t_z} = \frac{r_w^2 - y^2}{3}, \quad \frac{Q_y}{t_y} = \frac{r_w^2 - z^2}{3}.$$

In view of Remark K.2, the tractions acting on Section B in the local coordinate system are:

$$T_x^{(B)} = -T_x^{(A)}, \quad T_y^{(B)} = -T_y^{(A)}, \quad T_z^{(B)} = -T_z^{(A)}.$$

---

2 Adhémar Jean Claude Barré de Saint-Venant 1797–1886.

Bibliography

1 Actis RL, Szabó BA and Schwab C. Hierarchic models for laminated plates and shells. *Comput. Methods Appl. Mech. Engng.* **172**(1), 79–107, 1999.

2 Ainsworth M. Discrete dispersion relation for hp-version finite element approximation at high wave number. *SIAM Journal on Numerical Analysis.* 42, 553–575, 2004.

3 Andersson B, Falk U and Jarlås R. Self-adaptive FE-analysis of solid structures, Part 1: Element formulation and a posteriori error estimation. The Aeronautical Research Institute of Sweden. FFA TN 1986-27. 1986.

4 Antman SS. The theory of rods. In: *Handbuch der Physik*, Volume 6/2, pp. 641–703, Springer, 1972. Reprinted in *Mechanics of Solids*, Vol. II, Springer-Verlag, 1984, pp. 641–703.

5 Arnold D. Stability, consistency, and convergence of numerical discretizations. In: Björn Engquist, editor, *Encyclopedia of Applied and Computational Mathematics.*, Springer, 2015, pp. 1358–1364.

6 Arnold DN, Babuška I and Osborn J. Finite element methods: Principles for their selection. *Comput. Methods Appl. Mech. Engng.* 45(1–3), 57–96, 1984.

7 ASTM Standard E466. Standard Practice for Conducting Force Controlled Constant Amplitude Axial Fatigue Tests of Metallic Materials. ASTM International, West Conshohocken, PA, 2015.

8 ASTM E606/E606M-12. Standard Test Method for Strain-Controlled Fatigue Testing. ASTM International, West Conshohocken, PA, 2012.

9 Babuška I. Error bounds for finite element method. *Numerische Mathematik.* 16(4), 322–333, 1971.

10 Babuška I, Andersson B, Guo B, Melenk JM and Oh HS. Finite element method for solving problems with singular solutions. *Journal of Computational and Applied Mathematics.* 74, 51–70, 1996.

11 Babuška I, Andersson B, Smith PJ and Levin K. Damage analysis of fiber composites Part I: Statistical analysis on fiber scale. *Comput. Methods Appl. Mech. Engng.* 172(1), 27–77, 1999.

12 Babuška I and Aziz AK. Lectures on mathematical foundations of the finite element method (No. ORO–3443-42; BN–748). Institute for Fluid Dynamics and Applied Mathematics, University of Maryland, College park, MD 1972.

13 Babuška I and Dorr MR. Error estimates for the combined h and p versions of the finite element method. *Numerische Mathematik.* 37(2), 257–277, 1981.

14 Babuška I and Gui W. The h, p and hp versions of the finite element method in 1 dimension. Part I. The error analysis of the p-version. *Numerische Mathematik.* 49, 577–612, 1986.

15 Babuška I, d'Harcourt JM and Schwab C. Optimal shear correction factors in hierarchical plate modelling. *Math. Modelling and Sci. Computing.* 1, 1–30, 1993.

Finite Element Analysis: Method, Verification and Validation, Second Edition. Barna Szabó and Ivo Babuška.
© 2021 John Wiley & Sons, Inc. Published 2021 by John Wiley & Sons, Inc.
Companion Website: www.wiley.com/go/szabo/finite_element_analysis

16 Babuška I, Nobile F and Tempone R. A systematic approach to model validation based on Bayesian updates and prediction related rejection criteria. *Comput. Methods Appl. Mech. Engng.* 197, 2517–2539, 2008.

17 Babuška I, Sawlan Z, Scavino M, Szabó B and Tempone R. Bayesian inference and model comparison for metallic fatigue data. *Comput. Methods Appl. Mech. Engng.* 304, 171–196, 2016.

18 Babuška I and Silva RS. Numerical treatment of engineering problems with uncertainties. The fuzzy set approach and its application to the heat exchanger problem. *Int. J. Numer. Meth. Engng.* 87(1–5), 115–148, 2011.

19 Babuška I and Silva RS. Dealing with uncertainties in engineering problems using only available data. *Comput. Methods Appl. Mech. Engng.* 270, 57–75, 2014.

20 Babuška I, Soane AM and Suri M. The computational modeling of problems on domains with small holes. *Comput. Methods Appl. Mech. Engng.* 322, 563–589, 2017.

21 Babuška I and Strouboulis T. *The Finite Element Method and its Reliability.* Oxford University Press 2001.

22 Babuška I and Suri M. The *hp* version of the finite element method with quasiuniform meshes. *ESAIM: Mathematical Modelling and Numerical Analysis-Modélisation Mathématique et Analyse Numérique.* 21(2), 199–238, 1987.

23 Babuška I, Szabó B and Actis R. Hierarchic models for laminated composites. *International Journal for Numerical Methods in Engineering.* 33(3), 503–535, 1992.

24 Babuška I, Szabó B and Katz IN. The p-version of the finite element method. *SIAM Journal on Numerical Analysis.* 18(3), 515–545, 1981.

25 Babuška I and Szabó B. On the rates of convergence of the finite element method. *International Journal for Numerical Methods in Engineering.* 18(3), 323–341, 1982.

26 Bathias C and Paris PC. *Gigacycle Fatigue in Mechanical Practice.* Marcel Dekker, New York 2005.

27 Belytschko T and Tsay CS. A stabilization procedure for the quadrilateral plate element with one-point quadrature. *International Journal for Numerical Methods in Engineering.* 19(3), 405–419, 1983.

28 Brenner S and Scott R. *The mathematical theory of finite element methods.* Vol. 15, Springer Science & Business Media, 2007.

29 Brezzi F. On the existence, uniqueness and approximation of saddle-point problems arising from Lagrangian multipliers. *Publications mathématiques et informatique de Rennes,* S4, 1–26, 1974.

30 Chapelle D and Bathe KJ. *The Finite Element Analysis of Shells – Fundamentals,* Springer Science & Business Media, 2010.

31 Chen Q and Babuška I. Approximate optimal points for polynomial interpolation of a real function in an interval and in a triangle. *Comput. Methods Appl. Mech. Engng.* 128, 405–417, 1995.

32 Cottrell JA, Hughes TJR and Reali A. Studies of refinement and continuity in isogeometric structural analysis. *Comput. Methods Appl. Mech. Engng.* 196(41–44), 4160–4183, 2007.

33 Dauge M and Suri M. On the asymptotic behaviour of the discrete spectrum in buckling problems for thin plates. *Mathematical Methods in the Applied Sciences.* 29(7), 789–817, 2006.

34 Di Nezza E, Palatucci G and Valdinoci E. Hitchhiker's guide to the fractional Sobolev spaces. *Bulletin des Sciences Mathématiques.* 136(5), 527–573, 2012.

35 Düster A. *High order finite elements for three-dimensional thin-walled nonlinear continua.* Dissertation. Technische Universität München. Shaker Verlag, Aachen, 2002.

36 Düster A and Rank E. The p-version of the finite element method compared to an adaptive h-version for the deformation theory of plasticity. *Comput. Methods Appl. Mech. Engng.* 190, 1925–1935, 2001

37 Fuchs HO and Stephens RI. *Metal Fatigue in Engineering.* John Wiley & Sons, New York, 1980.

38 Gates N and Fatemi A. Notch deformation and stress gradient effects in multiaxial fatigue. *Theoretical and Applied Fracture Mechanics.* 84, 3–25, 2016.

39 Girkmann K. *Flächentragwerke.* 4th edition. Springer Verlag,Wien, 1956.

40 Grover HJ, Bishop SM and Jackson LR. Fatigue Strengths of Aircraft Materials. Axial-Load Fatigue Tests on Unnotched Sheet Specimens of 24S-T3 and 75S-T6 Aluminum Alloys and of SAE 4130 Steel. NACA Technical Note 2324, March 1951.

41 Grover HJ, Bishop SM and Jackson LR. Fatigue Strengths of Aircraft Materials. Axial-Load Fatigue Tests on Notched Sheet Specimens of 24S-T3 and 75S-T6 Aluminum Alloys and of SAE 4130 Steel with Stress Concentrations Factors of 2.0 and 4.0. NACA Technical Note 2389, June 1951.

42 Grover HJ, Bishop SM and Jackson LR. Fatigue Strengths of Aircraft Materials. Axial-Load Fatigue Tests on Notched Sheet Specimens of 24S-T3 and 75S-T6 Aluminum Alloys and of SAE 4130 Steel with Stress Concentrations Factors of 5.0. NACA Technical Note 2390, June 1951.

43 Grover HJ, Hyler WS and Jackson LR. Fatigue Strengths of Aircraft Materials. Axial-Load Fatigue Tests on Notched Sheet Specimens of 24S-T3 and 75S-T6 Aluminum Alloys and of SAE 4130 Steel with Stress Concentrations Factor of 1.5. NACA Technical Note 2639, February 1952.

44 Grover HJ, Hyler WS and Jackson LR. Fatigue Strengths of Aircraft Materials. Axial-Load Fatigue Tests on Notched Sheet Specimens of 2024-T3 and 7075-T6 Aluminum Alloys and of SAE 4130 Steel with Notched Radii of 0.004 and 0.070 inch. NASA Technical Note D-111, September 1959.

45 Gui WZ and Babuška I. The h, p and $h - p$ versions of the finite element method in 1 dimension. Part II. *The error analysis of the h- and hp-versions.* Numerische Mathematik. 49(6), 613–657, 1986.

46 Guo B and Babuška I. Direct and inverse approximation theorems for the p-version of the finite element method in the framework of weighted Besov spaces. Part III: Inverse approximation theorems. *Journal of Approximation Theory.* 173, 122–157, 2013.

47 Hashin Z and Rosen BW. The elastic moduli of fiber reinforced materials. *Journal of Applied Mechanics.* 31(2), 223–232, 1964.

48 Hinton MJ, Kaddour AS and Soden PD. *Failure Criteria in Fibre Reinforced Polymer Composites: The World-Wide Failure Exercise (WWFE).* Elsevier, 2004. ISBN: 0-08-044475-X

49 Hoff NJ and Soong T-C. Buckling of circular cylindrical shells in axial compression. *Int. J. Mech. Sci.* 7, 289–520, 1965.

50 Jacobus K, DeVor RE and Kapoor S.G. Machining-induced residual stress: experimentation and modeling. *J. Manuf. Sci. Eng.* 122(1), 20–31, 2000.

51 Juntunen M and Stenberg R. Nitsche's method for general boundary conditions. *Mathematics of Computation.* 78(267), 1353–1374, 2009.

52 Konishi S and Kitagawa G. *Information Criteria and Statistical Modeling.* Springer Science & Business Media, 2008.

53 Kozlov VA, Mazia VG and Rossmann J. *Spectral Problems Associated with Corner Singularities of Solutions of Elliptic Equations.* American Mathematical Society, Providence 2000.

54 Kuhn P and Hardrath HF. An Engineering Method for Estimating Notch-Size Effect in Fatigues Tests on Steel. NACA Technical Note 2805, October 1952.

55 Leoni G. *A First Course in Sobolev Spaces.* American Mathematical Society, 2017.

56 MacNeal RH. *The MacNeal-Schwendler Corporation. The First Twenty Years,* 1988.

57 McCombs WF, McQueen JC and Perry JL. Analytical design methods for aircraft structural joints. Report AFFDL-TR-67-184, Air Force Flight Dynamics Laboratory, Wright-Patterson Air Force Base, January 1968.

58 Metallic Materials Properties Development and Standardization (MMPDS) Handbook, Battelle Memorial Institute, 2012.

59 Miller KS. Derivatives of Noninteger Order. *Mathematics Magazine.* 68(3), 183–192, 1995.

60 Muskhelishvili NI. *Some Basic Problems of the Mathematical Theory of Elasticity.* English translation of the 3rd edition. P. Noordhoff Ltd., Groningen, Holland, 1953.

61 Naghdi PM. Foundations of elastic shell theory. In: *Progress in Solid Mechanics.* Vol. 4. North-Holland, Amsterdam, 1963, pp. 1–90.

62 Nazarov S and Plamenevsky BA. *Elliptic Problems in Domains with Piecewise Smooth Boundaries.* Vol. 13. Walter de Gruyter, 2011.

63 Nervi S and Szabó BA. On the estimation of residual stresses by the crack compliance method. *Comput. Methods Appl. Mech. Engng.* 196(37–40) 3577–3584, 2007.

64 Nervi S, Szabó BA and Young KA. Prediction of distortion of airframe components made from aluminum plates. *AIAA Journal.* 47(7) 1635–1641, 2009.

65 Neuber H. *Kerbspannungslehre.* Springer Verlag, Berlin, 1937.

66 Niemi AH, Babuška I, Pitkäranta J and Demkowicz L. Finite element analysis of the Girkmann problem using the modern hp-version and the classical h-version. *Engineering with Computers.* 28(2) 123–134, 2012.

67 Nitsche JA. Über ein Variationsprinzip zur Lösung von Dirichlet-Problemen bei Verwendung von Teilräumen, die keinen Randbedingungen unterworfen sind. *Abhandlungen aus dem Mathematischen Seminar der Universität Hamburg.* 36 9–15, 1971.

68 Novozhilov VV. *Thin Shell Theory.* P. Noordhoff Ltd., Groningen, 1964.

69 Oden JT, Babuška I and Faghihi D. Predictive Computational Science: Computer Predictions in the Presence of Uncertainty. In: Erwin Stein, René de Borst and Thomas J. R. Hughes, editors, *Encyclopedia of Computational Mechanics.* Volume 1: Fundamentals. John Wiley & Sons, Ltd., 2017.

70 Oden JT and Reddy JN. *An Introduction to the Mathematical Theory of Finite Elements.* Courier Corporation, 2012.

71 Olver FW, Lozier DW, Boisvert RF and Clark CW. *NIST Handbook of Mathematical Functions.* US Department of Commerce, National Institute of Standards and Technology, 2010.

72 Papadakis PJ and Babuška I. A numerical procedure for the determination of certain quantities related to the stress intensity factors in two-dimensional elasticity. *Comput. Methods Appl. Mech. Engng.* 122, 69–92, 1995.

73 Pascual FG and Meeker WQ. Estimating fatigue curves with the random fatigue-limit model. *Technometrics.* 41(4), 277–289, 1999.

74 Peterson RE. *Stress Concentration Design Factors.* John Wiley & Sons, Inc. New York, 1953.

75 Pilkey WD. *Peterson's Stress Concentration Factors.* 2nd edition. John Wiley & Sons, Inc. New York, 1997.

76 Pitkäranta J, Babuška I and Szabó B. The Girkmann problem. *IACM Expressions.* 22, January 2008.

77 Pitkäranta J, Babuška I and Szabó B. The problem of verification with reference to the Girkmann problem. *IACM Expressions.* 24, 14–15, January 2009.

78 Pitkäranta J, Babuška I and Szabó B. The dome and the ring: Verification of an old mathematical model for the design of a stiffened shell roof. *Computers & Mathematics with Applications.* 64(1), 48–72, 2012.

79 Popov EP. *Mechanics of Materials.* 2nd edition. Prentice-Hall, Inc. Englewood Cliffs, 1978.

80 Schijve J. Fatigue of structures and materials in the 20th century and the state of the art. *International Journal of Fatigue.* 25(8), 679–702, 2003.

81 Reddy JN. *An Introduction to Nonlinear Finite Element Analysis.* Oxford University Press, 2004.

82 Ramm E, Rank E, Rannacher R, Schweizerhof K, Stein E, Wendland W, Wittum G, Wriggers P and Wunderlich W. In: E. Stein, editor, *Error-controlled Adaptive Finite Elements in Solid Mechanics.* John Wiley & Sons Ltd., Chichester, 2003.

83 Sanchez-Palencia E. Boundary layers and edge effects in composites. In: E. Sanchez-Palencia and A. Zaoui, editors, *Homogenization Techniques for Composite Media.* Lecture Notes in Physics 272, Part III, Springer Verlag, Berlin, 1987.

84 Schwab C. *p- and hp-Finite Element Methods: Theory and Applications in Solid and Fluid Mechanics.* Clarendon Press, Oxford, 1998.

85 Schwab C, Suri M and Xenophontos C. The *hp* finite element method for problems in mechanics with boundary layers. *Comput. Methods Appl. Mech. Engng.* 157, 311–333, 1998.

86 Simo JC and Hughes TJR. *Computational Inelasticity.* Springer-Verlag, New York, 1998.

87 Sivia DS with Skilling J. *Data Analysis. A Bayesian Tutorial.* 2nd edition. Oxford University Press, Oxford, 2006.

88 Smith RN, Watson P and Topper TH. A stress-strain function for the fatigue of metals. *Journal of Materials* ASTM, 5(4), 767–778, 1970.

89 Strang G and Fix GJ. *An Analysis of the Finite Element Method.* Prentice Hall, Englewood Cliffs, 1973.

90 Szabó BA. Mesh design for the *p*-version of the finite element method. *Computer Methods in Applied Mechanics and Engineering.* 55(1–2), 181–197, 1986.

91 Szabó BA and Actis RL. Hierarchic models for laminated composites. *International Journal for Numerical Methods in Engineering* 33(3), 503–535, 1992.

92 Szabó B and Actis R. Simulation governance: Technical requirements for mechanical design. *Comput. Methods Appl. Mech. Engng.* 249–252, 158–168, 2012.

93 Szabó B, Actis R and Rusk D. Predictors of fatigue damage accumulation in the neighborhood of small notches. *International Journal of Fatigue.* 92, 52–60, 2016.

94 Szabó B, Actis R and Rusk D. Validation of notch sensitivity factors. *Journal of Verification, Validation and Uncertainty Quantification.* 4, 011004, 2019.

95 Szabó B, Actis R and Rusk D. Validation of a predictor of fatigue failure in the high-cycle range. *Computers and Mathematics with Applications.* 11, 2451–2461, 2020.

96 Szabó BA and Babuška I. Computation of the amplitude of stress singular terms for cracks and reentrant corners. In: T. A. Cruse, editor, *Fracture Mechanics: Nineteenth Symposium ASTM STP 969.* American Society for Testing and Materials, Philadelphia, 101–124, 1988.

97 Szabó B and Babuška I. *Introduction to Finite Element Analysis. Formulation, Verification and Validation.* John Wiley & Sons Ltd. Chichester, UK, 2011.

98 Szabó B, Babuška I and Chayapathy BK. Stress computations for nearly incompressible materials by the p-version of the finite element method. *International Journal for Numerical Methods in Engineering.* 28(9), 2175–2190, 1989.

99 Szabó B Babuška I Pitkäranta J and Nervi S. The problem of verification with reference to the Girkmann problem. *Engineering with Computers.* 26, 171–183, 2010. DOI 10.1007/s00366-009-0155-0.

100 Szabó BA and Mehta AK. *p*-Convergent finite element approximations in fracture mechanics. *International Journal for Numerical Methods in Engineering.* 12(3), 551–560, 1978.

101 Szabó BA and Sahrmann GJ. Hierarchic plate and shell models based on *p*-extension. *International Journal for Numerical Methods in Engineering.* 26(8), 1855–1881, 1988.

102 Timoshenko S. *History of strength of materials: With a brief account of the history of theory of elasticity and theory of structures.* Courier Corporation, 1953.

103 Timoshenko S and Gere JM. *Theory of Elastic Stability.* 2nd edition. McGraw-Hill Book Company, Inc., New York, 1961.

104 Timoshenko S and Woinowsky-Krieger S. *Theory of Plates and Shells.* 2nd edition. McGraw-Hill Book Company, Inc., New York, 1959.

105 Timoshenko S and Goodier JN. *Theory of Elasticity.* McGraw-Hill Book Company, inc., New York, 1951.

106 Todhunter I and Pearson KA. *A History of the Theory of Elasticity and of the Strength of Materials: From Galilei to the Present Time.* Vol. 1. Galilei to Saint-Venant, 1639–1850. Cambridge University Press, 1886.

107 Turner MJ, Clough RW, Martin HC and Topp LJ. Stiffness and deflection analysis of complex structures. *Journal of the Aeronautical Sciences.* 23(9) 805–823, 1956.

108 Vasilopoulos D. On the determination of higher order terms of singular elastic stress fields near corners. *Numer. Math.* 53, 51–95, 1998.

109 Wahl AM. *Mechanical Springs.* 2nd edition. McGraw-Hill Book Company, 1963.

110 Wollschlager JA. *Introduction to the Design and Analysis of Composite Structures.* Jeffrey A. Wollschlager, Troy, 2012.

111 Young W and Budynas R. *Roark's Formulas for Stress and Strain.* 8th edition. McGraw-Hill Companies, 2011.

112 Yosibash Z. *Singularities in Elliptic Boundary Conditions and their Connection with Failure Initiation.* Springer, 2012.

113 Zienkiewicz OC. *The Finite Element Method in Structural and Continuum Mechanics.* McGraw-Hill, London, 1967.

114 Zienkiewicz OC, Taylor RL and Too JM. Reduced integration technique in general analysis of plates and shells. *International Journal for Numerical Methods in Engineering.* 3(2), 275–190, 1971.

Index

Finite Element Analysis: Method, Verification and Validation, Second Edition. Barna Szabó and Ivo Babuška.
© 2021 John Wiley & Sons, Inc. Published 2021 by John Wiley & Sons, Inc.
Companion Website: www.wiley.com/go/szabo/finite_element_analysis

Printed and bound by CPI Group (UK) Ltd, Croydon, CR0 4YY

16/04/2025

14658589-0003